Power Electronics and Power Systems

Series Editors

Joe H. Chow, Rensselaer Polytechnic Institute, Troy, NY, USA
Alex M. Stankovic, Tufts University, Medford, MA, USA
David J. Hill, Department of Electrical and Electronics Engineering, University of Hong Kong, Pok Fu Lam, Hong Kong

The Power Electronics and Power Systems Series encompasses power electronics, electric power restructuring, and holistic coverage of power systems. The Series comprises advanced textbooks, state-of-the-art titles, research monographs, professional books, and reference works related to the areas of electric power transmission and distribution, energy markets and regulation, electronic devices, electric machines and drives, computational techniques, and power converters and inverters. The Series features leading international scholars and researchers within authored books and edited compilations. All titles are peer reviewed prior to publication to ensure the highest quality content. To inquire about contributing to the Power Electronics and Power Systems Series, please contact Dr. Joe Chow, Administrative Dean of the College of Engineering and Professor of Electrical, Computer and Systems Engineering, Rensselaer Polytechnic Institute, Jonsson Engineering Center, Office 7012, 110 8th Street, Troy, NY USA, 518-276-6374, chowj@rpi.edu.

More information about this series at http://www.springer.com/series/6403

Sarma (NDR) Nuthalapati
Editor

Use of Voltage Stability Assessment and Transient Stability Assessment Tools in Grid Operations

Editor
Sarma (NDR) Nuthalapati
Adjunct Professor
Department of Electrical and
Computer Engineering
Texas A&M University
College Station, TX, USA

Principal Engineer - Contractor
System Operations Engineering
Dominion Energy
Richmond, VA, USA

ISSN 2196-3185   ISSN 2196-3193   (electronic)
Power Electronics and Power Systems
ISBN 978-3-030-67481-6   ISBN 978-3-030-67482-3   (eBook)
https://doi.org/10.1007/978-3-030-67482-3

© Springer Nature Switzerland AG 2021
This work is subject to copyright. All rights are reserved by the Publisher, whether the whole or part of the material is concerned, specifically the rights of translation, reprinting, reuse of illustrations, recitation, broadcasting, reproduction on microfilms or in any other physical way, and transmission or information storage and retrieval, electronic adaptation, computer software, or by similar or dissimilar methodology now known or hereafter developed.
The use of general descriptive names, registered names, trademarks, service marks, etc. in this publication does not imply, even in the absence of a specific statement, that such names are exempt from the relevant protective laws and regulations and therefore free for general use.
The publisher, the authors, and the editors are safe to assume that the advice and information in this book are believed to be true and accurate at the date of publication. Neither the publisher nor the authors or the editors give a warranty, expressed or implied, with respect to the material contained herein or for any errors or omissions that may have been made. The publisher remains neutral with regard to jurisdictional claims in published maps and institutional affiliations.

This Springer imprint is published by the registered company Springer Nature Switzerland AG
The registered company address is: Gewerbestrasse 11, 6330 Cham, Switzerland

This book is dedicated to the memory of
Dr. Prabha Shankar Kundur
for this pioneering work in the area of Power System Dynamics and Voltage Stability

(March 18, 1939–October 9, 2018)

Foreword

Reactive power and *system stability*[1] rank low on the list of preferred topics in academia and power system industry. For students, they are difficult to grasp; for teachers, they are laborious to explain; and, for practitioners and operators, they symbolize a necessary evil that raises the head when system voltages decay and generation is in short supply. However, regardless of whether system stability and reactive power are, or, rather, are not "easy to catch" conceptually, the need to handle them online cannot be overestimated, especially in the current context of transcontinental electricity markets that encompass every conceivable form of generation, from the large scale, reliable and dependable nuclear to the intermittent and never-guaranteed renewables.

The real-time computation of the loadability limits, to the extent such limits are quantifiable and computable, is essential for the effective and reliable grid utilization in an open access environment. In the past, the calculation of these limits in power system control centers was relegated to off-line studies, and time-consuming simulations were performed to determine the maximum loadability limits resulting from stability constraints. Today, the complexity of online stability assessment has been mastered, and both simple, yet theoretically sound, applications that can quickly tell how far a given operating state is from instability, and sophisticated packages that perform comprehensive stability analysis have been successfully deployed and are currently used on a daily basis in system operations.

This edited volume is a compendium of articles written by experienced professionals. It reflects a world-wide agreed-upon state-of-the-art. It is intended to serve as both a textbook for students and teachers and a reference guide for

---

[1] Power systems may become unstable in various ways and for several reasons. The phenomena are complex and are handled in the realms of angle and voltage stability, transient and steady-state stability, load stability, and small-signal oscillation stability. Although unified, though not standardized, definitions have been formulated and are periodically updated by the IEEE PES Power System Dynamic Performance (PSDP) Committee. Nevertheless, alternate terminologies still coexist and are being used by practitioners.

practitioners. And it addresses a broad array of issues, from underlying techniques and technologies to actual implementations and lessons learned.

The first chapter familiarizes the reader with the real-time and study-mode network analysis functionality that nowadays is at hand in any modern power system control center, offers superb application integration opportunities that were not available until just a few years ago, and constitutes the socket where stability applications are plugged in. The next two chapters summarize the fundamentals and set the stage for the theoretical framework needed to understand the system stability issues covered in the book.

The subsequent sections document some of the online transient security assessment and voltage stability analysis approaches, out of a larger array of alternate solutions, that have been adopted to-date in large-scale control centers both in the USA and overseas. Considerations about software validation and integration, experience in handling the challenges created by wind generation, and details about solution techniques and implementation architectures are provided. Indispensable factors that can help improve situational awareness, such as WAMS and real-time VAR management are also addressed, thus enhancing the already broad array of topics tackled in the book.

A judicious balance between extensive theoretical details and practical implementation features permeates this volume from the first to the last page and reflects the editor's brilliant trajectory in academia and industry. Dr. Sarma Nuthalapati is one of those experts, unfortunately rather rare among us nowadays, who complement a thorough, in-depth understanding of theory and algorithms with the hands-on knowledge of how such tools are actually implemented and used in real life.

It is therefore a pleasure to welcome this effort by him especially when the electric power industry has undergone radical transformation with, on the one hand, the impact of WAMS, renewable forms of energy and transcontinental AC and DC interconnections and, on the other, the advent of the computational power and software wizardry needed to bring the real-time stability assessment to the operator's fingertips. This book is a worthy addition to an already valuable collection of earlier publications hosted by Springer in the field of real-time stability.

Energy Consulting International, Inc. Bayside, NY, USA   Savu C. Savulescu
November 24, 2020

Preface

When I started my career in Power Systems 30 years ago in the early 90s in India, we were involved in developing network advanced application functions for Energy Management Systems (EMS) but did not have stability assessment tools as a part of those functions. Later when I started working at Electric Reliability Council of Texas, Inc. (ERCOT) in the USA, I had the opportunity to work more closely with Advanced Network Applications for EMS and support operators in control centers for security assessment. We also had the opportunity to listen to Dr. Prabha Kundur in a three-day workshop on Voltage Stability. It was a significant learning experience for us to hear from the Guru on Voltage Stability Assessment. ERCOT uses both Voltage Stability and Transient Stability assessment tools in its control center.

The concept of this book started when I had an idea for an issue of IEEE PES Magazine focusing on "Use of Stability Assessment Tools in Control Centers." I shared it with Dr. Prabha Kundur and he agreed that such a compilation would benefit the readers and industry. In July 2017, we proposed the idea to IEEE PES Magazine Editorial Board. However, we were not successful in publishing the magazine. Later, I envisioned a book that aggregated the experiences of different utilities utilizing stability assessment tools in their grid operations. Dr. Kundur again encouraged me to pursue the idea and this is that book today. It is unfortunate that we lost Dr. Prabha Kundur, a pioneer in Power System Dynamics and Voltage Stability. I was constantly inspired by him and always felt supported when I approached seeking guidance. I am dedicating this book in his memory as a mark of my respect to him.

This book is an addition to an already valuable 2005 book on "Real-time Stability in Power Systems—Techniques for early Detection of the Risk of Blackout" by Springer that was edited by Dr. Savu C. Savulescu. The book was built around the panel session on Real-Time Stability Challenges hosted during IEEE Power Systems Conference and Exposition on October 13, 2004. The panel provided a forum for showcasing the progress achieved, identifying and discussing challenges to innovate on during future research. Though Dr. Savulescu's book was written in 2005, it is still appropriate and presents the need and importance of stability assessment tools in grid operations. The second edition of his book released in 2014 addresses some of the latest developments in this area. Dr. Savulescu kindly agreed to provide an overview of security assessment in Chap. 1 of this book. I am also grateful to him for writing the foreword.

This book discusses the use of stability assessment tools in grid operations by many utilities across the world. Initial chapters provide a basic introduction and background on security assessment, voltage stability assessment, and transient stability assessment. Remaining chapters present the unique usage of these tools in each of the utilities. Broad contributions for the book are from the following:

- Grid Operators/ Independent System Operators:
  - California ISO
  - ISO New England
  - PJM Interconnection
  - Electric Reliability Council of Texas (ERCOT)
  - PEAK Reliability (Former Reliability Coordinator for the majority in Western Interconnection)
  - Midwest ISO
  - California ISO
  - Australian Energy Market Operator (AEMO), Australia
  - Nordic Power System Operators, Norway

- Utilities:
  - BC Hydro
  - National Grid, UK
  - State Grid Corporation of China
  - Tennessee Valley Authority (TVA)
  - San Diego Gas & Electric (SDG&E)
  - Dominion Energy
  - Bonneville Power Authority (BPA)
  - New York Power Authority (NYPA)
  - Tokyo Electric Power Company (TEPCO), Japan

Initially, Chap. 1 by Savu C. Savulescu provides an overview of security assessment for Grid Operations and aims at familiarizing the reader with the state-of-the-art in real-time and study-mode network analysis as currently implemented in SCADA/EMS installations worldwide.

Chapters 2 and 3 provide the basic theoretical background for all the other chapters. In Chap. 2, authors describe the fundamentals of long-term voltage stability and short-term voltage stability phenomenon in power systems by studying the various parametric dependencies along with the respective assessment methods. Numerical techniques that enable scalable assessment of voltage stability in practical systems are described along with examples. In Chap. 3, the author provides the basics of Transient Stability Assessment.

In Chap. 4, the author presents an online transient security assessment tool (TSAT) that was successfully implemented in Peak Reliability (formerly, WECC

Reliability Coordinator) at its control rooms. The author introduces implementation experience, model validation findings, tool integration, and software enhancements in this chapter. The simulation and study results are provided for a few system events.

Chapter 5 presents a success story about how Peak Reliability (formerly, WECC Reliability Coordinator) collaborated with V&R Energy and Peak member entities to implement an online voltage stability analysis tool in a control room setting for real-time assessment of Interconnection Reliability Operating Limits (IROLs) in the Western System. Major technical challenges and the resolutions, including modeling aspects, software improvements, validation, and integration efforts are presented in the chapter. Lessons learned and future work are discussed in the end.

Chapter 6 looks at the voltage and transient stability considerations in the daily operations of the ERCOT system. The authors review the voltage stability assessments conducted as part of real-time operations and the need in ERCOT for a "system strength" metric. The Weighted Short Circuit Ratio (WSCR) proposed by ERCOT for this purpose is presented in detail. They also review the transient stability assessments conducted in real-time operations at ERCOT, starting with the West–North Interface and going through recent developments such as Transient Event Metrics in performing such assessments in daily operations.

Chapter 7 presents the details of voltage stability and transient stability tools at PJM Interconnection. Over the past few decades, PJM has moved voltage, transient, and dynamic stability analysis—historically the domain of planners and back-office engineers—into the control room. PJM utilizes the Real-Time Transfer Limit Calculator (RTTLC) to perform voltage stability analysis and determine operation limits. The Voltage Stability Analysis & Enhancement tool (VSA&E) application is used to further analyze the limits from RTTLC and find non-cost options (such as transform tap moves or capacitor switching) which can be used to increase the interface limit. This chapter provides more details of these tools used for voltage stability assessment at PJM. It also presents the details of Transient Stability Application (TSA) which is used to determine transient and dynamic stability limits for real time conditions allowing PJM to operate more efficiently than relying on off-line studies.

Chapter 8 presents the developments of the National Grid's Online and Off-line Stability Assessment (OSA/OFSA) systems. It presents business requirements, technical implementations, and gives some examples of use cases. OSA is capable of identification of unknown system stability issues during real-time operation; while OFSA provides what-if assessment in the operational planning phase. Performance of OSA meets the requirement for analyzing the full contingency set (more than 2000 cases) in each State Estimation cycle. OSA and OFSA together support National Grid's delivery of the Great Britain Security and Quality of Supply Standards in a changing and more uncertain environment.

Chapter 9 describes two power system monitoring applications that have been developed and introduced to operators at *Statnett* control centers in Norway. *Statnett* is the transmission grid owner and system operator (TSO) in Norway. Operators have identified online monitoring and assessment of stability properties as the most useful application of synchrophasor information that can be easily implemented in their control centers. The two applications for online voltage stability and transient stability monitoring provide critical information that until now has not been available to operators. The work presented in this chapter is a part of several efforts to introduce synchrophasors applications in the control room at Statnett.

Chapter 10 presents the details of real-time stability assessment tools used at Australian Energy Market Operator (AEMO) in Australia. The nature of the interconnected power system that dominates the eastern and southern regions of Australia, means that it has often dynamic stability concerns rather than thermal issues that dictate the operating boundary of this network. Adding in the ever-increasing penetration of intermittent asynchronous generation and the subsequent reduction in system strength has meant the operation of this network is increasingly challenging. This chapter examines some of the practical issues of assessing stability in a real-time operational time frame at AEMO. It provides some power system events which highlight the level of modelling detail required to properly understand these dynamic stability phenomena, especially in the context of significant penetration of asynchronous generation.

Chapter 11 presents the details of stability assessment tools at China State Grid. The increasing penetration of large-scale renewable energy resources, power electronics-based transmission equipment, and advanced protection and control systems all contribute to the even more complicated dynamics which are observed in the operation of today's China State Grid. In this chapter, online transient stability and voltage stability assessment tools are developed which input real-time EMS snapshots, perform dynamic contingency analysis under various conditions, and calculate real-time transfer limits. Some unique and innovative features are introduced to address the practical challenges, such as real-time integration of power flow information from both node/breaker and bus/branch models, online corrections, and enhancements for power flow and dynamic models. The developed online assessment system has been deployed in the unified dispatching and control system D5000 and has achieved satisfactory performance in improving situational awareness and safe operation of the power grid. Several examples focusing on transient stability and voltage stability are presented to illustrate the effectiveness of the online assessment system.

Chapter 12 discusses the implementation of online voltage and transient stability assessment tools at ISO New England. It presents details on how these tools were configured to achieve the best performance to meet operations' business requirements and attain a stable and robust solution and provides the steps on how these online tools are used to improve operating decisions, how transmission operating text guides are currently developed and will be improved in the future.

Chapter 13 describes the implementation of voltage and transient stability assessment tools at California Independent System Operator (CAISO). The architectural setup of the various real-time systems and data flow needed to achieve the implementation of real-time voltage and transient stability assessments that support the requirements laid out in the System Operating Limits Methodology are provided. The needed configurations of the scenarios voltage and transient to be monitored for stability assessments in real-time are described. In addition, various examples are provided on how the voltage and transient stability assessment tools can be utilized to provide additional information and assessments that can be helpful for better situational awareness of operators. Various challenges of implementing voltage and transient stability assessment tools and ensuring their continuous operation are also provided.

Chapter 14 provides details of stability assessment tools at Midcontinent Independent System Operator (MISO) which is an independent system operator across 15 US states and the Canadian province of Manitoba. To better prepare and maintain stability of the system MISO utilizes stability assessment tools in both real time horizon and operational planning horizon. This chapter provides the details of these tools and provides process flow on usage of these tools at MISO.

Chapter 15 provides the details of VAR Management Systems (VMSs) for monitoring and maintaining adequate reactive power reserves at Tennessee Valley Authority (TVA). This tool leverages algorithms for voltage stability analysis and steady-state power flow to deal with potential reactive power issues. The VMS tool seeks to define localized reactive power "zones," then monitor and ensure sufficient reactive reserves are available locally. In addition to monitoring and analysis, a control recommendation engine for mitigation is provided to enhance reactive reserve zones.

Chapter 16 presents the use of online voltage and transient security assessment tools at BC Hydro. It also discusses how these tools are configured in the EMS environment and integrated with other subsystems to meet the requirements of real-time operations and attain reliable, consistent, and robust solutions. It illustrates how these online tools are used to support operating decisions in conjunction with system operating guides in the form of operating orders that are implemented in an in-house application, Transient Stability Analysis Pattern Matching (TSAPM) in EMS environment.

Chapter 17 presents the details of stability assessment tools used at San Diego Gas & Electric (SDG&E). SDGE has been performing voltage stability studies for close to 30 years, developed various routines and methodologies to address voltage stability concerns. Being a participant of the Peak Reliability Synchrophasor Program (PRSP), SDG&E had access to the Peak-ROSE (Region Of Stability Existence) program/package. Since 2015, SDG&E has been running a real-time version of the Peak-ROSE program every 5 min. This chapter describes practical aspects, assumptions, and methodology for practical online implementation of a voltage stability analysis tool, ROSE. It also explains how certain system events lead to changes in voltage stability methodology and assumptions. A special focus is given to lessons learned after four years running real-time voltage stability analysis.

Chapter 18 provides the details of stability analysis at Dominion Energy. The design and implementation of a stability application in the control room requires both technical and procedural knowledge. Not only do the current processes and employee workflow need to be well understood, but there also must be a vision for how the stability analysis will fit into day-to-day tasks. With these prerequisites in mind, this chapter provides a discussion of the key concerns for Dominion Energy Virginia (DEV) and the lessons learned as the company introduces stability analysis in both real-time and system operations planning studies.

Chapter 19 presents a methodology that has been developed for Tokyo Electric Power Company (TEPCO) which takes the merit of both a direct method and fast time-domain method. The TEPCO-BCU is developed under this direction by integrating BCU method, improved BCU classifiers, and BCU-guide time domain method. The chapter presents methodology to perform transient stability assessment for planning purposes in which the contingency list is extensive to cover possible and yet credible contingencies including the network contingency list and the renewable contingency list.

Chapter 20 discusses a methodology to monitor voltage stability using synchrophasor technology. Voltage Stability Index (VSI) explained here is computed using the PMU measurements for a transmission corridor to determine if the system has any voltage stability issues. The method reduces a complicated transmission corridor to a single line equivalent. The index is computed at PMU measurement rate providing quick indication of voltage stability issues. Therefore, the method can be implemented in control rooms with the associated displays, alarming, and data recording features to support the application. The index limits are determined by studies to stress the system under various loading scenarios and system conditions to find a warning level and an alarm level that requires emergency action. Implementation of this method at two utilities as part of a research project was explained in this chapter.

All these chapters give a good overview of how voltage stability and transient stability assessment is performed at various utilities in the USA, Canada, Australia, China, Europe, and Japan. I am hoping that this book can serve as a useful reference to other utilities who are exploring these tools in their grid operations. It also serves as a textbook for teachers and students in understanding the implementation of stability assessment tools in real-world grid operations.

I would like to thank Dr. Joe Chow and other editors of Springer Power Electronics and Power System Series for publishing this book and through the process providing guidance on improving the contents of the book. I also want to thank all the authors for their time and efforts in preparing the chapters. I am grateful to Dr. Savu C Savulescu for kindly writing the foreword. I would like to thank all my IEEE and NASPI colleagues from whom I always learn.

I am what I am today because of my teachers at the Railway High School (Kazipet), National Institute of Technology Warangal (formerly known as Regional Engineering College Warangal), and the Indian Institute of Technology (IIT) at Delhi in India and my dear parents Sri N Hanumantha Rao and Late Smt. Kamala Devi. I am indebted to them for all their teachings which have made me a good engineer and a hardworking human being. I also appreciate my wife Vasudha and our daughter Sruti for their understanding and patience with my passion for my professional life.

Austin, TX, USA
30th November 2020

Sarma (NDR) Nuthalapati

Contents

1. **Overview of Security Assessment for Grid Operations** 1
   Savu C. Savulescu

2. **Basics of Voltage Stability Assessment** 25
   Amarsagar Reddy Ramapuram Matavalam, Alok Kumar Bharati,
   and Venkataramana Ajjarapu

3. **Basics of Transient Stability Assessment** 79
   U. D. Annakkage

4. **Implementation of the Online Transient Security Assessment
   Tool for RAS Real-Time Operation Monitoring** 99
   Hongming Zhang

5. **Implementing the Real-Time Voltage Stability Analysis Tool
   for the Western IROL Real-Time Assessment** 117
   Hongming Zhang

6. **Voltage and Transient Security Assessment in ERCOT Operations** .. 135
   Sidharth Rajagopalan, Jose Conto, Yang Zhang,
   and Sarma (NDR) Nuthalapati

7. **Use of Voltage Stability Assessment and Transient Stability
   Assessment Tools at PJM Interconnection** 145
   Dean G. Manno and Jason M. Sexauer

8. **Online and Offline Stability Assessment Development
   at National Grid, UK** 161
   Fan Li, Martin Bradley, and Frederic Howell

9. **Use of Voltage Stability Monitoring and Transient Stability
   Monitoring Tools at the Nordic Power System Operators:
   Introduction of Synchrophasor Applications in the Control Room** ... 181
   Kjetil Uhlen, Dinh Thuc Duong, and David Karlsen

10	**Voltage Stability and Transient Stability Assessment Tools to Manage the National Electricity Market in Australia** Stephen Boroczky and Lochana Perera	199
11	**Use of Online Transient Stability and Voltage Stability Assessment Tools at State Grid China** Shi Bonian, Yan Jianfeng, and Jin Yiding	217
12	**Online Voltage and Transient Stability Implementation at ISO New England** Omar A. Sanchez, Yuan Li, Xiaochuan Luo, Slava Maslennikov, and Song Zhang	247
13	**Stability Assessment at CAISO** Aftab Alam, Ruili Zhao, and Ran Xu	273
14	**Use of Stability Assessment Tools at Midwest ISO** Raja Thappetaobula	293
15	**Reactive Power Management in Real-Time at Tennessee Valley Authority (TVA)** Tim Fritch, Ulyana Pugina Elliott, Josh Shultz, Patrick Causgrove, and Gilburt Chiang	303
16	**Use of Voltage and Transient Security Assessment Tools for Grid Operations at BC Hydro** Ziwen Yao and Djordje Atanackovic	315
17	**Use of Voltage Stability Assessment Tools at San Diego Gas & Electric** Anita Hoyos, Kenneth Poulter, Robin Manuguid, Michael Vaiman, and Marianna Vaiman	325
18	**Stability Applications in the Dominion Energy System Operations Center** Katelynn Vance and Gilburt Chiang	351
19	**TEPCO-BCU for Transient Stability Assessment in Power System Planning Under Uncertainty** Ryuya Tanabe, Hsiao-Dong Chiang, and Hua Li	371
20	**Voltage Stability Assessment Using Synchrophasor Technology** Iknoor Singh, Ken Martin, Neeraj Nayak, Ian Dobson, Anthony Faris, and Atena Darvishi	385

Index 395

Contributors

Venkataramana Ajjarapu Department of Electrical and Computer Engineering, Iowa State University, Ames, IA, USA

Aftab Alam Operations Planning, California Independent System Operator, Sacramento, CA, USA

U. D. Annakkage Department of Electrical and Computer Engineering, University of Manitoba, Winnipeg, MB, Canada

Djordje Atanackovic Real-time Systems, BC Hydro, Vancouver, BC, Canada

Alok Kumar Bharati Department of Electrical and Computer Engineering, Iowa State University, Ames, IA, USA

Shi Bonian R&D Center, Beijing Sifang Automation Company, Beijing, China

Stephen Boroczky Australian Energy Market Operator, Melbourne, Victoria, Australia

Martin Bradley National Grid, St. Catherine's Lodge,, Wokingham, UK

Patrick Causgrove Bigwood Systems, Inc., Ithaca, NY, USA

Gilburt Chiang Bigwood Systems, Inc., Ithaca, NY, USA

Hsiao-Dong Chiang School of Electrical and Computer Engineering, Cornell University, Ithaca, NY, USA

Jose Conto Electric Reliability Council of Texas Inc. (ERCOT), Austin, TX, USA

Atena Darvishi New York Power Authority, White Plains, NY, USA

Ian Dobson Iowa State University, Ames, IA, USA

Dinh Thuc Duong Lede AS, Porsgrunn, Norway

Ulyana Pugina Elliott Tennessee Valley Authority (TVA), Chattanooga, TN, USA

Anthony Faris Bonneville Power Authority, Portland, OR, USA

Tim Fritch Tennessee Valley Authority (TVA), Chattanooga, TN, USA

Frederic Howell Powertech Labs, Inc., Surrey, BC, Canada

Anita Hoyos Electric Grid Operations Department, San Diego Gas and Electric, San Diego, CA, USA

Yan Jianfeng State Grid Simulation Center, China EPRI, State Grid Corporation of China, Beijing, China

David Karlsen Statnett SF, Oslo, Norway

Fan Li National Grid, St. Catherine's Lodge, Wokingham, UK

Hua Li Bigwood Systems, Inc., Ithaca, NY, USA

Yuan Li Real-Time Support, EMS Applications, Business Architecture and Technology, ISO New England, Holyoke, MA, USA

Xiaochuan Luo Real-Time Support, EMS Applications, Business Architecture and Technology, ISO New England, Holyoke, MA, USA

Dean G. Manno PJM Interconnection, Audubon, PA, USA

Robin Manuguid Electric Grid Operations Department, San Diego Gas and Electric, San Diego, CA, USA

Ken Martin Electric Power Group, Pasadena, CA, USA

Slava Maslennikov Real-Time Support, EMS Applications, Business Architecture and Technology, ISO New England, Holyoke, MA, USA

Neeraj Nayak Electric Power Group, Pasadena, CA, USA

Sarma (NDR) Nuthalapati Department of Electrical and Computer Engineering, Texas A&M University, College Station, TX, USA
System Operations Engineering, Dominion Energy, Richmond, VA, USA

Lochana Perera Australian Energy Market Operator, Melbourne, Victoria, Australia

Kenneth Poulter Electric Grid Operations Department, San Diego Gas and Electric, San Diego, CA, USA

Sidharth Rajagopalan Electric Reliability Council of Texas Inc. (ERCOT), Austin, TX, USA

Amarsagar Reddy Ramapuram Matavalam Department of Electrical and Computer Engineering, Iowa State University, Ames, IA, USA

Omar A. Sanchez Real-Time Support, EMS Applications, Business Architecture and Technology, ISO New England, Holyoke, MA, USA

Savu C. Savulescu Energy Consulting International, Inc., Bayside, NY, USA

Jason M. Sexauer PJM Interconnection, Audubon, PA, USA

Josh Shultz Tennessee Valley Authority (TVA), Chattanooga, TN, USA

Iknoor Singh Electric Power Group, Pasadena, CA, USA

Ryuya Tanabe System Planning Department, Tokyo Electric Power Company, Tokyo, Japan

Raja Thappetaobula Midcontinent Independent System Operator (MISO), Carmel, IN, USA

Kjetil Uhlen Norwegian University of Science and Technology, NTNU, IEL, Trondheim, Norway

Marianna Vaiman V&R Energy, Systems Research, Inc., Los Angeles, CA, USA

Michael Vaiman V&R Energy, Systems Research, Inc., Los Angeles, CA, USA

Katelynn Vance Dominion Energy Inc., Richmond, VA, USA

Ran Xu Operations Planning, California Independent System Operator, Sacramento, CA, USA

Ziwen Yao Real-time Systems, BC Hydro, Vancouver, BC, Canada

Jin Yiding System Operation Department, National Dispatching and Control Center, State Grid Corporation of China, Beijing, China

Hongming Zhang coreWSM Consulting LLC, Fort Collins, CO, USA

Song Zhang Real-Time Support, EMS Applications, Business Architecture and Technology, ISO New England, Holyoke, MA, USA

Yang Zhang Electric Reliability Council of Texas Inc. (ERCOT), Austin, TX, USA

Ruili Zhao Operations Planning, California Independent System Operator, Sacramento, CA, USA

Chapter 1
Overview of Security Assessment for Grid Operations

Savu C. Savulescu

1.1 Introduction

1.1.1 Background

The use of digital computers to evaluate[1] the impact of scheduled and potential transmission and/or generation outages emerged in the late 1950s and started to be performed routinely in the early 1960s when the load-flow calculations became settled science. At that time, power utilities owned and operated both generation and transmission facilities and the label "vertically integrated utility" had not been devised. Markets did not exist either, but the sheer size of the power pools, which had emerged in the 1930s to operate economically multiple utilities *without* owning

---

[1] At the outset, let us note that the techniques addressed in this chapter aim at the static, or steady-state, analysis of power system conditions that are reached long after the sub-transient and transient phenomena have subsided. In this context: "long after" is actually never longer than a couple of seconds; the after-contingency steady-states are determined with load-flow computations; and *dynamics, system oscillations, relay settings,* and other stability aspects are not taken into account. Originally, this approach was called "static security assessment" as opposed to "dynamic security assessment," which entailed primarily transient stability calculations. Since a thorough assessment of the power system operating reliability would not be complete unless dynamics would also be considered, a few stability concepts are briefly introduced in Sect. 1.2.3 and Sect. 1.4.2—and then, the main voltage and transient stability tools will be addressed extensively throughout the remaining chapters of this book

S. C. Savulescu (✉)
Energy Consulting International, Inc., Bayside, NY, USA
e-mail: one@eciscs.com

© Springer Nature Switzerland AG 2021
S. (NDR) Nuthalapati, *Use of Voltage Stability Assessment and Transient Stability Assessment Tools in Grid Operations*, Power Electronics and Power Systems,
https://doi.org/10.1007/978-3-030-67482-3_1

equipment, was already making it difficult to assess the reliability of energy transfers across vast system areas.

The real-time calculations entailed in system supervision and control, short-term scheduling, and post-operation were performed on SCADA platforms referred to as "master stations" and deployed in operational units known as "control centers." By contrast, medium- and long-term planning and forecasting belonged in separate quarters of the utility, known as "system planning," and used mainframes situated either locally, in the utility's "data processing" department, or remotely at large external data centers.

Dubbed either "real-time" or "general purpose," depending upon the type of data they were designed to process, the digital computers had completely replaced the analog systems and network analyzers of yesteryears but the applications were still aligned with the traditional landscape: *quality*, i.e., constant frequency and acceptable voltage levels, and *economy* of supply were handled in real-time; *operating reliability* was assessed off-line.

This state of affairs changed dramatically on Tuesday, November 9, 1965, when, at 5:29 p.m., approximately 80,000-square miles of the Northeastern United States and the Province of Ontario, Canada, fell into darkness. Toronto, the first city afflicted by the blackout, went dark at 5:15 p.m. Rochester followed at 5:18 p.m., then Boston at 5:21 p.m. New York, finally, lost power at 5:28 p.m. The failure affected four million homes in the metropolitan area and left between 600,000 and 800,000 people stranded in the city's subway system.

Later that evening, President Lyndon Johnson sent a memorandum (Fig. 1.1) to Joseph C. Swidler, Federal Power Commission Chairman, underlining *"the*

Fig. 1.1 The Northeast blackout of 1965—and President Lyndon Johnson's reaction to it

importance of the uninterrupted flow of power to the health, safety, and well-being of our citizens and the defense of our country," and directing the FPC to "*immediately and carefully investigate*" the incident and "*launch a thorough study of the cause*" [1].

The Federal Power Commission answered the call. That very evening it established an Advisory Board to assist in its subsequent investigation of the power failure. Then, on December 6, 1965, it reported the preliminary findings [1]. Various Study Groups were formed, including an Advisory Subcommittee, chaired by Glenn W. Stagg [2], which aimed at studying the role of digital computers in power system planning and operation.

Two years later, the Federal Power Commission issued its Final Report [3]. One of the key recommendations was to "*establish a real-time measurement system and develop computer-based operational and management tools,*" which triggered, among other developments, the advent of state estimation and power system security assessment—and the rest is history.

1.1.2 Context

It is not the purpose of this chapter to review the evolution of network analysis applications from what they were during those pioneering years to what they are today; a brief history of the contingency analysis tool in control centers is provided in [4]. Accordingly, we will jump directly to the current state-of-the-art in the assessment of power system operating reliability. But before addressing a number of key aspects of this sophisticated technology, it is important to realize that: the early *real-time control* systems evolved significantly and became comprehensive *information* systems (Fig. 1.2); the modern control center of today is supported by a complex structure of hardware, software, and communication components (depicted

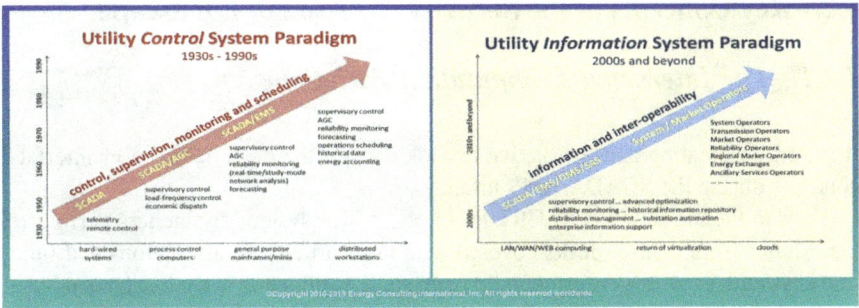

Fig. 1.2 Utility *control* system vs. utility *information* system

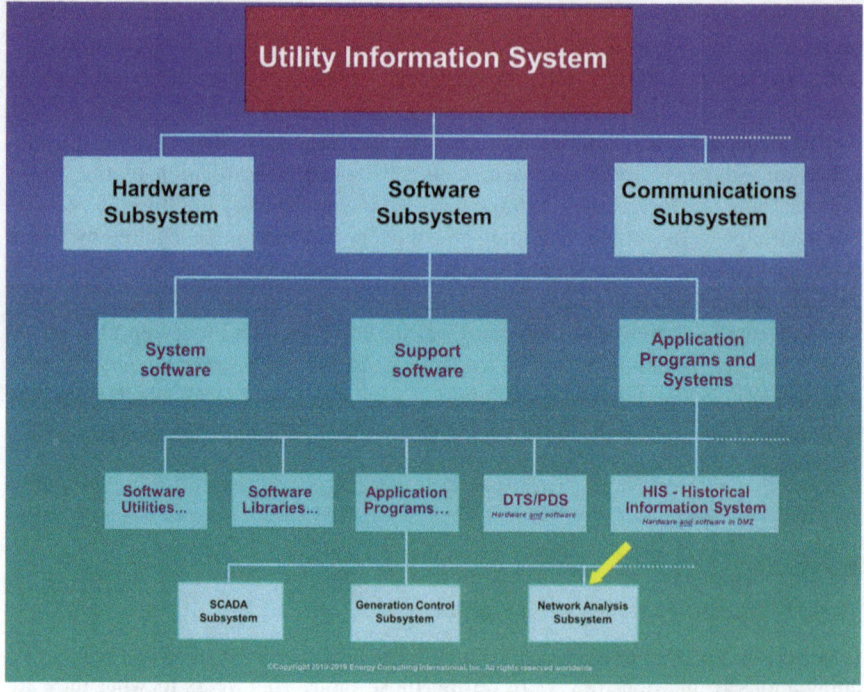

Fig. 1.3 Network Analysis functions in the hierarchy of a modern utility information system

in HIPO[2] format in Fig. 1.3); and that, in order to be appreciated correctly, the network analysis applications must be understood as being just one of the many modules of the utility information system hierarchy.

1.2 Key Concepts in the Security Assessment Landscape

1.2.1 The Interactive Computation Paradigm

It is both fascinating and instructive to look back at the evolution of user interface concepts during the SCADA/EMS infancy.

Prior to the introduction of CRT monitors, power system dispatchers performed their duties from control desks covered with large arrays of push buttons and function keys and, usually, got further help from static wallboards in the background. But

---

[2]HIPO (Hierarchy + Input-Process-Output) is a tool developed by IBM in the 1970s [5], which facilitates the planning, documentation, and specification of computer programs and complex systems that encompass both hardware and software

1 Overview of Security Assessment for Grid Operations

Fig. 1.4 Carolina Power and Light control room in the 1970s

even after the computer display became the main user interface tool, the supervisory control was still handled the old-fashioned way and most of the switching operations were still performed by pushing buttons [6] or, at best, by using track-balls to select and activate CRT poke-points (Fig. 1.4).

The advent of the *security assessment*[3] paradigm, which constituted one of the key corollaries of the technology breakthroughs that took place after the 1965 Northeast blackout, triggered further dilemmas:

- Is there a way to replace the static representation of the power system network with models that can be updated with data collected in real-time, and, if the answer is "yes," how to validate such models and, most importantly, how to use them to predict future states?
- Is it still adequate, or even possible, to conduct complex computational suites that involve multiple application programs in batch processing mode or a new software execution paradigm is needed instead?

---

[3] At that time: "system security" referred to the power system ability to withstand the impact of generation and/or transmission outages; "generation reliability," or just "reliability," designated the capability of the utility's generating units to cover the load duration curve within a specified Loss of Load Probability (LOLP); and "transmission reliability" belonged in long-range transmission studies and consisted of evaluating line and transformer contingencies in the context of planned network topologies. Nowadays, the term "system security" is normally used as a synonym of "cyber security" whereas the meaning of the early concept of "system security" is conveyed by the term "operating reliability".

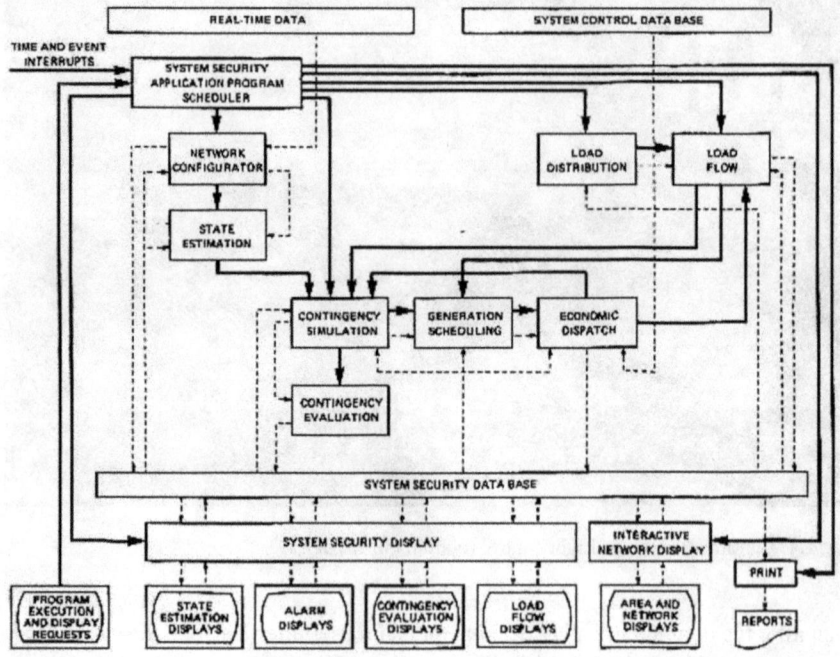

Fig. 1.5 Conceptual design of the first security assessment system deployed in the industry

The first question was brilliantly answered by compounding the *state estimation* with *load-flow* computations and *contingency simulations*. Figure 1.5 depicts the first security assessment system implemented in an actual control center in 1972 [7].

The answer to the second question came from the *Interactive Load-Flow* (ILF) [8], which opened the era of interactive computations. Today, of course, everything is "interactive" and we take it for granted, but at that time batch processing was king and the ILF represented a major change of paradigm.

A quick glance at the complex software interactions depicted in Fig. 1.5 can help explain why the interactive approach taken by the ILF was perceived as a major breakthrough when it was introduced—and can help grasp a better understanding of what today is referred to as *real-time and study-mode network analysis*.

1.2.2 Real-Time Vs. Study-Mode Processes

Another look at the security assessment flow-chart, this time through the lenses of the *hierarchy-input-process-output* paradigm, as shown in Fig. 1.6, tells us that:

1 Overview of Security Assessment for Grid Operations

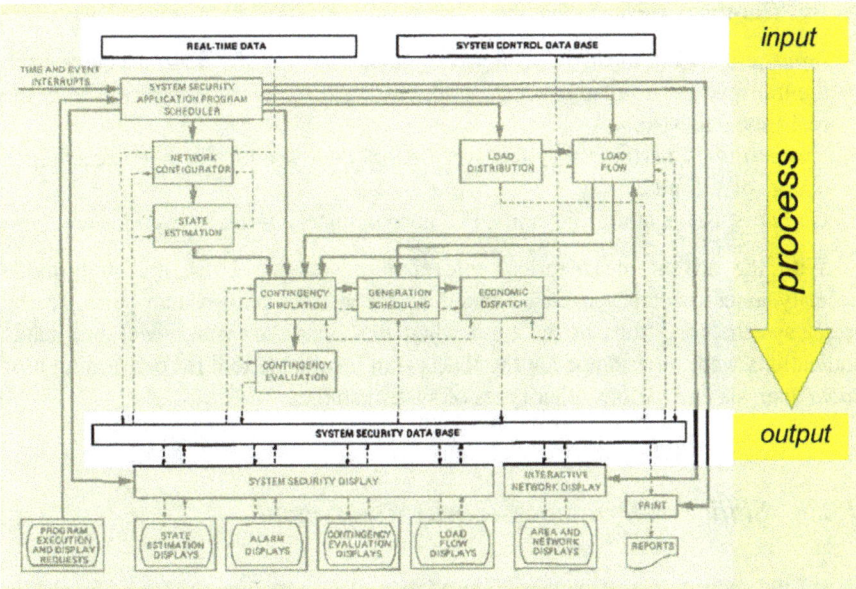

Fig. 1.6 HIPO view of the security assessment flow-chart

- The model entails using both real-time readings and off-line parameter data.
- The computations' elapsed times, no matter how short, imply that what we call "real-time model" is, at best, just a snapshot of what the system state was moments earlier.
- The process, in order to be meaningful, needs to be *executed periodically*, so that it would follow the evolution of the system load, and, also, to be *triggered* either automatically or manually upon the occurrence of system events.
- There is nothing in this paradigm that prohibits using conveniently stored data to perform certain functions in *study-mode* to assess alternate operating scenarios and/or events.

Indeed, once the power of interactive computations was unleashed, the next logical step was to execute *online*[4] and to *interactively control* the entire sequence of security assessment processes, from building the real-time network model to estimating the system state, assembling a base case, and evaluating the impact of potential contingencies—and to do it both in real-time and in study-mode for postulated system conditions.

---

[4] In the SCADA/EMS context, "online" implies that the calculation results are available to the operator in the SCADA/EMS system itself, as opposed to being available on some other separate system, which would be designated as "off-line." However, there is no guarantee that the online computational process will be fast enough to produce results that can be labeled "real-time." A detailed discussion of the "real-time" and "study-mode" paradigms is provided in reference [9]

By "real-time" we mean that the:

- *Input* reflects the most recent power system state—in the field, it comes from the transducers installed in RTUs; at the SCADA master, it is retrieved from the real-time database.
- *Processing* is performed within very short delays typically not exceeding a couple of seconds.
- *Output* is usable almost instantly, i.e., approximately 1–2 s, or even faster.

With the advent of Historical Information Systems (HIS), the study-mode security assessment has been extended to reconstruct past system states to the exact system conditions at the time when the analog measurements and status indications were timestamped at the RTUs—an invaluable tool for performing *post facto* analysis and various other types of system studies.

1.2.3 Static Vs. Dynamic Security Assessment

When the security assessment was introduced, it was recognized from Day 1 that the analysis it entailed was *static* because neither transient nor steady-state stability[5] checks were performed, which is why it was labeled S*tatic Security Assessment*. Many years and several blackouts later, it became obvious that the need to consider stability limits[6] also needed to be addressed in real-time, and so, the *Dynamic Security Assessment (DSA)* paradigm was born.

For network analysis systems, though, the dichotomy static vs. dynamic is not absolute: some SCADA/EMS specifications do include voltage stability and, sometimes, even transient stability assessment requirements *as part of* the standard network analysis subsystem, whereas most vendors offer the DSA capability separately, *in addition to* it. Likewise, the fast computation of the risk of blackout, which is discussed in Sect. 1.3.4, was offered in the past as a piggyback addition to the standard network analysis sequence[7] but today is seamlessly integrated as functionality with its own rights.

---

[5]In the old times, power system stability was classified as *transient*, or *dynamic*, and *steady-state*. Today, one of the earlier steady-state stability concepts known as *system loadability* was relegated to the field of voltage stability whereas the remaining ones are categorized as *small signal stability*

[6]The key concept of stability limit is briefly addressed in Sect. 2.4.2 and further expounded in [10] and related references

[7]Currently, this functionality is seamlessly integrated within SIGUARD®, which is a stand-alone product owned and marketed by Siemens AG, Nuremberg, Germany, and is deployed as a contingency analysis front-end computation to [11–13]

1.3 Anatomy of the Network Analysis Subsystem in a Modern SCADA/EMS

1.3.1 General Considerations

1.3.1.1 Purpose and Scope

The real-time power system monitoring and study-mode security analysis use both real-time information acquired through the SCADA and off-line parameter data. The software subsystem that provides this broad functionality is called *Network Analysis* and encompasses:

- *Standard Network Analysis* applications, which are incorporated and used on a continuing basis in virtually every single SCADA/EMS in the industry and encompass the Network Topology, State Estimation, Contingency Analysis, and Dispatcher's Power Flow programs.
- *Non-standard Network Analysis* applications that are sometimes requested on an optional basis and may include programs such as the Fast Computation of the Risk of Blackout and the Optimal Power Flow, as well as DSA add-ons including Voltage Stability and Transient Stability programs.
- *Service routines* and procedures that facilitate the interaction between user and software and/or between the applications themselves.

This is illustrated in HIPO format in Fig. 1.7.

Software advancements being what they are, the body of network analysis applications has been, and it will most probably be continuously evolving. In the early days, it included modules such as *Model Updating* and *Network Equivalencing* that were taken for granted but today are not even specified any longer—not because they would not be useful anymore, but because the services they provide have been seamlessly embedded in other functions. Likewise, certain applications, e.g., *Remedial Action*, were heavily promoted at their announcement but eventually all disappeared from sight. And things will certainly change again if and when the SCADA/EMS and WAMS[8] paradigms would eventually be merged.

1.3.1.2 Modeling Requirements

Typical Network Analysis modeling requirements include some or all of the following:

---

[8] In the realm of Wide Area Measurement Systems (WAMS), phasor data are collected by PMUs at 2 up to 5 cycle intervals, whereas the status and analog data handled by SCADA systems are gathered at much lower rates. The Sect. 2.4.1 provides a cursory review of using PMU data in SCADA environments

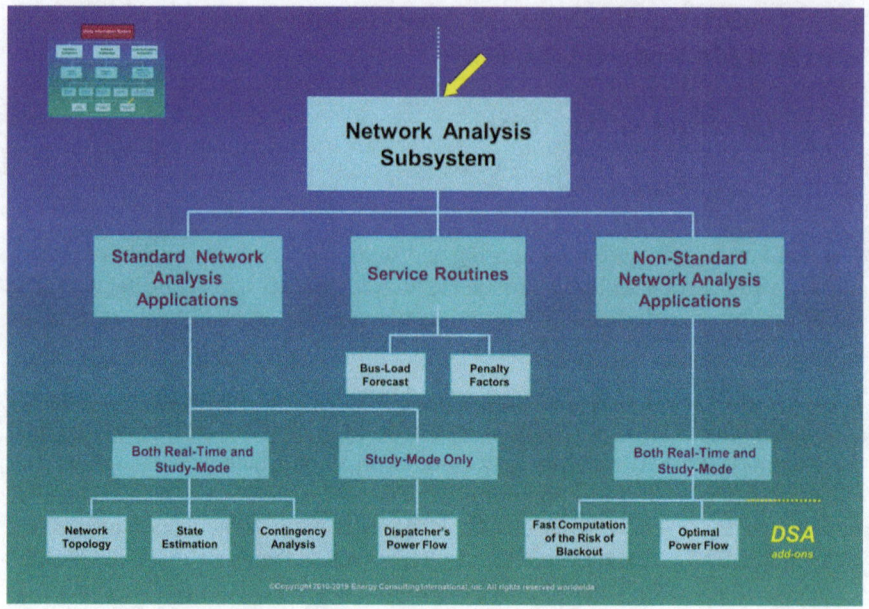

Fig. 1.7 HIPO view of the Network Analysis subsystem

- Power system network elements, such as overhead lines and underground cables, serial and shunt capacitors and reactors, Tap Changing Under Load Transformers (TCUL), Bus Injections (load and generation), Static VAr Compensators, DC lines and DC terminals.
- Control devices, such as breakers, disconnect switches, bus couplers, bus disconnect switches, and load disconnect switches, among others.
- Special modeling data such as equivalent branches, equivalent injections, and P-Q capability curves for the generating units.

System areas are usually defined in terms of operational jurisdiction, e.g., *internal* and *external* areas, but can also be classified as *observable* and, respectively, *unobservable*, depending upon the ability of the state estimator to converge for a given metering configuration and set of pseudomeasurements.

In the early days, non-observability happened frequently because not all of the substations were equipped with RTUs and, furthermore, not all of the installed RTUs were fully populated with transducers. But even if nowadays the power system metering facilities tend to be redundant, the power system networks may still split during major disturbances into electrically disconnected subnetworks, or islands, some of which may be unobservable. Therefore, the software should be able to dynamically identify the boundaries of observable/unobservable areas depending upon the available telemetry and/or set of pseudomeasurements.

1.3.1.3 Execution Modes

The industry-accepted practice is to execute the standard Network Analysis programs as follows:

- *In real-time mode*, the Network Topology, State Estimation, and Contingency Analysis functions: run automatically; do not require manual data entries; are typically scheduled for periodical execution; and can also be triggered by events or initiated upon request. This process is referred to as the *Real-time Network Analysis Sequence* and is monitored via displays that indicate: which module is currently running; what is the current execution status; and error conditions, if any.
- *In study-mode*, the Network Topology, State Estimation,[9]Contingency Analysis, and Dispatcher's Power Flow modules are executed upon request to analyze actual (current or past) and postulated (future or potential) system states.

Both in real-time and in study-mode, these applications work as a group of *seamlessly integrated* functions and the data exchanges between them need to be transparent. In real-time, this is also true for the service routines, which are automatically invoked if and when needed. In study-mode, a rich set of interactive capabilities enables the user to initiate studies from previously stored system conditions, completely reconstruct past states[10] starting from real-time database snapshots saved in HIS, and/or build future scenarios spanned by postulated system conditions.

As far as the nonstandard Network Analysis functions are concerned:

- The Fast Computation of the Risk of Blackout has been deployed both in real-time, to continuously monitor the distance to instability on trending charts, and in study-mode.
- The Optimal Power Flow typically comprises a voltage/reactive power component, which can be used as a study-mode advisory tool to identify controls that alleviate voltage and/or VAr violations, and, respectively, a real power component, which is said to provide the ability to remove MW flow violations within so-called remedial action procedures.

---

[9]The ability to run State Estimation in study-mode may be needed for a variety of reasons, e.g., to assess the system observability when building or upgrading the network analysis database

[10]Past system conditions are usually stored in savecases but can also be retrieved from the HIS as snapshots of the real-time database, in which case the capability to run Network Topology and State Estimation in study-mode is required

1.3.2 *Standard Network Analysis Applications*

1.3.2.1 Network Topology

The Network Topology program determines the topological configuration of the power system network and the related measurement topology, and identifies the energized or de-energized status of the network elements.

A set of typical input-process-output specifications for this application is depicted in HIPO format in Fig. 1.8.

1.3.2.2 State Estimation

The State Estimator processes the available real-time measurements, along with parameter and static data, develops the best state estimate of the power system conditions, and formulates it in terms of line loadings, bus injections, and bus voltage magnitudes and angles. If the process converges, the estimate of the operating state is assembled and saved in a *base case* that is subsequently used for performing security assessment. In addition, the State Estimator provides information about erroneous and missing data.

A set of typical input-process-output specifications for this application is depicted in HIPO format in Fig. 1.9.

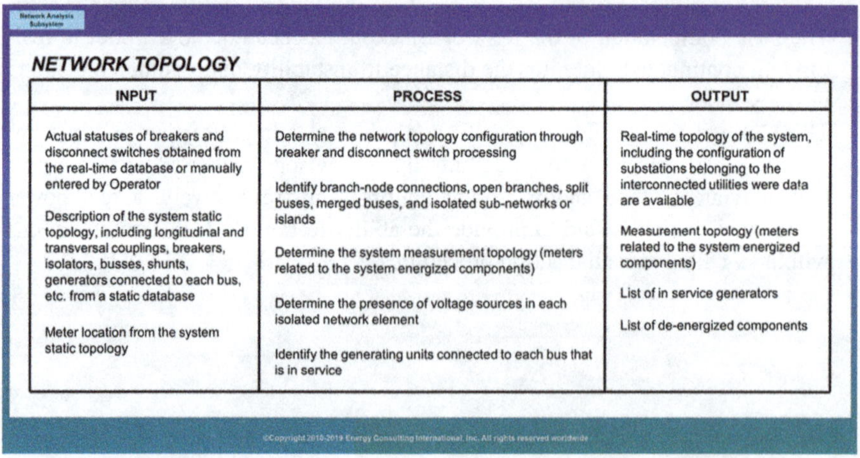
Network Analysis Subsystem
NETWORK TOPOLOGY
INPUT
PROCESS
OUTPUT
Actual statuses of breakers and disconnect switches obtained from the real-time database or manually entered by Operator
Description of the system static topology, including longitudinal and transversal couplings, breakers, isolators, busses, shunts, generators connected to each bus, etc. from a static database
Meter location from the system static topology
Determine the network topology configuration through breaker and disconnect switch processing
Identify branch-node connections, open branches, split buses, merged buses, and isolated sub-networks or islands
Determine the system measurement topology (meters related to the system energized components)
Determine the presence of voltage sources in each isolated network element
Identify the generating units connected to each bus that is in service
Real-time topology of the system, including the configuration of substations belonging to the interconnected utilities were data are available
Measurement topology (meters related to the system energized components)
List of in service generators
List of de-energized components
©Copyright 2010-2019 Energy Consulting International, Inc. All rights reserved worldwide

Fig. 1.8 Typical input-process-output specifications for the Network Topology program

STATE ESTIMATION

INPUT	PROCESS	OUTPUT
System topology assembled by the Network Topology program	Use a state-estimation algorithm adapted to the specifics of the input system, such as the complete or decoupled version of the Weighted Least Squares technique (other methods are also been deployed in the industry)	Estimated values of the bus voltage magnitudes and phase angles
Static parameters (line and trafo impedances, etc.) of the energized system elements		Estimated values of the reactive, real and apparent power at both sides of each transmission line and transformer
Measurement topology (location and type of meters related to the energized system elements)	Process active and reactive power measurements either separated or in pairs (no value should be left unused just because is single)	Estimated values of the bus injected reactive and real power
Redundant telemetered data such as bus voltage magnitudes, active and reactive bus injected powers, reactive powers of shunt capacitors and reactors, etc.	Perform limit checking (current, voltage and reactive power generation) for all the estimated values	Currents at both ends of the transmission lines
	Include a procedure to be executed off-line to determine the meters' standard deviation	Information about the: unobservable zones and/or electrical islands; equivalent injections at the boundary buses; total system load (real and reactive power); system losses (real and reactive power); quality of the estimated results
Tap position measurements from TCUL transformers	Perform a pre-processing analysis of the analog readings such as: coherence verification among measurements; plausibility checks, e.g., voltage values in relation to the node voltage level, etc.	
Manually entered data to inhibit measurements, control convergence characteristics, modify a measurement weight factor, etc.	Classify the measurements according to the pre-processing analysis as valid, suspicious or invalid	
	Detect, identify and eliminate from the calculation process, in a reliable manner, the measurements affected by gross error	
Pseudomeasurements, such as zero injections (passive nodes), calculated reactive power in compensation nodes, pseudo-injections from historical or statistic files, etc.	Replace the erroneous measurements with pseudomeasurements	
	Detect and correct system configuration errors	
Equipment limits that need to be checked	Perform parameter calculation for TCUL transformers with non-telemetered taps	
	Process and achieve convergence on islanded networks	

Fig. 1.9 Typical input-process-output specifications for the State Estimation program

1.3.2.3 Contingency Analysis

For a given base case of the power system, the Contingency Analysis program screens a set of potential contingencies and identifies and evaluates in detail those cases that would entail limit violations.

Historically, the set of contingencies to be screened[11] consisted of a list of line, transformer, and/or generator outages; eventually however the EMS Vendors simplified the process and introduced the so-called N-1 approach, whereby *all*

---

[11]Here is what Dr. Roland Eichler and his colleagues from Siemens say about *contingency screening* in the Sect. 9.1.1.4 of [4]: "Historically screening was utilized as a means of improving performance for contingency analysis. With modern CPU performance there is minimal time difference between screening and then fully simulating a subset of contingencies versus performing a full simulation without screening. While screening capability is available, there is a limited motivation to perform screening with the associated risk of missing contingency violations as a result of heuristic screening indexes. Full simulation without screening also eliminates maintenance effort spent tuning the screening algorithms. Evaluation of time savings with and without screening should be performed to determine the value of screening and its appropriate use to address certain cases."

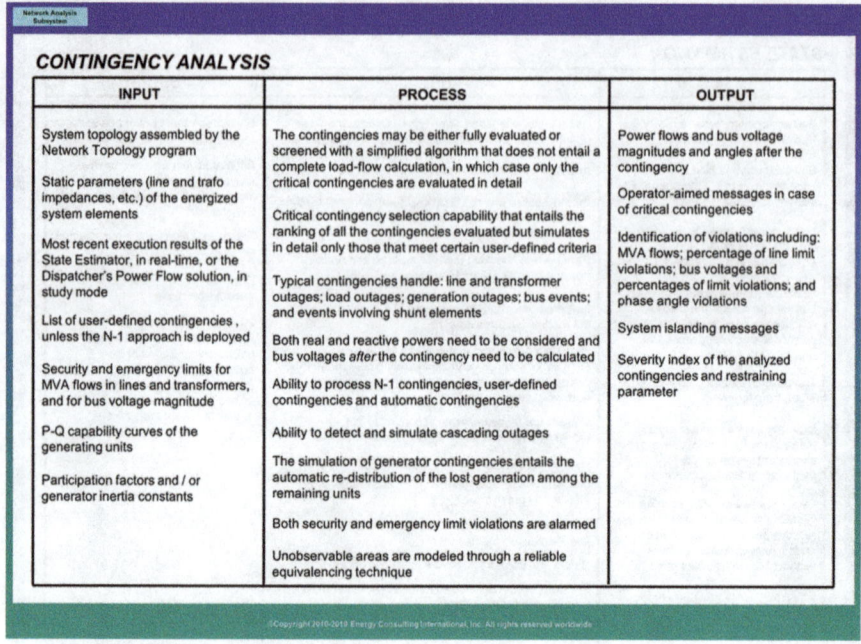

Fig. 1.10 Typical input-process-output specifications for the Contingency Analysis program

the cases that entail the outage of one single network element are processed combinatorially.

This eliminates the need to custom-define a contingency list, but, of course, implies the inconvenience of missing multiple contingencies, such as combined transmission and generation outages, which actually may be much more significant than the outage of just one single power system component.

Modern-day SCADA/EMS solutions handle both N-1, user-defined and automatic contingencies, and support the simulation of cascading outages as well.

A set of typical input-process-output specifications for the Contingency Analysis application is depicted in HIPO format in Fig. 1.10.

1.3.2.4 Dispatcher's Power Flow

The Dispatcher's Power Flow is used both to *simulate* the results of planned operating actions without actually implementing them, and as a general-purpose tool for power system analysis in study-mode.

Through the appropriate selection of input data, the Dispatcher's Power Flow program user is able to assess:

- System scenarios that are close or identical to real-time conditions
- Postulated cases developed for future scenarios, and

1 Overview of Security Assessment for Grid Operations

DISPATCHER'S POWER FLOW

INPUT	PROCESS	OUTPUT
System topology assembled by the Network Topology program	Ability to interactively select the location of loads, generation, interchanges, control actions, etc.	Bus voltage magnitudes an angles
Static parameters (line and trafo impedances, etc.) of the energized system elements	Facilities to set-up a base case	MW, MVAr, MVA and Ampere flows in lines and transformers
	Automatically switch the solution algorithm between a fast decoupled load-flow calculation and the full Newton-Raphson routine depending on the convergence or non-convergence of the case	Real and reactive bus injected powers
Most recent execution results of the State Estimator (real-time) or Initial conditions resulting from a previously conducted Dispatcher's Power Flow calculation (study mode)		Losses in lines, transformers and in the entire system
	Both real and reactive powers need to be considered and bus voltages *after* the contingency need to be calculated	Tap position values
Security and emergency limits for MVA flows in lines and transformers, and for bus voltage magnitude	Incorporate an external system equivalencing procedure, such as unreduced load-flow, Ward or REI equivalent, and so on	Reactive compensation in shunt devices and synchronous condensers
External equivalents information	Retrieve bus injections from the most-recent execution of the State Estimator or have them calculated by the Bus Load Forecast service routine	
P-Q capability curves of the generating units		

Fig. 1.11 Typical input-process-output specifications for the Dispatcher's Power Flow program

- Actual cases retrieved from the historical database for study and/or auditing purposes

A set of typical input-process-output specifications for the Dispatcher's Power Flow program is depicted in HIPO format in Fig. 1.11.

1.3.3 Service Routines

1.3.3.1 Bus-Load Forecast

In real-time, the Bus-load forecast generates bus-load values for the substations where metering is temporarily unavailable or just not implemented. These calculated load values are referred to as pseudomeasurements and are part of the input to the State Estimator. In study-mode, the Bus-load forecast service routine computes the individual bus loads to be used on input by the Dispatcher's Load-Flow when simulating postulated system load conditions.

The bus injected loads are calculated as a function of the total system load. Each bus load is construed as the sum of two components: a *conforming* and, respectively, a *non-conforming* load component. The MW part of the conforming load at each bus is modeled using load distribution factors that express the individual bus loads as a percentage of the total system load. The MVAr part is calculated from the MW component by applying the corresponding $\cos \phi$ power factor.

1.3.3.2 Transmission Losses Penalty Factors

The Transmission Losses Penalty Factors service routine uses the most recent State Estimate solution to compute the transmission losses penalty factors and store them in the real-time database for subsequent use by the Economic Dispatch (if implemented). The penalty factors are derived from the sensitivity factors computed from the transposed Jacobian matrix corresponding to a converged state estimate. The real-time calculation of penalty factors was useful at the time when utilities were vertically integrated and the economic dispatch capability was seamlessly integrated with the Automatic Generation Control (AGC) but the advent of electricity markets has changed all of this and rendered this functionality virtually obsolete.

1.3.4 Non-Standard Network Analysis Applications

1.3.4.1 Fast Computation of the Risk of Blackout

The concept of monitoring the risk of blackout by using a fast-computational tool piggybacked to the static security assessment system was introduced in an EPRI project [14] and validated by a US utility [15] in the early 1990s. The first production-grade installation came in 2002 [16], but the industry's real motivation to accept this approach was triggered by the August 14, 2003 blackout in the United States and Canada [17].

The Fast Computation of the Risk of Blackout was initially integrated with the standard network analysis functionality in Europe and US [18–22], but eventually was embedded into a DSA package by Siemens[12] [11–13]—perhaps because of the perception that a program that computes the distance to instability should be positioned as a DSA feature.

Be it as it may, the "distance" from the current state to a calculated point where voltages may collapse and generating units may get out of synchronism needs to be recomputed at each run of the network analysis sequence. This is because the value of this "distance to the stability limit" quantifies the risk of blackout, is not fixed, and changes each time the load, topology, and system voltages change—and this is why we are addressing this functionality in this section.

Simplified flow-charts that depict the Fast Computation of the Risk of Blackout in real-time and study-mode are shown in Fig. 1.12 and, respectively, Fig. 1.13. Summary input-process-output specifications for this program are illustrated in HIPO format in Fig. 1.14.

---

[12]The DSA functionality offered by Siemens is known commercially as SIGUARD®

1 Overview of Security Assessment for Grid Operations

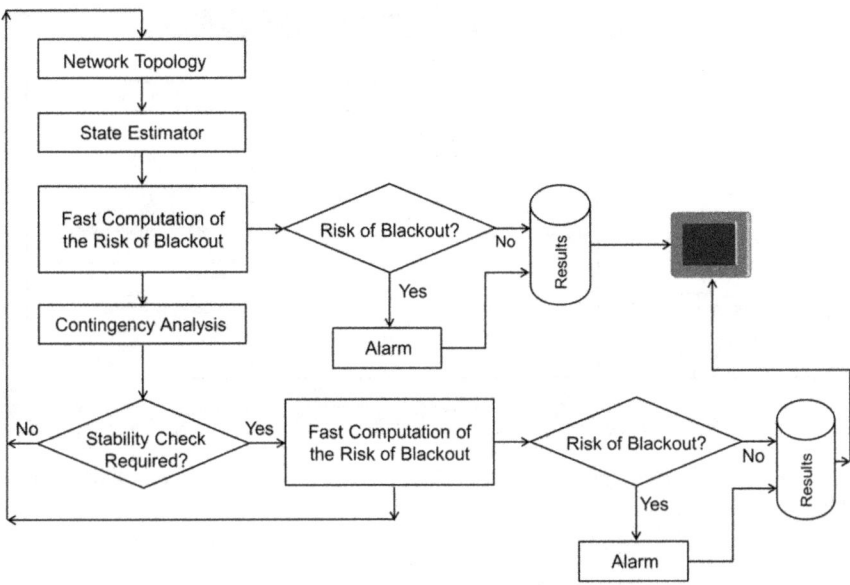

Fig. 1.12 Fast Computation of the Risk of Blackout in real-time

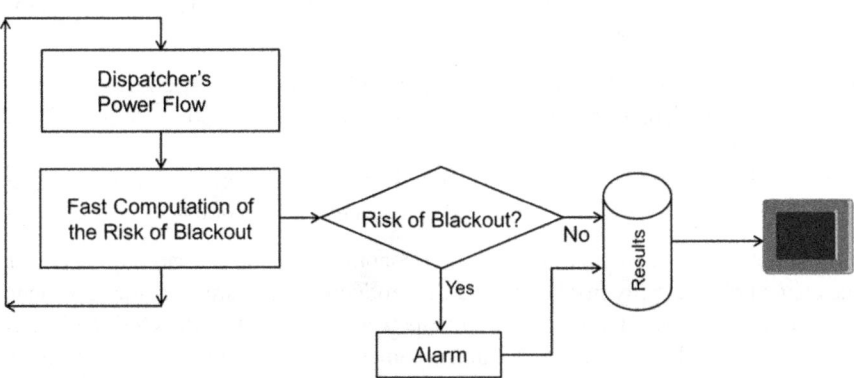

Fig. 1.13 Fast Computation of the Risk of Blackout in study-mode

1.3.4.2 Optimal Power Flow

The software commonly referred to as Optimal Power Flow emerged in the SCADA/EMS landscape in the mid-1980s. It aimed at enhancing the economy and security of power system operations while taking into account the equipment and network constraints.

The stated goal was magnificent, but, in real life, the calculations often diverged and convergence control variables had to be introduced in order to assist the

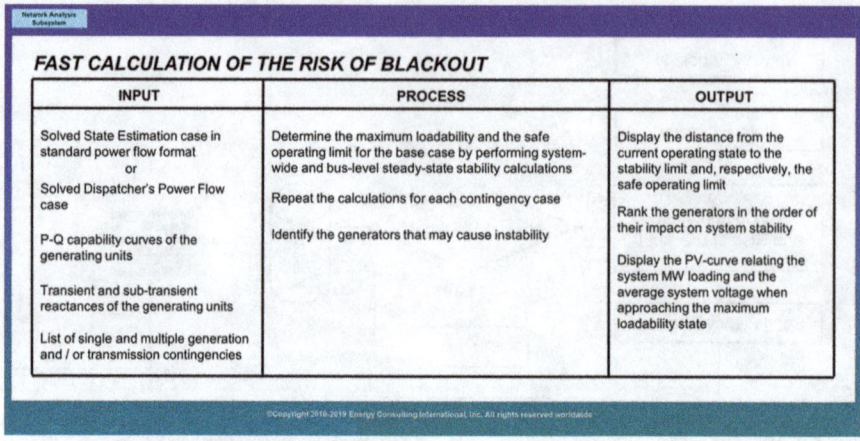

Fig. 1.14 Input-process-output specifications for the Fast Computation of the Risk of Blackout program

optimization algorithm to find a valid solution.[13] This, in turn, complicated this application's operational procedures and reduced the enthusiasm for its deployment. The currently available implementations are designed to segregate the optimization process into two separate capabilities as follows:

- Calculate an optimal state that minimizes the production cost by considering only real power variables and linearized constraints while attempting to find a load-flow solution.
- Compute voltage/reactive power controls that minimize transmission losses while observing t network constraints and reactive power limits.

The former functionality, just like the Economic Dispatch, became irrelevant in the context of electricity markets where the production costs are minimized through market mechanisms. The voltage/reactive power component of the Optimal Power Flow however did get some traction and is sometimes used to improve and maintain the system voltage profile.

---

[13]The Kuhn-Tucker Theorem, which sits at the foundation of nonlinear programming states that, when formulating a minimization problem, both the objective function and the domain of constraints have to be continuous, *convex* and twice differentiable. In reality, the domain of constraints in the optimum power flow problem is: *non-convex*, because of the nonlinearity of the complex voltage variables; and *non-continuous*, since many potential solution vectors are either unstable or physically unfeasible. This explains why, regardless of the technique deployed to solve the optimum power flow problem, there is no guarantee that a global optimum can be reached, assuming of course that a valid solution could be identified

1.4 Quick Glance at Additional Security Assessment Topics

1.4.1 The Impact of Phasor Measurements

Phasor measurements, which were introduced in the early 1980s mainly due to the visionary work of Arun Phadke [23] and are currently being deployed extensively in the industry, can be used to detect possible system separation and system oscillations in close to real-time. They give information on a millisecond time frame that can assist in the rapid detection of system separation—and can help enhance the modeling of external areas since the availability of synchronized phasor measurements can simplify and improve the external area models [9].

It has been advocated that PMUs can help prevent blackouts, but this a little bit of a stretch, to put it mildly, because phasor measurements are just another mechanism for monitoring system variables and do not incorporate the computational capabilities needed to *predict* states that are not yet there. In other words, one still has to perform some analysis to determine where the system will be at a future time, for the raw data, no matter how accurate and precise, are just raw data and cannot anticipate anything—and if they were collected when the system was collapsing, it would obviously be too late to do anything whatsoever. However, PMU data can be and are used for the *postmortem analysis* of system events that led to service interruptions.

In theory, also, the direct measurement of *all* the voltage and current phasors *throughout the entire network* can provide for a complete and error-free measurement set thus eliminating the need for state estimators and, in an ideal case, can give a complete picture of the system on a milliseconds (2 up to 5 cycles) time frame. But measurement errors and bad or missing data are still part of real-life and can only be identified by state estimation. Nevertheless, the latter would become simpler if PMU data were used since a linear model would suffice [24].

At the current time, phasor measurements are not fully integrated with the standard SCADA/EMS system; rather, a separate Phasor Data Concentrator is used to receive and process phasors and send the data archived to the SCADA/EMS database (see [25, 26] for some recent examples). Another potential hurdle comes from the fact that phasor measurements, on the one hand, and the SCADA data, on the other, are collected within time frames that differ by two orders of magnitude: milliseconds for the former, and seconds for the later.

There are other potential PMU applications in control centers that have not been realized in practice and remain a subject of active research. However, installation of phasor measurement units is becoming more common and their use can only increase, especially as the cost of these units continues to fall.

1.4.2 Stability Limits in Contingency Analysis

There are many stability tools that may be used for a broad range of purposes. The standard method is to run detailed transient stability checks for a set of contingencies and assess whether the post-contingency states are stable or not. This is the typical off-line approach and is being done across the industry on a standard basis.

In real-time however the objective is *not* to find out whether the system is stable or unstable, for instability means blackout and, quite obviously, the very *existence* of the current operating state implies that the system *is* stable[14]; rather, the objective in real-time is to ascertain whether the current operating conditions are *at risk of deteriorating* into a widespread failure. In other words, the immediate goal is to identify the stability limit[15] that corresponds to the current operating state—and to do it quickly enough so that the results could be used for online decision-making.

Due to a number of intrinsic algorithmic and modeling difficulties, which are extensively addressed in [10, 27, 28] and related references, the scope of DSA in system operations reflects a compromise between the:

- Depth and extent of the stability analysis.
- Level and granularity of the modeling details.
- Need and/or ability to seamlessly integrate the stability computational process, or processes, with the SCADA/EMS platform.
- Acceptable elapsed times for performing the calculations and presenting the results.

In terms of computational speed and implementation complexity, the methods range between:

- At one end of the spectrum, the fast-computational approach discussed earlier in Sect. 1.3.4, which is the fastest and the easiest to deploy, and,

---

[14] Let us mention *en passant* the ill-advised, yet relatively widespread, practice of "assessing stability" by running load-flows at successively increased load levels and stopping when the load-flow diverged. While it is true that Newton-Raphson load-flows diverge near instability, they may diverge for many other reasons and the state of maximum power transfer has probably been reached *before* the load-flow diverged. Sauer and Pai [33] demonstrated conclusively that "*for voltage collapse and voltage instability analysis, any conclusions based on the singularity of the load-flow Jacobian would apply only to the voltage behavior near maximum power transfer. Such analysis would not detect any voltage instabilities associated with synchronous machines characteristics and their controls.*"

[15] Conceptually, the "stability limit" is a function of the system state vector: for each new system state, there is a new stability limit. But not even the stability limit associated with the current or post-contingency operating state is unique, for it depends upon the trajectory followed throughout the computational search. Simply stated, "stability limits" exist; are not fixed; change with the system's loading, topology, and voltage profile; and depend upon the procedure used to stress the system conditions until instability has been reached. It is precisely this dynamic nature of the "stability limits" that makes it necessary to recompute and track them online. An extensive theoretical discussion of this topic is provided in [10] and related references

- At the other end, the comprehensive transient *and* voltage stability solutions documented in [29–32], among others, which are time and resource intensive and may require the deployment of dedicated servers and workstations *in addition* to the existing SCADA/EMS equipment.

The above considerations about stability limits are, of course, just a few cursory remarks.

The remaining chapters of this book are solely dedicated to DSA and address implementation details of, and practical experience results with a number of solutions that have been developed to date in the industry.

References

1. Report to the President by the Federal Power Commission on the Power Failure in the Northeastern United States on November 9–10, 1965, Federal Power Commission, December 6, 1965, http://blackout.gmu.edu/archive/pdf/fpc_65.pdf
2. A.G. Phadke, A.F. Gabrielle Half a Century of Computer Methods in Power System Analysis, Planning and Operations. Part I: Glenn W. Stagg: His Life and His Achievements—paper PSE-11 PSCE0208 presented at the IEEE Power Systems Conference & Expo PSCE'11, Phoenix, AZ, March 20–24, 2011
3. "Prevention of Power Failures, An Analysis and Recommendations Pertaining to the Northeast Failure and the Reliability of US Power Systems", A Report to the president by the Federal Power Commission, July, 1967., http://blackout.gmu.edu/archive/pdf/fpc_67_v1.pdf
4. IEEE PES Task Force on Real-time Contingency Analysis, "Real-time Contingency Analysis", Final Report, Power System Operation, Planning and Economics (PSOPE) Committee, Bulk Power System Operations Subcommittee, August (2019)
5. IBM Corporation, *"HIPO—A Design Aid and Documentation Technique", Publication Number GC20–1851* (IBM Corporation, White Plains, NY, 1974)
6. T.E. Dy Liacco, Design elements of the man-machine interface for power system monitoring and control, in *Computerized Operation of Power Systems*, ed. by S. C. Savulescu, (Elsevier Publishing, Amsterdam, 1976), pp. 20–33
7. R. Rice and G.W. Stagg, "Application Program Design Criteria for the Southern Company Control System", 8th Power Industry Computer Applications (PICA) Conference Proceedings, June 1973, Minneapolis, MN, pp. 128–134
8. L.W. Coombe, M.K. Cheney, D.C. Wisneski, and G.W. Stagg, "Interactive Load Flow System", 9th Power Industry Computer Applications (PICA) Conference Proceedings, June 1975, New Orleans, LA, pp. 96–104
9. S.C. Savulescu, S. Virmani, The real-time and study-mode data environment in modern SCADA/EMS, in *Real-Time Stability Assessment in Modern Power System Control Centers*, ed. by Savulescu, (John Wiley & Sons and IEEE Press, New York, NY, 2009)
10. S.C. Savulescu, Overview of key stability concepts applied for real-time operations, in *Real-Time Stability Assessment in Modern Power System Control Centers*, ed. by Savulescu, (John Wiley & Sons and IEEE Press, New York, NY, 2009)
11. R. Eichler, R. Krebs and M. Wache, "Early Detection and Mitigation of the Risk of Blackout in Transmission Grid Operation", CIGRE International Symposium "The Electric Power System of the Future—Integrating Supergrids and Microgrids" that will be held in Bologna, Italy, September, 2011
12. B.O. Stottok, R. Eichler, "Visualizing the Risk of Blackout in Smart Transmission Grids", CIGRE International Symposium Smart Grids: Next Generation Grids for New Energy Trends, Lisbon, Portugal, April 2013

13. R. Eichler, C.O. Heyde, B.O. Stottok, Composite approach for the early detection, assessment and visualization of the risk of instability in the control of smart transmission grids, in *Real-Time Stability in Power Systems. Techniques for Early Detection of Blackouts*, 2nd edn., (Springer, New York, NY, 2014)
14. S.C. Savulescu, M.L. Oatts, J.G. Pruitt, F. Williamson, R. Adapa, Fast steady-state stability assessment for real-time and operations planning. IEEE Trans. Power Syst. **8**, 1557–1569 (1993) T-PWRS
15. R.S. Erwin, M.L. Oatts, S.C. Savulescu, Predicting steady-state instability. IEEE Comput. Appl. Power **7**(3), 10–15 (1994)
16. L.A. Gonzalez, "Post-Facto Analysis of a Near-Blackout Event", Presented on behalf of ETESA, Panama, at the 7th International Workshop on Electric Power Control Centers, May 25–28 2003, Ortisei, Italy
17. U.S.-Canada Power System Outage Task Force, "Final Report on the August 14, 2003 Blackout in the United States and Canada: Causes and Recommendations", Energy.gov - Office of Electricity Delivery & Energy Reliability (Report). U.S./Canada Power System Outage Task Force. United States Department of Energy. 2004
18. H.S. Campeanu, E. L'Helguen, Y. Assef N. Vidal, "Real-time Stability Monitoring at Transelectrica", Paper PSCE06–1288 presented at the "Real-time Stability Applications in Modern SCADA/EMS" Panel, IEEE Power Systems Conference & Exposition 2006 (IEEE PSCE'04), Atlanta, GA, October 29–November 2, 2006
19. S. Virmani, D. Vickovic, "Real-time Calculation of Power System Loadability Limits", IEEE Powertech 2007 Conference, Lausanne, Switzerland, July 1–5, 2007
20. D. Vickovic, R. Eichler, Real-time stability monitoring at the independent system operator in Bosnia and Herzegovina, in *Real-Time Stability Assessment in Modern Power System Control Centers*, (John Wiley & IEEE Press, New York, NY, 2009)
21. L.E. Arnold, J. Hajagos, "LIPA Implementation of Real-time Stability Monitoring in a CIM Compliant Environment", Paper PSE-09PSCE0253 presented at the "Real-time Stability Assessment in Modern Power System Control Centers" Panel, IEEE Power Systems Conference & Exposition 2009 (IEEE PSCE'09), Seattle, WA, March 15–18, 2009
22. L.E. Arnold, J. Hajagos, S.M. Manessis, A. Philip, LIPA implementation of real-time stability monitoring in a CIM compliant environment, in *Real-Time Stability Assessment in Modern Power System Control Centers*, (John Wiley & IEEE Press, New York, NY, 2009)
23. A.G. Phadke, Synchronized phasor measurements in power systems. IEEE Comput. Appl. Power **6**(2), 10–15 (1993)
24. M. Zhou, V.A. Centeno, J.S. Thorp, A.G. Phadke, An alternative for including phasor measurements in state estimators. IEEE Trans. Power Syst. **21**(4), 1930–1937 (2006)
25. D. Atanackovic, J.H. Clapauch, G. Dwernychuk, J. Gurney, H. Lee, "First Steps to Wide Area Control", both in IEEE Power and Energy Magazine, pp 61–68, IEEE Power and Energy Magazine, January/February 2008
26. D. Novosel, V. Modani, B. Bhargava, K. Vu, J. Cole et al., "Dawn of Grid Synchronization", pp 49–601, IEEE Power and Energy Magazine, January/February 2008
27. S.C. Savulescu, *Real-Time Stability Assessment in Modern Power System Control Centers (Editor)* (John Wiley & IEEE Press, New York, NY, 2009)
28. S.C. Savulescu, *Real-Time Stability in Power Systems. Techniques for Early Detection of Blackouts (Editor)*, 2nd edn. (Springer, New York, NY, 2014)
29. J. Jardim, Online security assessment for the Brazilian system—a detailed Modeling approach, in *Real-Time Stability Assessment in Modern Power System Control Centers*, ed. by Savulescu, (John Wiley & Sons and IEEE Press, New York, NY, 2009)
30. S.J. Boroczky, Real-time transient security assessment in Australia at NEMMCO, in *Real-Time Stability Assessment in Modern Power System Control Centers*, ed. by Savulescu, (John Wiley & Sons and IEEE Press, New York, NY, 2009)
31. K. Morison, L. Wang, F. Howell, J. Viikinsalo, A. Martin, Implementation of online dynamic security assessment at Southern Company, in *Real-Time Stability Assessment in Modern Power System Control Centers*, ed. by Savulescu, (John Wiley & Sons and IEEE Press, New York, NY, 2009)

32. G. Beissler, O. Ruhle, R. Eichler, Dynamic network security analysis in a load dispatch Center, in *Real-Time Stability Assessment in Modern Power System Control Centers*, ed. by Savulescu, (John Wiley & Sons and IEEE Press, New York, NY, 2009)
33. W.P. Sauer, M.A. Pai, Relationships between power system dynamic equilibrium, load-flow, and operating point stability, in *Real-Time Stability in Power Systems*, ed. by Savulescu, (Springer Verlag, Norwell, MA, 2006), pp. 1–30

Chapter 2
Basics of Voltage Stability Assessment

Amarsagar Reddy Ramapuram Matavalam, Alok Kumar Bharati, and Venkataramana Ajjarapu

2.1 Introduction

From the advent of power generation and induction machines in the 1800s, the power system has been evolving. The power system at large consisted of bulk power generation, transmission system that transported the bulk electrical energy to the load centers, and distribution systems that distributed the bulk energy to individual loads. Power system is probably the biggest machine invented by man. The power grid is a complex interconnected system that spans the geographies of countries and continents. One can trace an electrical path between any two devices connected in the electric grid of these countries or continents. Ensuring the entire grid operates in stable operating conditions under large or small disturbances involves assessment of the operating conditions of the power grid. Since the power system is so large and complex, the power system stability or instability is not a simple classification. Power system instability can occur due to various reasons and can be controlled in various ways. The power system studies used some assumptions like the passive distribution systems: these were pure consumers of electrical energy and that their behavior was well known based on the seasons and the types of loads connected in the system. Until recently, the system had not changed much, and hence the assumptions made about the system and the individual components worked well.

More recently, over the past few decades, with the introduction of distributed generation, increased renewable generation, and fast-changing nature of load is forcing the power system community to reevaluate these assumptions for the various power system studies. In this chapter, we will address the basics of voltage stability

---

A. R. Ramapuram Matavalam · A. K. Bharati · V. Ajjarapu (✉)
Department of Electrical and Computer Engineering, Iowa State University, Ames, IA, USA
e-mail: amar@iastate.edu; alok@iastate.edu; vajjarap@iastate.edu

assessment in power systems and some novel methods that are being proposed in the literature that are important under the recently changing distribution systems.

2.1.1 Power System Stability Classification

Stability of a system is a condition of equilibrium between internal and external forces in the system. The ability of the system to return to an equilibrium after a small or large disturbance is often embedded while defining the stability of the system. The stability in power system deals with the various phenomena and components of the power systems that can drive the power system to instability which means drive the system to an operating point from where the system cannot return to its normal operating conditions. The major classification of the power system stability is addressed by the subject matter experts from around the globe that form the IEEE-CIGRE joint task-force has classified the power system stability into *voltage stability, frequency stability, and rotor angle stability* [1]. This classification is classical and is continuously evolving with the newer additions of power system components like distributed energy resources (DERs) leading to newer kinds of instabilities manifesting in the power system operations. The definitions and classification were recently updated with the converter/inverter control instabilities that arise from high penetration of DERs or inverter-based resources (IBRs) [2]. The definitions and classification detailed in reference [1] were updated in the IEEE taskforce report on "Stability definitions and characterization of dynamic behavior in systems with high penetration of power electronic interfaced technologies" published in [3].

2.1.2 Voltage Stability

Voltage stability in power systems is the ability of the system to maintain voltages at normal acceptable values at all the nodes in the system at a given operating condition or after a disturbance.

This chapter deals with the fundamentals of voltage stability assessment. This chapter will address some fundamentals of what voltage stability means and how the voltage instability can manifest due to various reasons. Voltage stability assessment in power systems is done differently for long-term and short-term voltage stability. Both, large and small disturbances can result in either short-term or long-term voltage instability in power systems. Voltage instability in a system begins to manifest when there is a continuous drop in voltages or a progressive droop in the bus voltages caused due to a disturbance or increase in the load or a change in operating condition.

We know from fundamental power flow equations that in power systems, the voltage and reactive power are closely related and are coupled. Therefore, the voltage instability problem is usually due to the inability of the system to meet the reactive power demand. Of course, the real power, the network impedances and many other parameters affect the voltage stability of the system, but, fundamentally it is the reactive power demand with losses that are the most crucial when addressing voltage stability problems or voltage instability. The voltage stability of a system is classified into long-term voltage stability and short-term voltage stability. The terms long-term and short-term are defined with respect to the time taken by the system to enter voltage instability after a disturbance or a change in operating point. Further sections address these two kinds of voltage stabilities and the fundamentals of voltage stability assessment for both these kinds of voltage instability.

2.2 Long-Term Voltage Stability

The long-term voltage stability is defined as time scales of a few minutes to few tens of minutes (rarely it can extend to few hours if the voltage instability goes undetected). Voltage collapse is a classic case of long-term voltage instability. The main reason for voltage collapse is the loss of generation or increase in load. The voltage collapse is usually caused due to the saddle-node bifurcation in the system. To understand this phenomenon mathematically, we will look at the basics of the bifurcation theory applied to voltage stability.

2.2.1 Maximum Loading (P-V Curve and Q-V Curve)

Long-term voltage instability or voltage collapse usually occurs due to saddle-node bifurcation. Let us consider a 2-bus system as shown in Fig. 2.1.

The generator terminal voltage and the line reactance is constant. The resistance of the line is assumed to be negligible. The power flow equations for this system can be written as follows:

$$P = \frac{EV}{X} \text{Sin}(\theta) \tag{2.1}$$

$$Q = \frac{EV}{X} \text{Cos}(\theta) - \frac{V^2}{X} \tag{2.2}$$

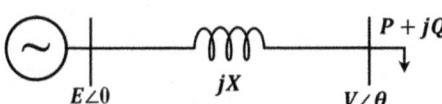

Fig. 2.1 A 2-Bus system with load connected to generator through a transmission line

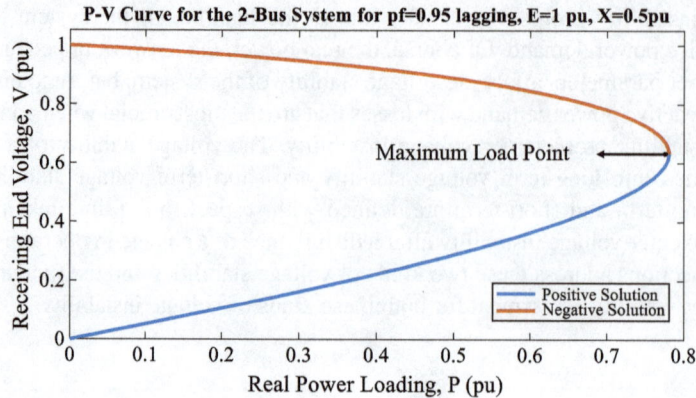

Fig. 2.2 Voltage stability curve ($\lambda - V$ or P-V Curve) for the 2-Bus case

We can write these two equations as functions of the state variables and parameters. The load P, Q can be related with the power factor of the load.

$$\beta = \tan \Phi; \text{ where, } \phi \text{ is the load power factor angle} \tag{2.3}$$

$$\Rightarrow Q = \beta P \tag{2.4}$$

Using this relation, we square Eqs. (2.1) and (2.2) and simplify to eliminate θ. We get a bi-quadratic equation in V as shown in Eq. (2.5)

$$\left(V^2\right)^2 + \left[2P\beta X - E^2\right]V^2 + P^2 X^2 \left[1+\beta^2\right] = 0 \tag{2.5}$$

This can be used to plot the P-V curve, that is using the solution of Eq. (2.6)

$$V = \sqrt{\frac{-\left[2P\beta X - E^2\right] \pm \sqrt{\left[2P\beta X - E^2\right]^2 - 4\left[1+\beta^2\right]}}{2}} \tag{2.6}$$

By varying P and β we can plot P-V curves. A specific case is shown here for power factor = 0.95 lagging, $X = 0.5$ pu and $E = 1$ pu. The positive solution corresponds to the "+" and the negative solution corresponds to the "-" of the "±" in Eq. (2.6). The variation of P and V is shown in Fig. 2.2.

From a given point to the point of saddle-node bifurcation, the power is called the voltage stability margin or the loading limit of the system for the given operating conditions. Researchers have established that voltage instability is mainly caused due to the saddle-node bifurcation in the system [4–9].

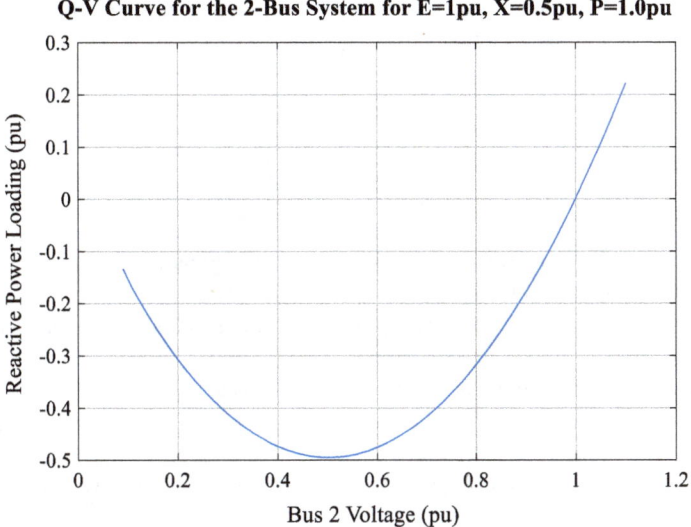

Fig. 2.3 Q-V curve for the 2-Bus case for $P = 0.1$, $E = 1$ pu, $X = 0.5$ pu

Using the solution from Eq. (2.6), for a given P, the angle θ can be calculated from Eq. (2.1) and using that, we can calculate the Q from Eq. (2.2). This results in a Q-V curve shown in Fig. 2.3.

The Q-V curves are typically used for a particular location of interest for locating a synchronous condenser or other reactive power sources. These can be typically drawn with a power flow program. These are typically easier than plotting P-V curves, because in case of P-V curves, at the saddle-node bifurcation point, the power flow solution cannot be solved. The Q-V curve needs one power flow solution and the variation in voltage establishes the amount of reactive power injection required.

2.3 Power Flow Divergence and Instability

The power system is a nonlinear dynamical system. The details of the dynamics and stability are dealt with in great detail by researchers in [4–9]. A bifurcation is an acquisition of a new quality by the motion of a dynamical system, caused by small and smooth changes in its parameters. A power system when undergoes a bifurcation, generally evolves into undesirable states. A saddle-node bifurcation occurs when there is disappearance of an equilibrium caused due to a zero eigenvalue, i.e., an eigenvalue at the origin.

Consider the dynamical power system representations in the mathematical form of differential algebraic equations given by Eq. (2.7)

$$\dot{x} = \underline{F}\left(\underline{x}, \underline{y}, \underline{\lambda}\right)$$
$$0 = \underline{G}\left(\underline{x}, \underline{y}, \underline{\lambda}\right) \tag{2.7}$$

In Eq. (2.7),

x represents the state variables of the system like generator rotor angle, speed, dynamic load variables, etc.

y represents the algebraic state variables like voltages and angles at each bus in the system.

λ represents the real and reactive power injections at each bus.

The function F represents the differential equations for the dynamic components in the power systems.

The function G represents the power flow equations and few other algebraic equations in the power system.

The unreduced Jacobian of the system represented by Eq. (2.7) is given as:

$$J_{DAE} = \begin{bmatrix} \underline{F}_X & \underline{F}_Y \\ \underline{G}_X & \underline{G}_Y \end{bmatrix} \tag{2.8}$$

$$\begin{bmatrix} \Delta \dot{x} \\ 0 \end{bmatrix} = J_{DAE} \begin{bmatrix} \Delta x \\ \Delta y \end{bmatrix} \tag{2.9}$$

Assuming $\underline{G}_Y$ is non-singular, we may reduce Eq. (2.9) by eliminating $\underline{\Delta y}$, which results in the reduced Jacobian matrix and is a Schur's complement. Eq. (2.10) and (2.11) are a result of this elimination of $\underline{\Delta y}$ in the $\underline{\Delta \dot{x}}$ expression:

$$\underline{\Delta \dot{x}} = \left[\underline{F}_X - \underline{F}_Y \underline{G}_Y^{-1} \underline{G}_X\right] \underline{\Delta x} \tag{2.10}$$

$$\underline{A} = \left[\underline{F}_X - \underline{F}_Y \underline{G}_Y^{-1} \underline{G}_X\right] \left(\text{Schur's Complement}\right) \tag{2.11}$$

From Eq. (2.10) we can clearly see that singularity of $\underline{G}_Y$ causes bad things to happen. Therefore, singularity of $\underline{G}_Y$ is directly associated to instability. The power flow Jacobian J_{LF} is part of $\underline{G}_Y$; $\underline{G}_Y = \begin{bmatrix} D_1 & D_2 \\ D_3 & J_{LF} \end{bmatrix}$. Reference [5] explains under special cases, the $\underline{G}_Y$ is reduced to the power flow Jacobian J_{LF} and singularity of the power flow Jacobian directly indicates the instability of power system.

Section 8 of reference [5] provides a detailed explanation that the singularity of the power flow Jacobian is an indication of the instability of the system. The determination of the point of singularity of the power flow Jacobian is not a trivial problem and continuation methods were applied to determine the point of instability. Development of continuation power flow method helped to determine the point of maximum loading [6].

2.3.1 Continuation Power Flow Applied to Determine Voltage Stability Margin

There are many methods applied for voltage stability margin dependent, and one of the most standard methods is the continuation power flow method.

To understand the continuation method, let us consider the following system equation:

$$g(x, \lambda) = 0 \qquad (2.12)$$

In Fig. 2.4, we see if we want to move from solution 1 (x_1, λ_1) to solution 2 (x_2, λ_2), there are multiple ways to accomplish this:

1. We can use a simple straight-line predictor by changing λ to λ_2 and use Newton's method to compute the value of x_2 with x_1 as the initial solution.

$$\begin{aligned} g_x\left(x^i, \lambda_2\right) \times \Delta x^i &= -g\left(x^i, \lambda_2\right) \\ x^{i+1} &= x^i + \Delta x^i \end{aligned} \qquad (2.13)$$

Geometrically, this amounts to approximating the curve first by a straight-line predictor and then correcting it at $\lambda = \lambda_2$.

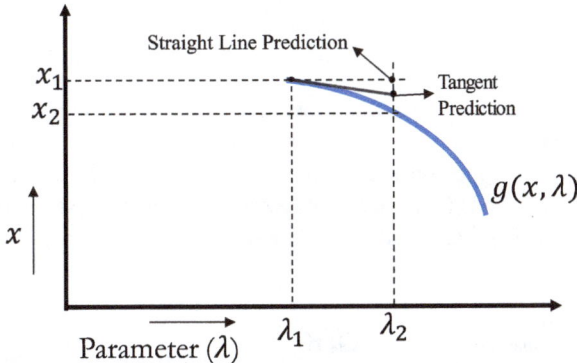

Fig. 2.4 Methods for prediction of the next solution for a change in parameter λ

2. We can use a tangent predictor shown in Fig. 2.4 to arrive at a tangent prediction at $\lambda = \lambda_2$. And then correct it. In this case, as seen, the correction is much smaller than the correction in the straight prediction case.

The continuation power flow method for voltage stability margin assessment uses the second method to determine the maximum load increase parameter λ_{max}.

2.4 Parameter Sensitivity in VSM Assessment

The main parameters that affect the voltage stability margin are the nature of load, losses in the system, and generator limits (generator capability). The fundamental purpose of the power system is to ensure there is electrical energy/power delivered to the load. Due to voltage instability, the power transferred to the load becomes limited and at the voltage collapse, the system is unable to supply power to the load. The long-term voltage stability assessment is directly related to the transfer of power from the generator to the load end. The loads in the power system are located at the far end of the distribution feeders and so it is important to account for the distribution system for this analysis. Let us Consider a 2-bus system with a transmission line, a generator, and a load as shown in Fig. 2.5.

We will vary all the parameters to see how they affect the voltage stability margin of this simple extended:

$$R_T + j\, X_T = 0.03 + j\, 0.3 \text{ pu}$$

$$\text{Base Load} = 30 + j\, 10 \text{ MVA}$$

2.4.1 Static Load Models

The nature of load is one of the most important aspects in voltage stability assessment. The load models for long-term voltage stability assessment are mainly static load models like the constant power loads (P), constant current loads (I), and constant impedance loads (Z). There can be combination of these load types as well which are in the form of ZIP loads.

Fig. 2.5 2-Bus system with a load connected to a generator

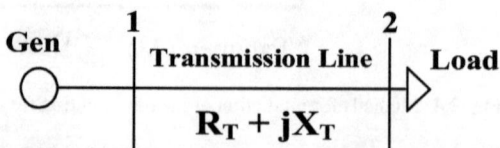

Equation (2.14) represents the ZIP load models. It can be seen from the ZIP load models that the constant impedance loads are proportional to the square of the voltage fraction, the constant current load is proportional to the voltage fraction, and the constant power is not dependent on the voltage. As the voltage decreases, the constant impedance load reduces maximum and then the constant current load and the constant power does not vary.

$$P_{ZIP} = P_0 \left(P_Z \left(\frac{V}{V_0}\right)^2 + P_I \left(\frac{V}{V_0}\right) + P_P \right)$$
$$Q_{ZIP} = Q_0 \left(Q_Z \left(\frac{V}{V_0}\right)^2 + Q_I \left(\frac{V}{V_0}\right) + Q_P \right)$$
(2.14)

Where,
$P_0, Q_0 \rightarrow$ base real and reactive powers of the load
$P_Z, Q_Z \rightarrow$ constant impedance fraction of real and reactive power
$P_I, Q_I \rightarrow$ constant current fractions of real and reactive power
$P_P, Q_P \rightarrow$ constant power fractions of real and reactive power

$$P_Z + P_I + P_P = Q_Z + Q_I + Q_P = 1$$

$$[ZIP] = [P_Z \ P_I \ P_P] = [Q_Z \ Q_I \ Q_P]$$

Therefore, as the load is increased, the voltage drops, and this has an impact on the voltage-dependent loads in turn. However, for the constant power load, as the line voltage drop increases due to higher load, it results in lower voltage at the load which means the line current increases causing the losses to increase further. Therefore, it is expected that the constant power load results in the lowest voltage stability margin (VSM) keeping all other parameters constant and the highest margin will be for the constant impedance load. Under ideal case, the constant impedance will have an infinite margin. To understand this better, let us consider the system shown in Fig. 2.5. We will model the load as ZIP load and consider three ZIP Profiles: ZIP1 = [0.8 0.1 0.1]; ZIP2 = [0.1 0.8 0.1]; and ZIP3 = [0.1 0.1 0.8]. We will use the continuation power flow to understand the influence of nature of load on VSM. Figure 2.6 shows the P-V curves for the load modeled with different ZIP profiles and the corresponding maximum load increase parameter λ_{max}.

2.4.2 Network Impedance

The impedance is responsible for the losses in the system and the transfer of power is directly influenced by the losses. Higher losses imply lower capability of transfer of power from the generator to the load. We will vary the transmission line impedance

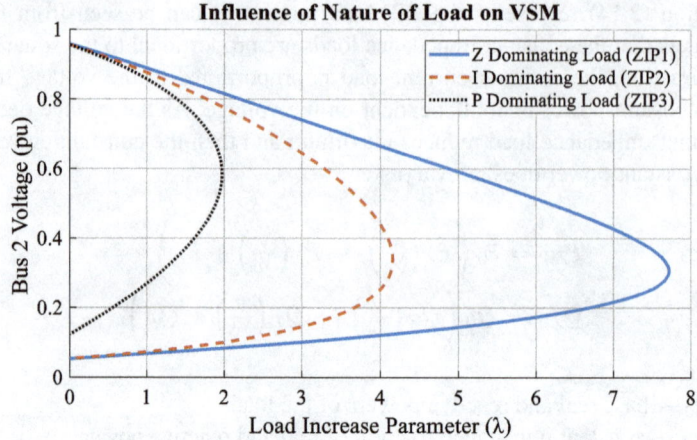

Fig. 2.6 Influence of nature of load on VSM for the extended 2-Bus system

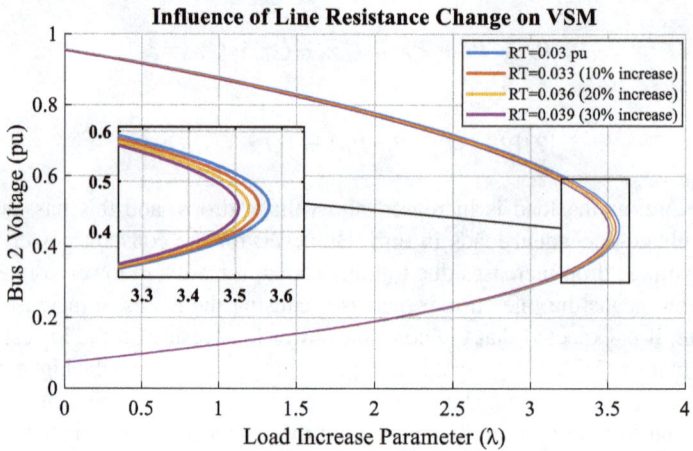

Fig. 2.7 Influence of line resistance on VSM for the extended 2-Bus system

by varying the resistance and reactance separately. And see how they affect the VSM of the system. We will model the load with a ZIP profile of [0.4 0.3 0.3] for all the cases.

We will now see the influence of change in Line reactance on VSM.

From Figs. 2.7 and 2.8, we can see that the change in the line reactance has more influence on the VSM. The reactive power losses are important for voltage stability and since transmission lines usually have low $\frac{R}{X}$ ratios, the real loss impact is lower in the transmission systems compared to the reactive power loss. However, it is important to capture the real and reactive power losses in the systems with significant $\frac{R}{X}$ ratios.

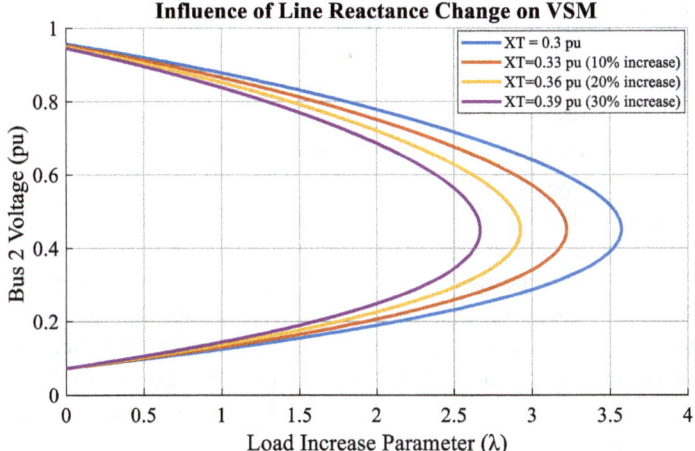

Fig. 2.8 Influence of Line reactance on VSM for the extended 2-Bus system

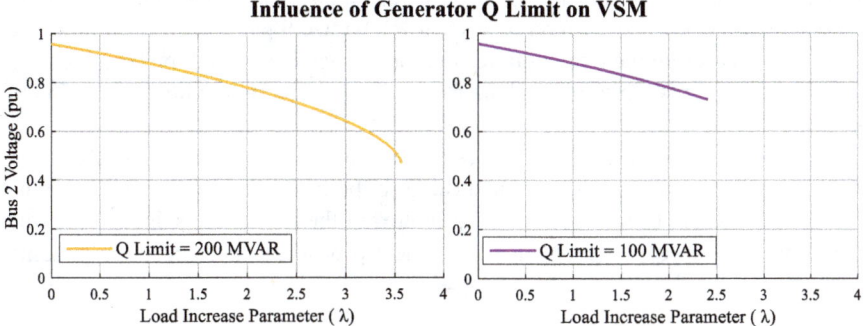

Fig. 2.9 Influence of generator Q_{Limit} on VSM for the extended 2-Bus system

2.4.3 Generator Limits

So far, the reactive power limits of the generator were kept at ±500MVAR. We will modify the reactive power limits of the generator in this section to understand how this influences the VSM assessment. The reactive power limits of the generator directly affect the loading limit as the reactive losses have to be met as the loading increases, the reactive losses increase and if the generator reactive power limit is hit, it is unable to meet the demand of the load any more. The generator reactive power limit is reduced to 200MVAR 100MVAR and the corresponding $\lambda - V$ curves are shown in Fig. 2.9. We can clearly see that the VSM of the system is lower for the case with lesser Q_{limit} in the generator.

2.4.4 Effect of Transformer Taps on VSM Assessment

For the simple 2-Bus system shown above, we add a transformer with taps before the load. The transformer secondary is equipped with taps to help restore the voltage in the distribution system as the load increases for the *P-V* curve tracing. Figure 2.10 shows the modified system:

For the system shown in Fig. 2.10 the load is increased to draw the *P-V* curve for VSM assessment. In this case, the load is modeled as ZIP load with ZIP profile [ZIP] = [0 0 1]. The response of the Tap changer depends on many aspects of the system. A simple case is shown here where the transformer taps are located on the load side and it is similar to an on-load tap changer (OLTC) that helps to restore the voltage on the distribution system side. The taps have a dead-band for voltages and if the voltage is going beyond the dead-band, the taps operate to maintain the voltage within the dead-band.

For this test case, the dead-band of voltages on the load side is 0.9–1.1 pu. The transformers usually have limited taps. In this case as the load is increased, the voltage decreases and since the taps are located on bus 3 and after a certain amount of load increase, the voltage tends to go below 0.9 pu. 0.9 pu is the lower limit of the dead-band, therefore the taps try to restore the voltage back to 0.9 pu by increasing the number of turns. And this is done for further load increase until the maximum number of turns are reached.

Figure 2.11 shows the $P - V$ curves for the cases with tap changer enabled and disabled with the voltages at Bus 2 and Bus 3 for both the cases. The response of the taps is important to be understood in the context of the load models, location for the taps, and location of the controlled bus. The tap positions and the corresponding

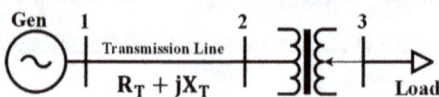

Fig. 2.10 Modified 2-Bus system with taps on the secondary of the substation transformer

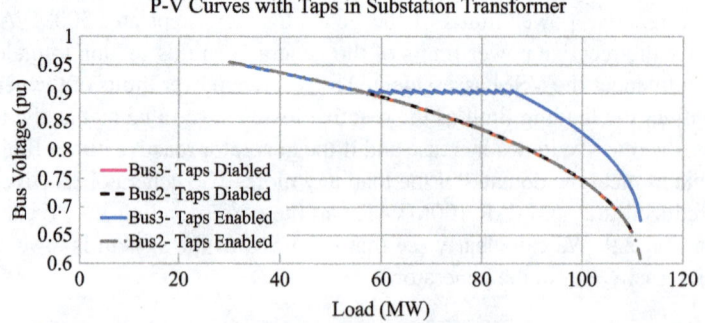

Fig. 2.11 Effect of OLTC tap change on VSM

variations can be captured only through accurate simulation models and correct forms of representing the power flow equations.

As it is seen from Fig. 2.11, even though the taps are enabled, they do not start operating until the voltage has reached the lower limit of the dead-band of the voltage, i.e., 0.9 pu in this case. The taps trying to restore the voltage that enables a higher amount of load increase in the system in this configuration. The taps play an important role in VSM assessment. Reference [8] provides lots of detailed case studies that address the importance of tap changers in VSM assessment.

2.5 Effects of T&D Interactions on Voltage Stability

The recent decades have seen many changes in the way consumers interact with the power grid. The distribution system has seen the integration of various distributed energy resources (DERs). These include electric vehicles, roof-top solar installations, small capacities of distributed wind in the distribution system, battery storage, flexible load, price responsive demand response, etc. as shown in Fig. 2.12.

It is important to understand that modeling distribution system can significantly impact the voltage stability of the system. Traditionally, the distribution systems have been aggregated as a simple load at the load bus in the bulk power system. This is a very drastic method to model the complete distribution system especially for voltage stability studies. Since voltage issues are usually local and require local control, it is important to understand when we model the load and increase the load for voltage stability margin assessment; What does it mean in the real physical world?

The load is located at the terminals of the distribution system feeders. Integration of the DERs in the distribution system has led to situations where there can be two-way power flow in the distribution systems. This is a concern not only for the distribution system operators but also a concern for the transmission system operators as this is largely driven by the renewable uncertainty that causes significant errors in the net-load seen by the transmission system compared to the predicted net-load. This can cause serious issues that are unexpected in the transmission and distribution systems.

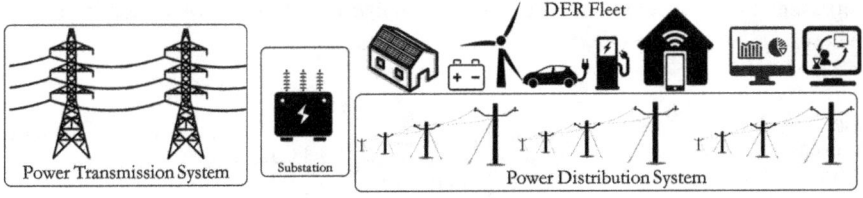

Fig. 2.12 Illustration of various grid-edge technologies (DER Fleet) in the power distribution system

2.5.1 Importance of Modeling Distribution Networks for VSM Assessment

Majority of the DER technologies are integrated into the power system at the distribution level. As discussed earlier, traditionally, the bulk power system studies aggregated the distribution system in the form of a simple constant power load. This is a drastic assumption considering the large number of changes that have occurred in the distribution. North American Reliability Corporation (NERC), a regulating body, also recommends modeling the DERs with as much detail as possible. Lumping the DERs or distributed generation (DG) as negative load is not recommended according to NERC [10]. Federal Energy Regulatory Commission (FERC) has also recently mentioned the need for representing details of the distribution system along with the transmission system for various studies to ensure accurate results and conclusions for planning and operations in power systems [11].

A simple method of modeling the distribution system is to add an equivalent distribution feeder before the load bus at the transmission system. Let us consider extending the 2-bus system with a substation transformer and an equivalent distribution feeder. The equivalent distribution system feeder impedance is calculated based on the IEEE 4-bus distribution system. The load is modeled as a constant power load. The eq. feeder parameters are computed to be $R_D + jX_D = 0.046 + j0.095$ pu. The extended 2-bus system is shown in Fig. 2.13. The load is slightly changed according to the load of the IEEE 4-Bus distribution system. The load is modeled as constant power load of $20 + j10\ MVA$.

Figure 2.14 shows the CPF results for the cases with and without the equivalent distribution feeder (Eq. D-Feeder) and the impedance added due to the D-Feeder clearly has an impact on the VSM of the system. It is important to model the distribution systems for accurate VSM assessment.

Reference [12] summarizes the different methods of representing distribution system and the respective trade-offs for VSM assessment. The main differences between the transmission and distribution systems are the significant real losses in the distribution systems and three-phase unbalanced operation. Table 2.1 shows the main physiognomies of the distribution system that should be accounted for along with the transmission system models for accurate VSM assessment.

It can be seen from reference [12] that the distribution system unbalance is also an important parameter that needs to be accounted for and for this, three-phase representation is important. The load unbalance effect on VSM is demonstrated below:

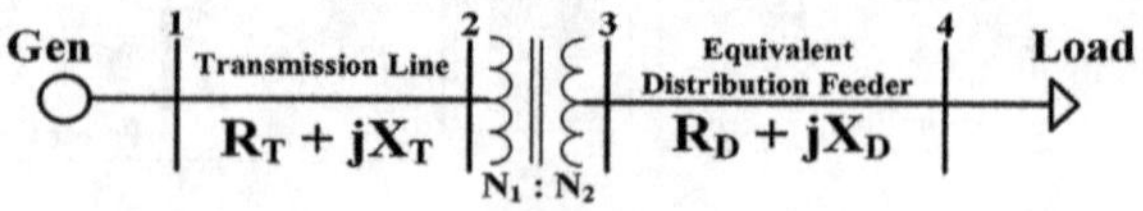

Fig. 2.13 2-Bus equivalent with an equivalent distribution system

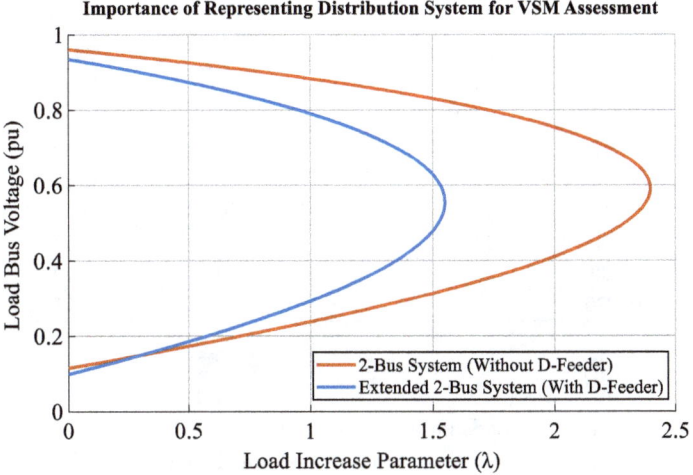

Fig. 2.14 2-Bus equivalent with an equivalent distribution system

Table 2.1 Methods of representing distribution system and the trade-off for VSM assessment [12]

Distribution system Physiognomies ↓	Distribution system physiognomies captured ↓		
	No D-System	Eq. D-Feeder	T&D co-simulation
D-losses	No	Yes (with error)	Yes
D-feeder voltage drop	No	Yes (with error)	Yes
D-feeder segment drop	No	No	Yes
Dist. Unbalance	No	Yes (with error)	Yes
Impact of T on D	No	Yes	Yes

2.5.2 Influence of Load Unbalance on VSM of a System

This is a more recently determined parameter that affects VSM. It directly affects the losses and hence affects the loading limit of the system. To account for the load unbalance in the system, we consider the IEEE 4-Bus distribution system and its losses transferred to the transmission system through the equivalent distribution feeder impedance. Reference [12] provides details of how the unbalance affects the VSM. Like voltage and current unbalance, let net-load unbalance (NLU) be defined as follows:

$$S_{avg} = \frac{S_A + S_B + S_C}{3} \quad (2.15)$$

$$U_i = \frac{S_i - S_{avg}}{S_{avg}} \quad \forall i = A, B, C \quad (2.16)$$

$$\text{NLU} = \max(|U_i|) \times 100\% \quad \forall i = A, B, C \quad (2.17)$$

Table 2.2 Variation of loss as the load unbalance (NLU) increases (same amount of load)

NLU%	Total P loss (W)	P loss/P load (%)	Total Q loss (VAR)	Q loss/Q load (%)
0	419,380	7.77	861,370	32.94
10	428,640	7.94	882,430	33.74
20	460,040	8.52	954,310	36.49
30	514,810	9.53	1,079,410	41.27
40	598,710	11.09	1,270,430	48.58
50	729,460	13.51	1,567,110	59.92
55	830,870	15.39	1,796,630	68.70
60	1,016,220	18.819	2,215,500	84.71

Where,

S_A, S_B, S_C → The net-loads on phases A, B, C.

NLU % → Percentage of maximum net-load unbalance.

The Loss is computed for the standard IEEE 4-Bus system with constant power loads for various load unbalance levels and the results demonstrate how the losses increase with increase in load unbalance. The increase in loss is further extended to show the increase in the effective impedance of the equivalent distribution feeder and thereby its effect on the VSM of the system. The load on the system is modeled as constant power loads to ensure the variation in losses observed is due to the load unbalance and no other parameters influence the increase in the losses.

The load is varied by varying the load on phase 'A' and 'C' to create an unbalance in the IEEE 4-Bus distribution system. Care is taken to ensure the total three-phase load is kept the same. The power factor of the load is also kept same ensuring the reactive power is also constant and the unbalance in the real and reactive powers are the same. Table 2.2 shows the results of the IEEE 4-Bus system loss for various percentages of NLU. We can see that the real and reactive losses increase with the load unbalance. The loss is expressed as percentage of load also and it can be seen for higher NLU, loss percentage is much higher than that for a lower NLU%. It can also be seen that the reactive loss % for a higher NLU are much higher than the real power losses and this is very important as this has a significant impact on the overall voltage stability margin of the system and this can be effectively captured by representing the distribution system in detail.

The Eq. D-Feeder parameters, $R_D + jX_D$, are calculated for the IEEE 4-bus system for various NLU. The results from Table 2.2 are used for determining the corresponding Eq. Feeder Parameters of the eq. distribution system. For each of the NLU case, the Eq. D-Feeder parameters are computed in pu for the IEEE 4-Bus system and are shown in Table 2.3.

We performed the continuation power flow on the extended 2-bus system for the various NLU cases and the results are shown in Fig. 2.15 for some of the cases. The results in Fig. 2.15 shows that the effect of increased eq. D-Feeder impedance is the reduction in margin. In addition, as the unbalance increases the amount of reduction in VSM increases further.

Table 2.3 Eq. D-Feeder parameters for the extended 2-Bus system

NLU%	Real loss (kW)	Reactive loss (kVAR)	$R_D + X_{Di}$
0	419.4	861.4	0.0463 + 0.0951i
10	428.6	882.4	0.0473 + 0.0974i
20	460.0	954.3	0.0508 + 0.1053i
30	514.8	1079.4	0.0568 + 0.1191i
40	598.7	1270.4	0.0661 + 0.1402i
50	729.5	1567.7	0.0805 + 0.1729i
55	830.9	1796.6	0.0917 + 0.1983i
60	1016.2	2215.5	0.1121 + 0.2445i

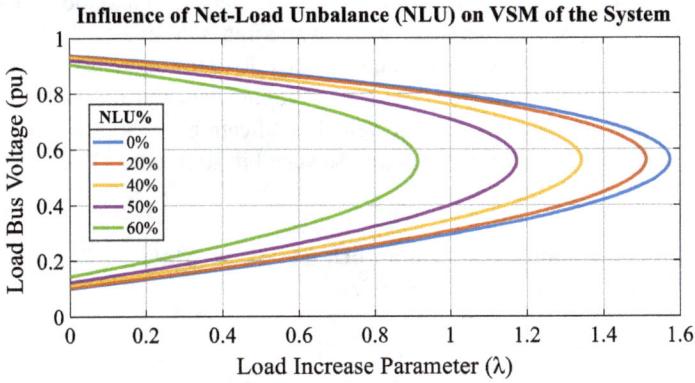

Fig. 2.15 Voltage stability curves for the extended 2-Bus system for various NLU% [12]

The three-phase continuation power flow discussed in [13–14] also demonstrates this for larger transmission system cases where the complete system is modeled in three phases. Much of this unbalance in real life is creeping in from the distribution systems. The distribution system operation is varying fast with the introduction of various grid-edge technologies. The load unbalance coupled with the integration of various distributed generation (DG) or distributed energy resources (DERs) can interact in ways that the net-load unbalance can be significantly high to affect the voltage stability of the system.

Table 2.1 shows that T&D co-simulation is an effective tool to capture both the transmission and distribution system physiognomies for voltage stability assessment and we briefly introduce some preliminary results of using T&D co-simulation for VSM assessment. Reference [12] provides details of the importance of modeling distribution system along with the transmission system for VSM assessment.

2.5.3 T&D Co-Simulation and its Application for VSM Assessment

T&D Co-simulation based on the method of solving coupled systems in a decoupled way. Researchers have been experimenting with different methods of performing T&D co-simulation and some are more practical and seamless than the others. The most commonly used method for T&D co-simulation is based on Master-Slave Splitting where the distribution system substation is the point of coupling between the T&D systems and convergence of solutions from both, the T-System and the D-System is measured based on the Substation voltages and powers. The assumption is that at the substation the voltages are balanced and the unbalance in the distribution is not transferred to the transmission side due to the use of load balancing equipment and reduction in unbalance due to the aggregated effect of the multiple distribution system feeders. Ideally, the unbalance effects are seen in the parts of sub-transmission systems as well. Identification of the boundary bus for T&D co-simulation is very important as also stated in Reference [12].

2.5.3.1 T&D Co-Simulation Framework for Steady-State and Quasi-Steady-State Studies

T&D co-simulation enables detailed modeling of both: the transmission and the distribution systems. This method of simulating the power systems captures all the details of the distribution system along with the inter-dependent nature of the transmission and the distribution systems. The trade-off however is the computational complexity and burden is increased leading to a longer time of simulation. T&D co-simulation employed in large systems requires to use commercial grade solvers for the transmission and distribution systems. The co-simulation method does not require development of new transmission system solvers or distribution system solvers, but, it efficiently integrates existing solvers that can be scaled to large systems in an easy way.

The methodology leveraged for T&D co-simulation is the "Master-Slave" method described by the authors of [15] and [16]. The master-slave method for transmission and distribution (T&D) system co-simulation considers the transmission network as the "Master" and the distribution networks as the "Slave" systems, respectively. Reference [15] is a textbook that discusses the detailed mathematical fundamentals necessary to establish the distributed method of solving coupled problems. Reference [15] discusses various implications of optimizations, dynamic co-simulation, and steady-state co-simulation formulations and the numerical stability of a T&D co-simulation framework that works on "Master-Slave Splitting" (MSS) method. A simple representation of T&D co-simulation framework is shown in Fig. 2.16.

2 Basics of Voltage Stability Assessment

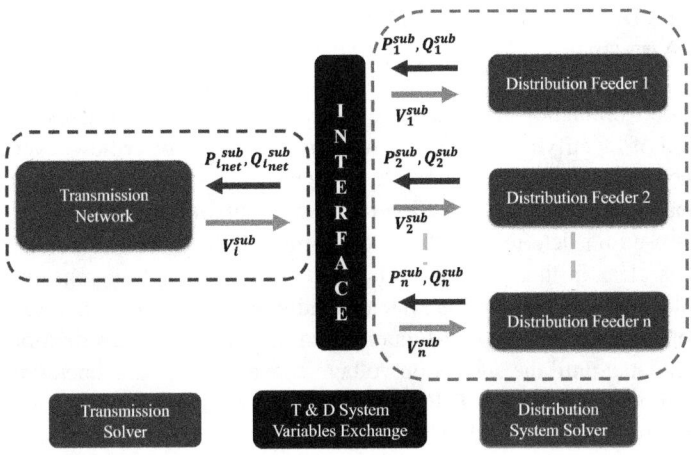

Fig. 2.16 Transmission and distribution co-simulation framework

Reference [12] provides an interface written in python to co-simulate opensource transmission and distribution solvers like Pypower and GridLAB-D. The interface code is written in python and is responsible for exchanging the variables between the transmission and distribution systems. The interface execution does not need any additional software and has built-in features for plotting and can be extended to generate reports as well. Effective T&D co-simulation tightly couples the T&D systems while performing the co-simulation, i.e., the transmission and distribution systems solutions are solved till the substation voltage reaches convergence for a given operating point. The T&D co-simulation framework is also extended to co-simulate commercial solvers on the transmission systems as many of the utilities use the commercial software for their system studies. Reference [12] mentions the use of co-simulating PSSE and GridLAB-D for T&D co-simulation. The salient features of the developed co-simulation interface to co-simulate PSSE and GridLAB-D for steady-state and quasi-steady-state simulations are:

1. The interface is developed using Python which is opensource and provides a simple way to operate GridLAB-D or OpenDSS which are also opensource.
2. The interface is parallel computing compatible and can interface multiple load buses to different distribution systems.
3. The interface can seamlessly integrate with other solvers as well but might need small tweaking to ensure its functionality with the solver. The solver is python version independent.
4. The interface developed can be effectively used for plotting and post simulation report generation.

2.5.3.2 T&D Co-Simulation Tool for Long-Term Voltage Stability Margin Assessment

The T&D co-simulation tool can be effectively used for VSM assessment using the method of identifying the loading limit by means of power flow divergence by slowly increasing the load in the system. Figure 2.17 shows the flowchart for the application of the T&D co-simulation for VSM assessment. The P-V curve tracing method is used for determining the loading limit on the system. The load increase direction is clear in this method as the load increase is not aggregated but each individual load in the distribution system is individually increased. For each loading point or operating point on the *P-V* curve, the transmission and distribution systems are co-simulated until the substation voltage converges, and this operating point is recorded to be plotted on the *P-V* curve. After the operating point is recorded, the load is slightly increased further by a small step and T&D co-simulation is carried out. This is continued till either system reaches its loading limit, i.e., the power flow diverges.

Figure 2.18 provides the preliminary results on a simple test system where the IEEE 9-Bus transmission system is co-simulated without distribution system representation, with eq. feeder and T&D co-simulation for balanced and unbalanced load as per the IEEE 4-Bus feeder datasheet.

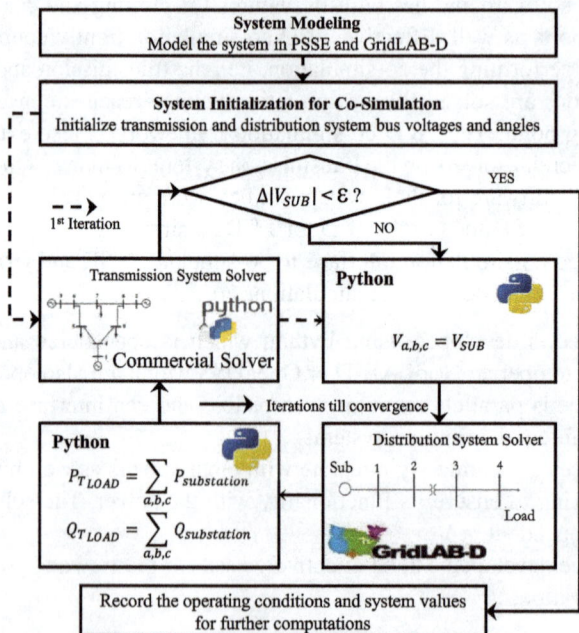

Fig. 2.17 Flow chart for one operating point on *P-V* curve to determine VSM [12]

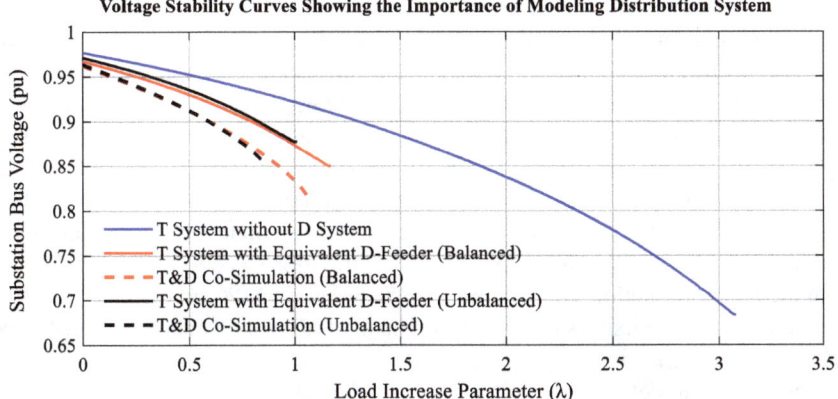

Fig. 2.18 Voltage stability curves with various method of simulations: IEEE 9-Bus transmission + IEEE 4-Node distribution system [12]

Figure 2.18 shows voltage stability curves corresponding different forms of representing the distribution system along with the transmission system for the voltage stability margin assessment. The blue curve is for the case with the transmission system where the distribution system losses are modeled as a part of the load and no distribution system network is represented (The loss is not modeled accurately in the lumped load + loss model). As discussed earlier, the distribution system can be represented as an equivalent feeder as it captures the distribution system physiognomies to a reasonable extent if there is not much unbalance in the system. For the equivalent distribution system feeder (D-Feeder) representation, the D-Feeder parameters are computed for balanced and unbalanced IEEE 4-Bus distribution system. The voltage stability curves for the balanced and unbalanced cases with equivalent feeder method are shown by the red and black curves, respectively. As described in Table 2.1, T&D co-simulation is an effective method to capture the distribution system physiognomies. The voltage stability curves with the T&D co-simulation method with the balanced and unbalanced IEEE 4-Bus distribution system are shown by the red and black dotted curves, respectively. The difference between λ_{max} for the balanced case is not much for the balanced case but for the unbalanced case, there is a significant error. Therefore, for unbalanced distribution systems, using T&D co-simulation is an effective method for VSM assessment.

2.5.3.3 Influence of DER on VSM

The distribution system is changing fast with the integration of various kinds of distributed energy resources (DERs) or distributed generation (DG). The DG when added in the distribution system can aggravate the unbalance in the distribution system if they are not added in a planned manner to ensure there is not high amounts of net-load unbalance. To study this further a simple case study is systematized. In this case study, DG in the form of solar PV inverters are added in the distribution system. The DG is added in different proportions, some extreme three-phase distribution of DG is chosen for 60% of DG added in the distribution system. The system considered is the IEEE 9-bus and IEEE 4-bus distribution system for T&D co-simulation. The DG added in the system is operating in two operating modes: unity power factor (UPF) and volt-VAR control (VVC) mode. The load is modeled with a ZIP profile $[ZIP] = [0.4\ 0.3\ 0.3]$. The distribution system load is unbalanced with the load distribution as: $A = 28.05$ MW; $B = 39.6$ MW; and $C = 52.25$ MW seen at the transmission system load bus. The total MW of DG added $=71.94$ MW (60% of total load).

The distribution of this ~72 MW of DG is different in the different phases in the three-phase system. In one case the DG is distributed in equal proportion of load, i.e., 60% of load on each phase, and, in other cases, there is low DG penetration in one of the phases and the other two phase % is adjusted to have the total DG added in the system to be ~72 MW. The DG penetration is computed with respect to load in that phase. For example, DG penetration in A-phase- %A = 10% means, amount of DG is 10% of phase-A load.

The *P-V* curves for all the cases of DG are shown in Fig. 2.19 which shows that for the same amount of DG added in different proportions, the VSM of the system can be different. This is primarily caused by different amount of NLU in the system, in turn resulting in different amounts of losses. Figure 2.19 also shows that for each case, the VSM for DG in VVC mode is higher than the VSM of DG in UPF mode because in the VVC mode, the smart inverter supplied additional reactive power to maintain the voltage set-point of the inverter. The additional reactive power helps to allow for a further load increase resulting in a larger VSM.

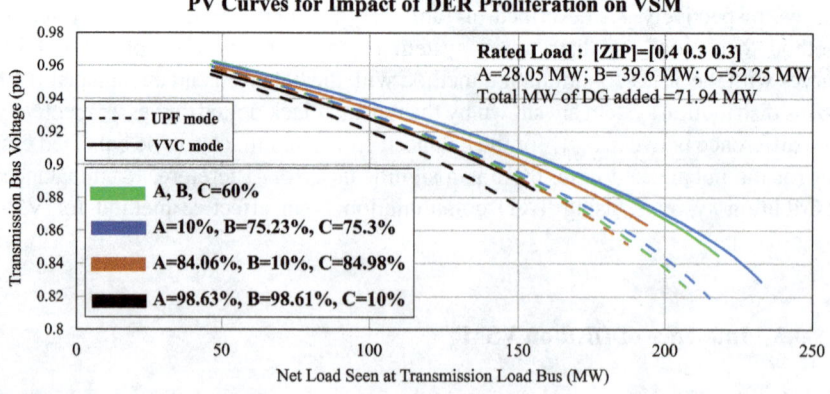

Fig. 2.19 Voltage stability curves for DG proliferation in various proportions—IEEE 9-bus transmission and IEEE 4-bus distribution systems [12]

DG helps increasing the margin but, to determine by how much, depends on how the DG proliferation occurs in the system with respect to the three phases. With the recent amendments to the IEEE 1547 standard, it is important to understand the impact of volt-var control (VVC) on the VSM and the influence of DG distribution on VSM compared to UPF.

2.6 Data-Driven Methods for Long-Term Voltage Stability Assessment

There have been recent efforts done to utilize the online measurements to estimate the margin or an index that can be used as a proxy for the long-term voltage stability. They can be split into methods requiring local measurements and centralized measurements. As the Thevenin methods are of interest in this dissertation, they are discussed in more detail. The main idea behind the Thevenin methods is to estimate an equivalent circuit for the system at the critical load and utilize the ratio between the load impedance and Thevenin impedance as an indicator of long-term voltage stability.

2.6.1 Local Thevenin Equivalent-Based Methods

The early Thevenin methods used only local PMU measurements and independently calculated the Voltage Stability Index (VSI) at each monitoring bus [17, 18]. The VSI was either used to initiate local control actions or transmitted to a centralized location for visualization or control applications. These techniques exploit the high sampling rate of the PMUs (30 samples per second) to capture small variations in the bus voltage at a quasi-steady state operating point and calculate a Thevenin equivalent circuit at each monitored bus. The estimated Thevenin equivalent parameters are then used to calculate the VSI at a bus. To improve accuracy, a multi-bus equivalent is proposed for load areas with several tie-lines [10] and an analytical derivation of the maximum power is used to monitor voltage stability.

One drawback of the local approaches is the reliance on the quasi-steady-state nature of the system. The small variations could be due to a specific phenomenon (forced oscillations, etc.) that skew the measurements and provide a false equivalent. Furthermore, measurement noise in the PMU can cause the LTI to oscillate wildly. This is a well-documented problem and [18] use multiple measurements over a time window to smooth out the errors by mathematical techniques. However, these methods assume a certain noise profile and might not work in presence of certain system behavior. Despite these drawbacks, the simplicity and local nature of these methods make them attractive to utilities and they have been implemented commercially in the field and can trigger emergency corrective actions [20].

2.6.2 Centralized Thevenin Equivalent-Based Methods

The centralized Thevenin methods are calculated at the EMS where the state estimation results and PMU measurements are available for the entire system. Since these methods do not utilize any quasi-steady-state nature of the system for the Thevenin equivalent calculation, they are more robust to noise compared to the local methods. However, the centralized nature means that these methods cannot be used for corrective schemes and instead are best used for preventive schemes. The initial method utilized a simplifying assumption to define the L-index [21], without explicitly calculating a Thevenin equivalent. This idea was formalized by the concept of coupled single-port circuit model [22] which is used to explicitly define a Thevenin equivalent. By utilizing the network equations relating the voltages and currents, the entire system can be equivalently described by an extended Thevenin circuit which includes an extra component (source, load, or impedance) to reflect the coupling with current injections at other load buses and generators. The more recent methods have included the reactive limits into the method by fitting a cubic curve and estimating the generators reaching the limit [23]. [24] presents a method to estimate the maximum power transfer in a transmission corridor utilizing the line admittances. A different paradigm by using the system Jacobian along with the admittance matrix to calculate the Thevenin impedance is proposed in [25]. In the next section, we demonstrate how the sensitivities calculated from the Jacobian are related to the Thevenin Index [26].

2.6.3 Sensitivity-Based Thevenin Index

Fig. 2.20 shows the 2-bus equivalent at a load bus where the rest of the system is reduced into an equivalent voltage E_{th} and an equivalent impedance Z_{th}. At low loading, $|Z_L| > |Z_{th}|$ and at critical loading, $|Z_L| = |Z_{th}|$. Thus, the ratio between $|Z_L|$ & $|Z_{th}|$ can be used as an indication of voltage stability and is referred to as Local Thevenin index (LTI) [17, 18] as shown in (2.20).

In principle, two subsequent phasor measurements of the pair V & I can be used to compute Z_{th} under the assumption that the equivalent parameters do not change

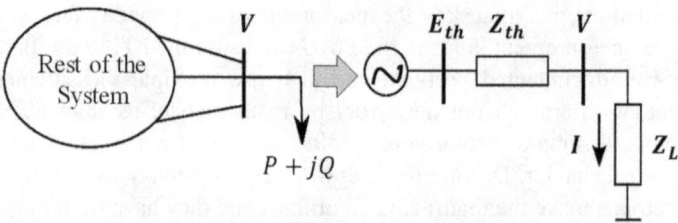

Fig. 2.20 The reduction of the rest-of-the-system into an equivalent Z_{th} and E_{th}

during the time interval between the two subsequent measurements [17]. This assumption is valid when the load increment ($\Delta\lambda$) between the two measurements is as close to 0 as possible. In practice, $\Delta\lambda$ between subsequent measurements is very small (~0.1%) and thus we can use this assumption but theoretically, the ideal value of the Z_{th} is determined when the load increment is as close to 0 as possible (i.e., $\Delta\lambda \to 0$). The conventional Local Thevenin index (LTI) at a load Bus i, uses the Thevenin Impedance and the Load Impedance and can be determined by Eq. (2.18), using two distinct operating points [17].

$$Z_{th_i} = -\frac{V_i^{(2)} - V_i^{(1)}}{I_i^{(2)} - I_i^{(1)}} = -\frac{\Delta V_i}{\Delta I_i} \tag{2.18}$$

$$Z_{L_i} = \left(\frac{V_i^{(1)}}{I_i^{(1)}}\right) \tag{2.19}$$

$$\text{LTI}_i = \left|\frac{Z_{th}}{Z_L}\right| = \left|\frac{\Delta V_i}{V_i^{(1)}}\right| \cdot \left|\frac{I_i^{(1)}}{\Delta I_i}\right| \tag{2.20}$$

The closer the operating points are, the better the estimate of the Thevenin impedance and the accuracy of the LTI. For simplicity, the loading direction is assumed to be proportional to the initial load implying that the $\Delta\lambda$ at all the buses is same. Let the load voltage at the first instance be $(V)e^{j(\theta)}$ and at the second instance can be expressed as $(V + \Delta V)e^{j(\theta + \Delta\theta)}$. As the LTI depends on the $\Delta\lambda$ chosen, it is explicitly written as a function of $\Delta\lambda$, using the expression LTI($\Delta\lambda$) [26].

$$\text{LTI}(\Delta\lambda) = \left|\frac{Z_{th}}{Z_L}\right| = \sqrt{\frac{\left(\frac{\Delta V}{\Delta\lambda} \cdot \frac{1}{V}\right)^2 + \left(\frac{\Delta\theta}{\Delta\lambda}\right)^2}{\left(1 - \frac{\Delta V}{\Delta\lambda} \cdot \frac{1}{V}\right)^2 + \left(\frac{\Delta\theta}{\Delta\lambda}\right)^2}} \tag{2.21}$$

The ideal value of LTI occurs by evaluating the limit of the expression in (2.21) as $\Delta\lambda \to 0$ and the terms $\Delta V/\Delta\lambda$ and $\Delta\theta/\Delta\lambda$ become $dV/d\lambda$ and $d\theta/d\lambda$, respectively. The terms $dV/d\lambda$ and $d\theta/d\lambda$ are the sensitivities of the voltage magnitude and the phase angle with respect to the load scaling factor. Hence, the proposed index is termed as the Sensitivity-based Thevenin Index (STI), to indicate that it connects sensitivity and the Local Thevenin Index. The expression of the STI is presented in Eq. (2.22) [26].

$$\text{STI} = \lim_{\Delta\lambda \to 0} \text{LTI}(\Delta\lambda) = \sqrt{\frac{\left(\frac{dV}{d\lambda} \cdot \frac{1}{V}\right)^2 + \left(\frac{d\theta}{d\lambda}\right)^2}{\left(1 - \frac{dV}{d\lambda} \cdot \frac{1}{V}\right)^2 + \left(\frac{d\theta}{d\lambda}\right)^2}} \tag{2.22}$$

The terms $dV/d\lambda$ and $d\theta/d\lambda$ are well known in industry and academia and similar sensitivities have been conventionally used as voltage stability indicators at the control center, before the widespread deployment of PMUs [27]. As the above derivation shows, there is a direct connection between the LTI and the sensitivities and hence the LTI can also be used as an indicator of static long-term voltage stability. Intuitively, the reason for using the sensitivities can be understood using Fig. 2.21 which shows a *P-V* curve with three operating points Point A, B, and C. Point A is the present operating point, point B corresponds to a negative load increment ($\Delta\lambda < 0$), and point C corresponds to a positive load increment ($\Delta\lambda > 0$). The LTI derived using the $\Delta\lambda$ is directly related to the slope of the secants AB or AC. As the ideal value of the LTI occurs when the $\Delta\lambda \to 0$, this corresponds to the slope of the tangent at point A (which is the same as the sensitivity). Thus, the sensitivities at an operating condition can be used to calculate the ideal LTI at a particular bus.

The calculation of the sensitivities in power systems is a standard procedure and requires the Jacobian at an operating point [28]. Let $f(\overline{V}, \overline{\theta})$ be the set of expressions for the active power injection at all PV and PQ buses and let $g(\overline{V}, \overline{\theta})$ be the set of expressions for reactive power injection at all PQ buses. The sensitivities are determined by solving the linear system of equations given in (2.23).

$$\begin{bmatrix} f_{\overline{\theta}} & f_{\overline{V}} & \overline{P}_\lambda \\ g_{\overline{\theta}} & g_{\overline{V}} & \overline{Q}_\lambda \\ 0 & 0 & 1 \end{bmatrix} \cdot \begin{bmatrix} d\overline{\theta} \\ d\overline{V} \\ d\lambda \end{bmatrix} = \begin{bmatrix} 0 \\ 0 \\ 1 \end{bmatrix} \quad (2.23)$$

The submatrices $f_{\overline{\theta}}$ & $f_{\overline{V}}$ are the partial derivatives of the active power flow injection expressions with respect to the angles and voltages and can be extracted

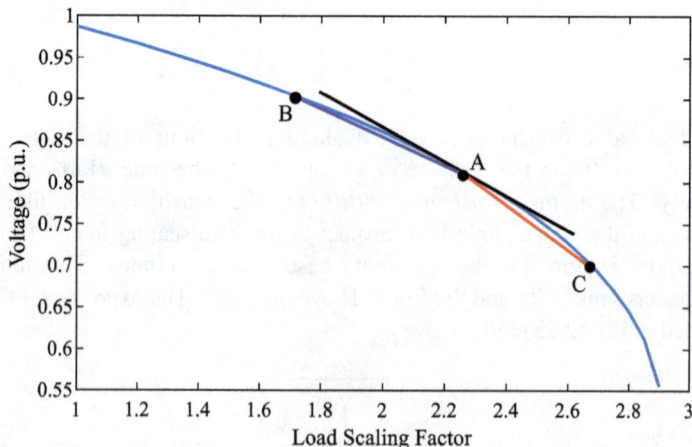

Fig. 2.21 A *P-V* curve indicating that the slope of tangent at a Point, A, is between the slope of secants, AB ($\Delta\lambda < 0$) and AC ($\Delta\lambda > 0$)

directly from the power system Jacobian at that operating point. Similarly, $g_{\overline{\theta}}$ & $g_{\overline{V}}$ correspond to the partial derivatives of the reactive power flow and are similarly extracted from the Jacobian. P_λ and Q_λ are column vectors and correspond to how the active and reactive power injections vary as a function of $\Delta\lambda$. As described before, the voltage sensitivity at an operating point is essentially the slope of the tangent of the *P-V* curve at that point and this method to determine sensitivities is numerically robust to noise, compared to numerically computing $\Delta V/\Delta\lambda$.

2.6.4 Incorporating the Distribution Network in Thevenin Index

One of the key assumptions in the Thevenin Equivalent-based methods using PMU measurements [17–20, 22–26] is that the load is connected to the transmission system. In reality, the loads are located in the sub-transmission and distribution networks and this has to be incorporated into the Thevenin Equivalent. This is conceptually done in the modified Thevenin equivalent represented in Fig. 2.22 where the impedance Z_{eq_D} represents an aggregation of the distribution feeders in a load area and the equivalent load impedance is given by Z_{L_D}. The parameters of the modified Thevenin Equivalent can be estimated from quasi-steady-state voltage and current phasor measurements in the transmission and distribution system. More details about the parameter estimation can be found in [29].

Comparing the two equivalents in Figs. 2.20 and 2.22, it can be seen that $Z_L = Z_{L_D} + Z_{eq_D}$. As the load is present at the distribution node, at the critical loading $|Z_{L_D}| = |Z_{eq_T} + Z_{eq_D}|$. Combining this information with it can be deduced that the LTI calculated at the transmission bus is the Thevenin equivalent including the distribution network equivalent is less than 1. The new voltage stability index (VSI$_D$) that accounts for the distribution network is given in Eq. (2.24) [29]. It is shown in [29] that this index successfully identifies the critical loading for T&D co-simulated systems [12] while the previous index LTI cannot identify it due to the absence of a distribution network representation in Fig. 2.20. The Thevenin equivalent in Fig. 2.22 can be used to represent three-phase unbalanced circuits in which the equivalent parameters ($E_{eq}, Z_{Eq_T}, Z_{eq_D}$ & Z_{L_D}) are all in three-

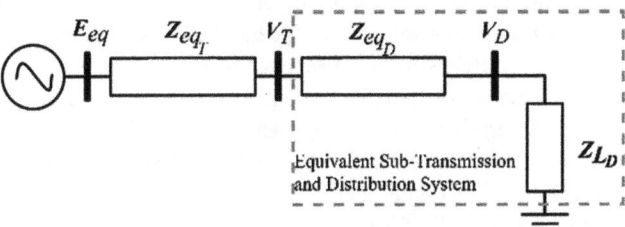

Fig. 2.22 Structure of the modified Thevenin Equivalent including the distribution network

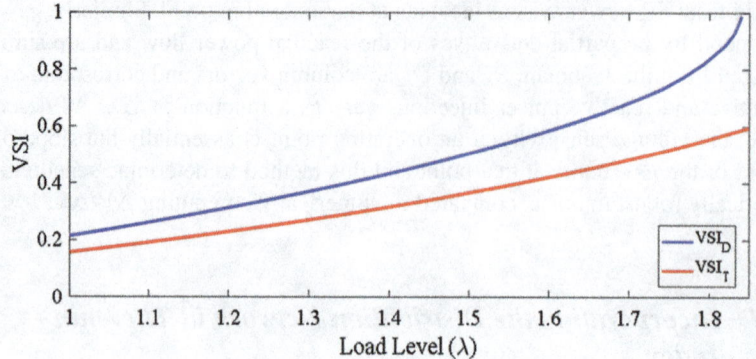

Fig. 2.23 $VSI_{D-3\phi}$ at critical node in distribution network and VSI_T at corresponding transmission node vs. load scaling (extract from [29])

phase representation [29]. The VSI in (2.24) can be extended for the three-phase equivalents ($VSI_{D-3\phi}$) to identify the critical loading for three-phase circuits by comparing the losses in the networks with the load power and is shown in Eq. (2.25). More details on the estimation of the three-phase Thevenin Equivalent parameters using three-phase quasi-steady-state voltage and current phasor measurements in the transmission and distribution system can be found in [29].

$$VSI_D = \frac{|Z_{eq_T} + Z_{eq_D}|}{|Z_{L_D}|} \quad (2.24)$$

$$VSI_{D-3\phi} = \frac{|S_{loss_{T-3\phi}} + S_{loss_{D-3\phi}}|}{|S_{L_{D-3\phi}}|} \quad (2.25)$$

To validate the methodology, co-simulation is used for identifying the critical loading on the test system. The test system has IEEE 9 bus as the transmission network with the loads at all three load buses (5, 7, and 9) replaced with the IEEE 13 bus distribution test feeders [29]. The critical nodes are in the distribution system connected to Bus 5 with a critical loading $\lambda = 1.85$. Figure 2.23 plots the $VSI_{D-3\phi}$ at the critical node in distribution network and the VSI at the transmission bus as the loading in the system increases. It can be seen that the value of the $VSI_{D-3\phi}$ reaches the critical value of 1 while the value of the VSI at the transmission bus only reaches a value of 0.6. Thus, only using the PMU measurements from the transmission system can lead to situations where the voltage instability is not detected/identified. This drawback can be mitigated by using measurements from distribution system are used and by accounting for the three-phase unbalanced nature of the distribution system in the Thevenin Equivalent [29].

The modified Thevenin Equivalent including the distribution network along with the $VSI_{D-3\phi}$ can also be used for offline studies with a co-simulation setup

[12] to identify the critical nodes in the distribution system and the overall T&D system [29]. Further, there have been scenarios when the overall system has been distribution limited [30]. The modified Thevenin Equivalent is able to distinguish between the transmission limited and distribution limited system [29].

2.7 Short-Term Voltage Stability

The phenomenon of voltage stability in the time scale of up to 30 s is referred to as short-term voltage stability as the dynamics involved are very different from the dynamics and components involved in long-term voltage stability. Short-term voltage stability involves dynamics of fast-acting load components such as induction motors, electronic loads, HVDC links, and inverter-based generator resources. The study period of interest is in the order of several seconds and so detailed models that can represent power system transient dynamics are critical. For short-term voltage stability, the dynamic modeling of loads is essential, and short circuit faults near loads are the main concern [1].

The typical case of short-term voltage instability is the stalling of induction motors (IM) after a large disturbance (such as a fault of a loss of generation) either due to the loss of equilibrium between electromagnetic and mechanical torques in the induction motor or due to escaping the region of attraction of the stable equilibrium due to delayed fault clearing [1]. During a fault, induction motors decelerate due to decreased electromagnetic torque, which makes them draw higher current and much larger reactive power, causing further voltage depression. After fault clearing, the load voltage partially recovers and the electromagnetic torque improves. If the motor has not decelerated below a critical speed, it reaccelerates towards the normal operating rotational speed and the load voltage returns to the nominal value. If the motor has decelerated below a critical speed, it cannot reaccelerate and the motor decelerates to a stop (stalls). Stalled motors can either be disconnected by undervoltage protections or remain connected, drawing a large (starting) current until they are disconnected by thermal overcurrent protections. If the stalled motors remain connected, the voltage remains depressed for a longer time (>10 s), possibly inducing a cascade of stalling on nearby motor loads. This mode of short-term voltage instability also applies to induction generators. The difference is that induction generators accelerate instead of stalling during faults and, if unstable, they are disconnected by overspeed relays instead of undervoltage relays.

2.7.1 Fault Induced Delayed Voltage Recovery

Single-phase induction motors that are often used in residential air-conditioners are more susceptible to stalling than large three-phase induction motors due to their smaller inertia [31, 32]. The stalling of large numbers of single-phase induction

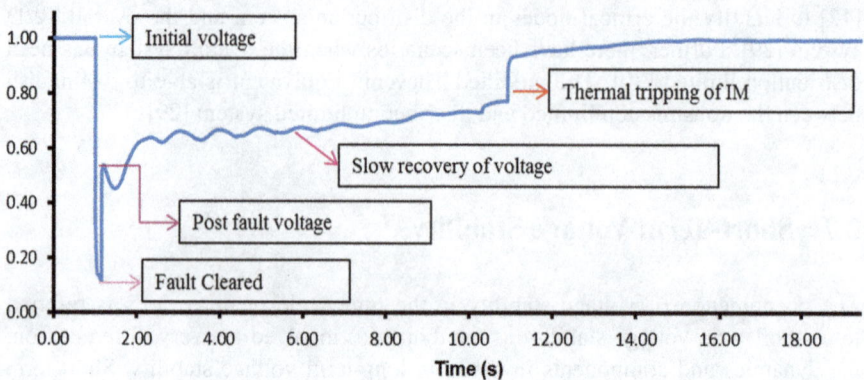

Fig. 2.24 Conceptual FIDVR waveform at a bus

motors leads to a phenomenon referred to as Fault Induced Delayed Voltage Recovery (FIDVR) [31, 33]. In FIDVR events, the eventual tripping of the stalled motors is also through the thermal protection of the individual motors. FIDVR is also a potential cause of cascading and/or instability depending on the network topology and available reactive support close to the event.

FIDVR is mainly caused in systems with a moderate amount of single-phase (1ϕ) induction motor loads (2530%). After a large disturbance (fault, etc.), these motors, that are connected to mechanical loads with constant torque, stall and typically draw 5–6 times their nominal current and this leads to the depression of the system voltage for a significant amount of time. The low voltages in the system inherently lead to some load being tripped by protection devices close to the fault. However, even after this, the concern is that the sustained low voltages (>10 s) can lead to cascading events in the system steering towards a blackout. A typical delayed voltage response after a fault along with the various features is shown in Fig. 2.24. In this particular example, all the IMs are tripped at the same time, leading to a sudden voltage recovery. There can also be scenarios in which the thermal tripping is more gradual, leading to a gradual voltage recovery.

Most single-phase induction motor are used in residential air-conditioners and so the FIDVR phenomenon has been historically observed in systems where a large number of residential AC's are operational at the same time (e.g., summer in California or Arizona). Most of these devices do not use undervoltage protection schemes and are only equipped with the thermal protection with an inverse time-overcurrent feature, delaying the tripping up to 20 s.

2.7.2 FIDVR Events Observed

Description of several FIDVR events observed in the field are listed in [34] and almost all of them occur in high residential load areas during a period of high temperature. As an example, Fig. 2.25 shows an FIDVR event on a 115 kV bus in Southern California on July 24, 2004. The sustained low voltage is likely caused by stalled AC IM's and the voltage finally recovered to pre-contingency voltage around 25 s after the fault. Out of the substation load of 960 MW, 400 MW of load was tripped by protection devices in residential and commercial units to recover the voltage.

FIDVR can also occur in distribution feeders due to lightning strikes on the feeders. An FIDVR event occurred on July 10, 2012, in the Southern California Edison System and lasted approximately 9 s [35]. This FIDVR event occurring in a single distribution feeder was detected using micro-PMUs [36] in the distribution feeder at Valley substation. The nominal power is 30 kW and so this micro-PMU essentially monitors the behavior of around 10–15 households. The purpose of the micro-PMUs is to capture load events and to enable proper load modeling. Lightning strikes caused multiple distribution faults and reduced voltage to 60% causing some loads to stall. The stair-shaped profile for real power indicates that several loads disconnected approximately 6 s after the FIDVR event was initiated and is due to thermal protection schemes tripping off residential A/C units. The voltage profile shown in Fig. 2.26 (Fig. 6.1 in [35]) is not so flat and there are several voltage sags (e.g., at 5 s) which make it hard to quantify FIDVR just from voltage.

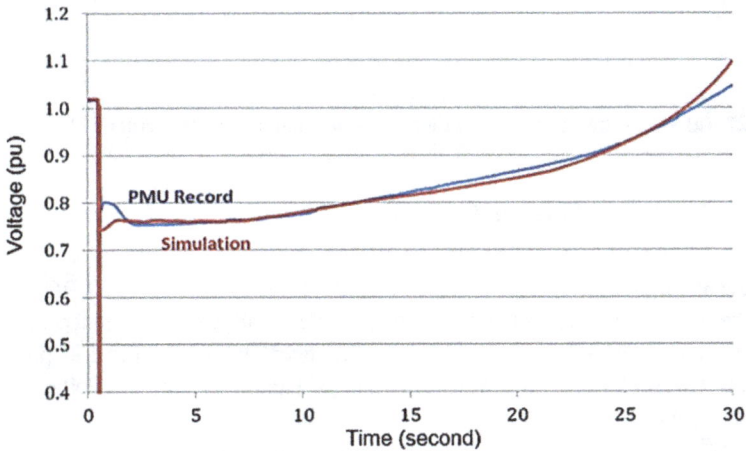

Fig. 2.25 Recorded delayed voltage recovery waveform at a 115kv bus in southern California on July 24, 2004 [34]

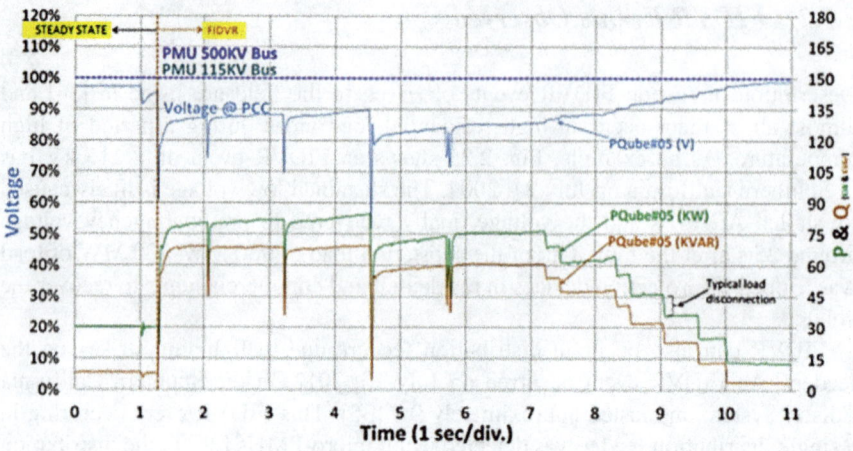

Fig. 2.26 Voltage, active power, and reactive power for the SCE FIDVR event on July 10, 2012. (extract from [35])

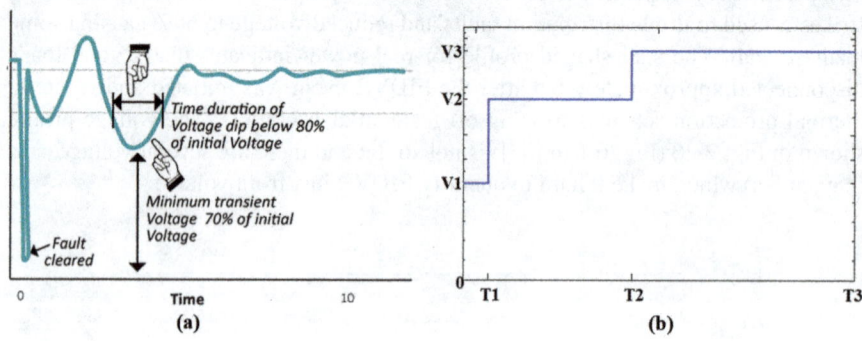

Fig. 2.27 (a) WECC transient voltage criteria [37] (b) simplified voltage criteria [38]

2.7.3 Transient Voltage Criteria

To prevent uncontrolled loss of load in the bulk electric system, NERC, WECC, and other regulatory bodies have specified transient voltage criteria that utilities and system operators need to satisfy after a fault has been cleared. Figure 2.27 provides a pictorial representation of the WECC criteria [37] and the simplified voltage criteria (PJM criteria [38]).

The WECC transient criteria is defined as the following two requirements [37]

1. Following fault clearing, the voltage shall recover to 80% of the pre-contingency voltage within 20 s of the initiating event.
2. Following fault clearing and voltage recovery above 80%, voltage at each applicable bulk electric bus serving load shall neither dip below 70% of

pre-contingency voltage for more than 30 cycles nor remain below 80% of pre-contingency voltage for more than 2 s.

A simplified voltage criteria is used generally by utilities and the trajectory of the recovering voltage must be above the curve in Fig. 2.3b where $V_1 = 0.5$, $V_2 = 0.7$ & $V_3 = 0.95$ and $T_1 = 1\ s$, $T_2 = 5\ s$ & $T_3 = 10\ s$. The ERCOT criteria for transient voltage response requires that voltages recover to 0.90 p.u. within 10 s of clearing the fault [39]. The utilities ensure that the voltage recovery satisfies the guidelines specified by their regulatory authority during their planning phase and operational phase by either installing VAR devices (STATCOM, SVC, etc.) in critical regions and by ensuring that sufficient dynamic VARS are available during operation. In order to study the phenomenon of short-term voltage instability in practical systems, power system time domain simulators are used along with the appropriate load models that can accurately model the phenomenon. This is described in the next section.

2.8 Dynamic Composite Load Model by WECC

In order to enable the utilities and system operators to simulate the transient voltage phenomenon and estimate the amount of VAR support required to prevent short-term voltage stability issues, a dynamic load model has been developed recently by WECC called as the Dynamic Composite Load Model [40]. The composite model essentially aggregates the various kinds of dynamic loads in the sub-transmission network into several 3-ϕ IM (representing high, medium, and low inertias) and an aggregate 1-ϕ IM (representing the AC loads). Furthermore, the protection schemes that trip a proportion of the loads are also implemented for each of the motor representing the Undervoltage and Underfrequency protection policies. An equivalent feeder is also present that tries to emulate the impact of voltage drop in the distribution system when a large current is drawn. The overall structure of the composite load model is shown in Fig. 2.28.

This model has 132 parameters and has been implemented by vendors in commercial software such as PSSE, PSLF, and PowerWorld. More details along with descriptions of the various parameters can be found in [40]. The various components of the composite load model are explained in the next subsections.

2.8.1 Substation and Feeder Model

The substation transformer is modeled as an on-load tap changing (LTC) transformer that can regulate its low-side voltage. A compensating impedance is used to represent line-drop compensation. A single shunt capacitor is represented on the low-side bus with a susceptance (Bss) to account for the capacitors in the distribution

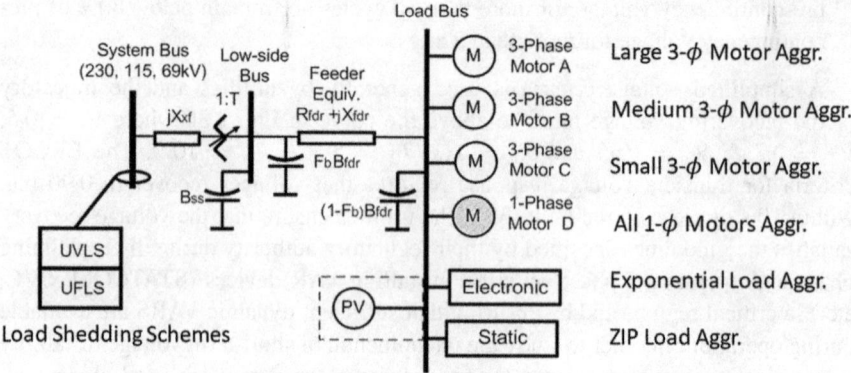

Fig. 2.28 Structure of the composite load model [40]

and sub-transmission systems. The feeder equivalent includes series resistance and reactance and shunt capacitors at both ends to capture the line charging of the individual feeders in the distribution and sub-transmission system.

2.8.2 Static and Electronic Load Models

The static load model is represented as either ZIP loads or as exponential load models [40] whose parameters are input by the user. The electronic load model is represented as a constant power load with an additional logic to reduce the load when the load voltage falls below a user-defined setpoint. These loads are not the main contributors in the short-term voltage stability phenomenon.

2.8.3 Three-Phase Induction Motor Load Model

The three-phase (3ϕ) induction motor (IM) is a highly dynamic load, and therefore it needs to be properly represented with detailed differential and algebraic equations. A standard way to model the $3\text{-}\phi$ IM is by an equivalent circuit [41] where the stator and rotor impedances along with the mutual inductances are specified. A summary of the governing equations of a single cage three-phase IM is given in Eq. (2.26).

$$\begin{aligned}
\frac{d\phi_{ds}}{dt} &= \omega_b \left[V_{ds} - R_s \cdot \frac{X_r}{X_e^2} \cdot \phi_{ds} - \phi_{qs} + R_s \cdot \frac{X_M}{X_e^2} \cdot \phi_{dr} \right] \\
\frac{d\phi_{qs}}{dt} &= \omega_b \left[V_{qs} - R_s \cdot \frac{X_r}{X_e^2} \cdot \phi_{qs} + \phi_{ds} + R_s \cdot \frac{X_M}{X_e^2} \cdot \phi_{qr} \right] \\
\frac{d\phi_{dr}}{dt} &= \omega_b \left[-R_r \cdot \frac{X_s}{X_e^2} \cdot \phi_{dr} - (1 - \omega_r) \phi_{qr} + R_r \cdot \frac{X_M}{X_e^2} \cdot \phi_{ds} \right] \quad (2.26)\\
\frac{d\phi_{qr}}{dt} &= \omega_b \left[-R_r \cdot \frac{X_s}{X_e^2} \cdot \phi_{qr} + (1 - \omega_r) \phi_{dr} + R_r \cdot \frac{X_M}{X_e^2} \cdot \phi_{qs} \right] \\
\frac{d\omega_r}{dt} &= \frac{1}{H} \left[\frac{X_M}{X_e^2} \left(\phi_{ds} \cdot \phi_{qr} - \phi_{dr} \cdot \phi_{qs} \right) - T_L \cdot \omega_r^{T_e} \right]
\end{aligned}$$

where the states of the dynamic model are $\phi_{ds}, \phi_{qs}, \phi_{dr}$ & ϕ_{qr} which correspond to the flux linkages along the d and q axis of the stator and rotor and the rotor speed (ω_r). R_r, R_s, X_r, X_s, X_M and H are the parameters of the induction motor with $X_e = \sqrt{X_s X_r - X_M^2}$ and ω_b is the synchronous rotor speed. T_L is the mechanical load torque coefficient and T_e is the mechanical load torque exponent (In practice $T_e < 2$). The input to this model are the voltages that are on the stator side and are V_{ds} & V_{qs} which correspond to the d- and q-axis components of the grid voltage. The active and reactive power consumed by the IM model can be written in terms of the states and the inputs and is given by Eq. (2.27).

$$\begin{aligned}
P &= \frac{V_{ds}}{X_e^2} \cdot (X_r \cdot \phi_{ds} - X_M \cdot \phi_{dr}) + \frac{V_{qs}}{X_e^2} \cdot (X_r \cdot \phi_{qs} - X_M \cdot \phi_{qr}) \\
Q &= \frac{V_{ds}}{X_e^2} \cdot (X_r \cdot \phi_{qs} - X_M \cdot \phi_{qr}) - \frac{V_{qs}}{X_e^2} \cdot (X_r \cdot \phi_{ds} - X_M \cdot \phi_{dr})
\end{aligned} \quad (2.27)$$

In most practical systems, the induction motor is a double cage model of either type-1 or type-2. The equivalent circuits of the two types of three-phase double cage induction motors are shown in Fig. 2.29. In order to apply the Eqs. (2.26) and (2.27) to the double cage model, the equivalent rotor resistance (R_r) and reactance (X_r) is calculated as a function of the rotor speed (ω_r) and the rotor parameters. More details can be found in [41].

The 3-ϕ IM components in the composite load model are specified by the transient and sub-transient parameters and not the impedances. The transient and sub-transient parameters (Ls, Lp, Lpp, Tp0, and Tpp0) can be determined from

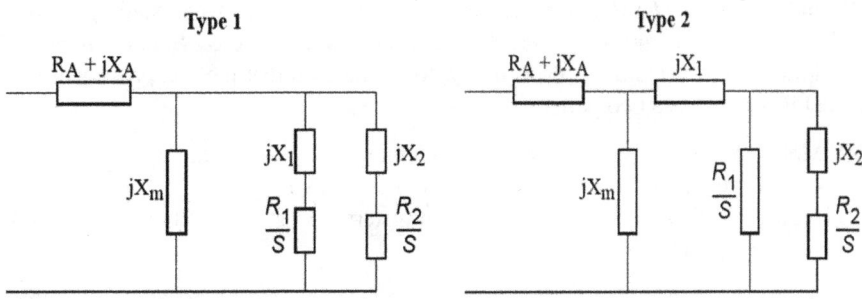

Fig. 2.29 Equivalent circuits of the two types of three-phase double cage induction motors

the impedance parameters of the standard model using the relations in Eq. (2.28) [42] where the terms X_m, X_A, X_1, X_2, R_A, R_1 & R_2 correspond to the components indicated in Fig. 2.29.

$$L_s = (X_A + X_m)/\omega_{base}$$
$$L_p = (X_A + (X_1 \cdot X_m)/(X_1 + X_m))/\omega_{base}$$
$$L_{pp} = (X_A + (X_1 \cdot X_2 \cdot X_m)/(X_1 \cdot X_2 + X_2 \cdot X_m + X_m \cdot X_1))/\omega_{base} \quad (2.28)$$
$$T_{p0} = (X_1 + X_m)/(\omega_{base} \cdot R_1)$$
$$T_{ppo} = (X_2 + (X_1 \cdot X_m)/(X_1 + X_m))/(\omega_{base} \cdot R_2)$$

The 3-ϕ IM model is also equipped with two undervoltage relays that are activated cumulatively based on the user settings. Appropriate settings of the undervoltage relays prevent short-term instability due to the stalling of the 3ϕ IM.

2.8.4 Single-Phase Induction Motor Load Models

The 1ϕ IM A/C performance-based model was developed by Western Electricity Coordinating Council (WECC) Load Modeling Task Force members based on extensive laboratory testing of a variety of A/C units. The model represents the combined positive sequence phasor behavior of several individual single-phase A/C compressors and can represent complex behavior such as

1. Stalling the compressor motors when the node voltage is below a threshold value (V_{stall}) for more than a pre-specified time (t_{stall}).
2. Restarting a fraction of the A/C load if the voltage recovers above a set value (V_{rst}) for more than a pre-specified time (t_{rst}), i.e., these motors are no longer stalled. This fraction is set by a parameter F_{rst}.
3. Disconnection of the stalled motors due to thermal protection after a few seconds. This is controlled by a thermal relay.

Figure 2.30 shows the block diagram of the 1ϕ IM A/C model and demonstrates how the active and reactive power demanded is dependent on the node voltage (V) and frequency (F). The thermal relays are only activated when stalling occurs and are inactive in normal operation. The circles with Π & Σ correspond to multiplication and addition, respectively. It can be seen that the compressor motor model is divided into two parts:

(a) Motor A—Those compressors that cannot restart after stalling and remain stalled even after the voltage rises above the stall voltage.
(b) Motor B—Those compressors that can restart after stalling if the voltage rises above a certain setpoint.

The motors A and B are represented by algebraic Eqs. (2.29) and (2.30) that describe the variation of the active power and reactive power load with the load bus

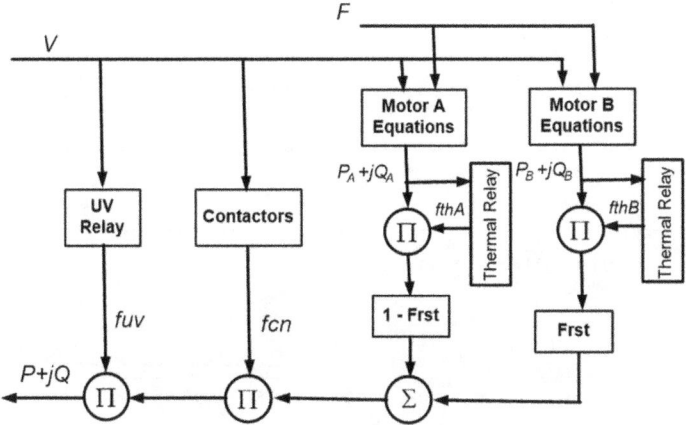

Fig. 2.30 The block diagram of the 1ϕ IM A/C model with various components [40]

voltage (V) and the frequency (F) per unit. The active and reactive powers of the motors A and B are scaled by $(1 - F_{rst})$ and F_{rst}, respectively before being added to get the final power of the 1ϕ IM A/C model. The various symbols in the Eqs. 2.29 and 2.30 such as $V_{brk}, K_{p1}, N_{p1}, G_{stall}$ are all parameters of the composite load model and are provided by the user.

$$P = \begin{cases} P_0 + K_{p1}\left(1 + (F-1)\right)(V - V_{brk})^{N_{p1}} & V > V_{brk} \\ P_0 + K_{p2}\left(1 - 3.3(F-1)\right)(V - V_{brk})^{N_{p2}} & V_{stall} < V < V_{brk} \\ G_{stall} \cdot V^2 & V < V_{stall} \end{cases} \quad (2.29)$$

$$Q = \begin{cases} Q_0 + K_{q1}\left(1 + (F-1)\right)(V - V_{brk})^{N_{q1}} & V > V_{brk} \\ Q_0 + K_{q2}\left(1 - 3.3(F-1)\right)(V - V_{brk})^{N_{p2}} & V_{stall} < V < V_{brk} \\ B_{stall} \cdot V^2 & V < V_{stall} \end{cases} \quad (2.30)$$

The motors stall when the voltage at the node becomes less than V_{stall} for more than t_{stall} seconds and this activates the thermal relay. The fraction of motors A and B connected after stalling is determined by the fraction f_{th} which is the output of the thermal relay. The power demanded by the motors are then scaled by the fractions fcn and fuv that correspond to the reduction in power due to contactors and undervoltage protection.

The 1-ϕ induction motor is the main reason why the FIDVR is observed. The 1-ϕ IM model has representations of the AC compressor motor, compressor motor thermal relay, undervoltage relays, and contactors. Depending upon the input voltage, the motor operates either in "running" or "stalled" state. The behavior of the motor as a function of the voltage can be understood based on the power consumption of the motor. Figure 2.31 plots the active and reactive power demand as a function of the voltage for the normal operation and stalled operation. From

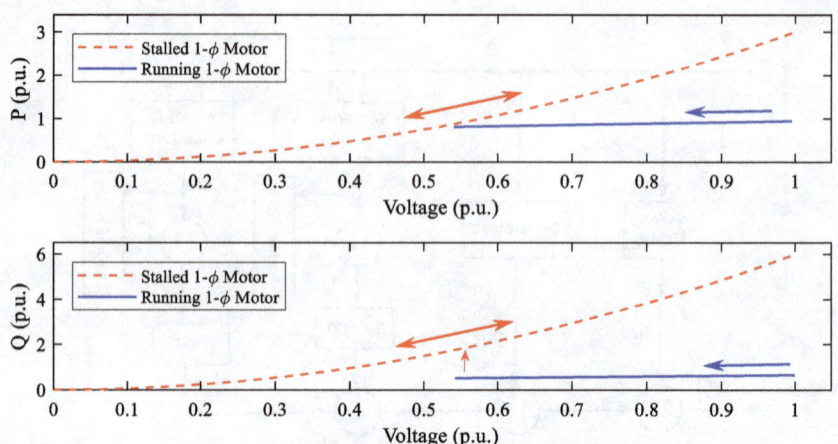

Fig. 2.31 Active power (Top) and reactive power (Bottom) versus the voltage for the normal operation (Blue) and stalled operation (Red) for the 1-Φ induction motor [43]

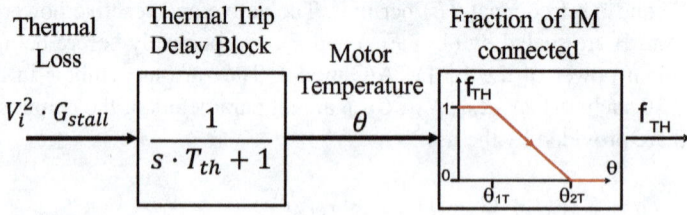

Fig. 2.32 The structure of thermal relay in the 1φ IM A/C model [40]

Fig. 2.31, it can be seen that in the stalled state, the active power demand is three times the nominal amount and the reactive demand is six times the nominal amount compared to the normal "running" state. This large demand is the reason why the voltage reduces at the substation causing FIDVR. This demand naturally is reduced via thermal protection that takes around 10–15 s.

The thermal relay block diagram is shown in Fig. 2.32, where $V_i^2 \cdot G_{stall}$ is the thermal power dissipated in the motor. T_{th} is the thermal relay time constant and θ is the motor temperature estimated by the relay. Initially, the internal temperature is zero and the thermal loss is zero. As the stalling condition occurs suddenly, the input to the thermal delay block can be approximated by a step function and the temperature (θ) rises in an exponential manner. The f_{th} fraction remains 1 till the temperature reaches θ_1 after which the fraction reduces linearly with the temperature until the temperature reaches θ_2 when all the motors are disconnected. A more analytical description of the rise of motor temperature can be found in [43, 44].

It can be seen from the equations and the description that the 1φ IM A/C model is highly nonlinear and has complex dynamics (stalling, restarting, thermal disconnection) with conditional arguments. These complex dynamics make the study and control of FIDVR challenging.

2.8.5 Key Parameters for the Composite Load Model

One of the key challenges for the utilities in the simulation of the composite load model by utilities is a large number of parameters (130+) of the model. To aid the utilities in their dynamic studies WECC has provided default parameters that are derived from the study of several feeders in their footprint [32, 40]. Furthermore, the US Department of Energy in conjunction with the WECC's Load Modeling Task Force has developed a tool to identify the parameters of the composite load model based on the geographic location of the feeder. The Load Model Data Tool [44] is available for utilities to create the composite load model dynamics file either in PSSE or PSLF compatible format based on the local weather and loading of the feeder.

Another method to estimate the model parameters is to use recorded FIDVR events at a substation and then utilize parameter estimation techniques [46] as the structure of the model is known. However, a large number of parameters make the problem ill-defined. To solve this issue, most of the parameters are fixed to be the same as the default values (or the values from the Load Model Data Tool) and only a few key parameters that impact the load dynamics behavior are chosen for the parameter estimation. The key parameters of the composite load model have been identified from sensitivity studies [47] and are the following:

- Stall voltage in p.u. (Vstall).
- Stall time delay in sec. (Tstall).
- Motor D fraction of load power (FmD).
- Fraction of load with undervoltage relay protection (Fuvr).
- Motor D thermal time constant in sec. (Tth).
- Motor D thermal protection trip start level (Th1t).
- Motor D thermal protection trip completion level (Th2t).

The report [47] also provides the key parameters of the three-phase motors in the composite load model and recommends Transmission Owners and Transmission Planners focus data collection on the key parameters.

2.9 Data-Driven Methods to Assess and Monitor Short-Term Voltage Stability

Once the modeling of the power system components is complete, the time domain power system simulators are used to perform various contingency studies under different plausible operating scenarios and load behaviors. The results of these simulations are then used for identifying regions in the power system that are susceptible to short-term voltage instability [48]. One challenge in directly utilizing the time series data is that the characterization of the stability/instability of a particular time domain simulation from the resulting data is not trivial and needs

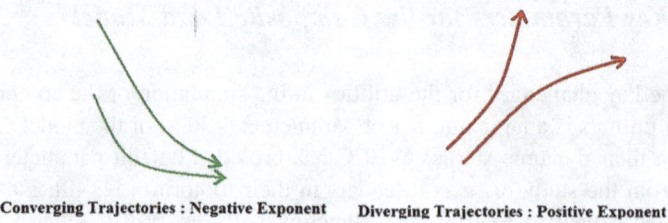

Fig. 2.33 Convergent and divergent trajectories and corresponding sign of Lyapunov exponents

to be done in a systematic manner using methods from control system theory. The Lyapunov Exponent (LE) [49, 50] has been shown to be the appropriate tool for the identification of stability/instability from time series data.

The Lyapunov exponent (LE) is an idea that is adapted from the ergodic theory of dynamical systems. The maximum Lyapunov exponent is a measure of the rate of separation of two trajectories in the system and is used to ascertain the system stability. If the maximum Lyapunov exponent is negative, the trajectories of the system converge to a stable equilibrium. However, if the maximum Lyapunov exponent is positive, the trajectories of the system diverge; this suggests a possibly unstable and chaotic system. This is illustrated in Fig. 2.33.

2.9.1 Computation of Lyapunov Exponent from Time Series Data

The algorithm for the computation of the maximum LE based on the voltage time series data is outlined below [49, 50]:

1. Let $V(t)$ be a vector of voltages at different buses at time t. The voltages are sampled at a constant sampling frequency Δt. Thus, $t = 0, \Delta t, 2\Delta t, \ldots$.
2. The values of ϵ_1 and ϵ_2, which are fixed in advance, determine when the algorithm is initialized. Choose integer N such that $\epsilon_1 < \|V(m\Delta t) - V((m-1)\Delta t)\| < \epsilon_2$ for $m = 1, 2, \ldots, N; 0 < \epsilon_1 < \epsilon_2$.
3. The maximum LE of the system at time $k\Delta t$ can be calculated using the following formula for $k = 1, 2 \ldots$.

$$\Lambda(k\Delta t) = \frac{1}{Nk\Delta t} \sum_{m=1}^{N} \log \frac{\left\| V\left((k+m)\Delta t\right) - V((k+m-1)\Delta t) \right\|}{\left\| V\left((m)\Delta t\right) - V((m-1)\Delta t) \right\|} \quad (2.31)$$

The basic idea behind the equation for calculating the LE is that the above equation measures the separation on the voltage trajectories with respect to separation present at the N initial conditions. If the separation of the measurements at a particular instant is lesser than the initial separation then it will result in negative LE, implying converging behavior. If the separation of the measurements at a particular instant is greater than the initial separation then it will result in positive LE, implying diverging behavior. Since the value of the initial separation is used at all the instants to calculate the LE, the selection of the initial points is critical for well-behaved behavior of the algorithm.

The initial points, that determine the initial separation, depend on the values of ϵ_1 and ϵ_2 and so these need to be selected appropriately. These quantities depend on the time difference between consecutive time series data. If the rate at which the measurements are obtained is high, then the change in the value between two consecutive measurements is low. Hence, the values of ϵ_1 and ϵ_2 have to be small. On the other hand, if the measurements are obtained at a low rate, then the values of ϵ_1 and ϵ_2 have to be chosen relatively larger. For simulation purposes, we have measurements at a frequency of 120 Hz. Therefore we chose $\epsilon_1 = 0.002$ and $\epsilon_2 = 0.01$.

The equation to calculate the system-wide LE can be slightly modified to computing the Lyapunov exponent of individual buses to determine the stability/instability contribution of individual buses to the overall system stability/instability. The Lyapunov exponent for the bus will be computed using the following equation.

$$\lambda_i(k\Delta t) = \frac{1}{Nk\Delta t} \sum_{m=1}^{N} \log \frac{\left\| V_i\left((k+m)\Delta t\right) - V_i((k+m-1)\Delta t) \right\|}{\left\| V_i\left((m)\Delta t\right) - V_i((m-1)\Delta t) \right\|} \quad (2.32)$$

Where $V_i((m)\Delta t)$ is the mth sample of voltage measurement at the ith bus and λ_i is the Lyapunov exponent at the ith bus. This is a useful concept as the bus where the exponent is largest is the main contributor to the instability and control actions taken at this bus will have a large impact on the system.

PMU RMS voltage measurements can also be used for the online computation of the Lyapunov exponent using the above formulas is relatively straightforward. The proposed Lyapunov exponent computed, using the voltage measurements from PMU devices at n buses, will provide stability information for all buses whose states can be estimated by using PMU measurements.

2.9.2 Simulation Results for Lyapunov Exponent in PSSE of the 9-Bus System

The WECC 9-Bus system is simulated in PSSE in order to test the Lyapunov exponent methodology described above. A three-phase fault is applied at bus location 5 and the fault is cleared by opening the line 7–5. Two cases of the fault clearing time (t_{cr}) are used—one with a stable scenario (0.05 s) and another with an unstable scenario (0.2 s). The voltage at the various buses for the two cases and the corresponding Lyapunov Exponents are shown in Fig. 2.34.

Simulation results of the WECC 9-Bus system and the IEEE 162-Bus system with different load models and their corresponding LE are described in detail in [49]. It can be observed from the results that the LE settles to a value less than 0 (around −0.5 to be specific) for the scenario when the voltage settles to a steady-state suggesting that the system is stable while the LE settles to a value greater than 0 (around 0.5 to be specific) for the scenario when the voltage is oscillating suggesting that the system is unstable. Thus, the LE is able to correctly predict the stability of the system. Another observation is that the LE sometimes crosses the zero line, changing the estimate of the stability. This is because of the fact that the algorithm presented correctly estimates the actual LE asymptotically, i.e., the estimation becomes better as the time increases. Thus, there is a trade-off between the simulation time and the accuracy of the stability characterization. It is important to note that only the stable cases are affected by this. The unstable case has a positive LE from the start at Bus 5. Thus, there may be situations where the stable cases may be detected as unstable but not the other way around.

2.10 Data-Driven Methods to Assess and Monitor FIDVR

In order to assess and quantify FIDVR from time series data, using only the voltage data might not always be appropriate as the voltage is the result of motor stalling. Thus, quantifying the severity of the stalling is more appropriate for assessing the FIDVR event. One challenge is that the composite load model is too complex for analysis and needs to be simplified. As the thermal relay dynamics is much slower compared to the dynamics of the 3-ϕ IM, the fast dynamics of the 3-ϕ IM can be neglected and only the dynamics of the 1-ϕ IM thermal relay determines the overall behavior of the FIDVR phenomenon. Since the 1-ϕ IM are represented as an admittance after stalling, the 3-ϕ IM and the static loads can also be represented as a voltage-dependent admittance. These observations and modeling assumptions lead to the admittance-based representation of the composite load model.

As a demonstration that the load admittance can indeed capture the load behavior during FIDVR, Fig. 2.35 plots the voltages and Fig. 2.36 plots the load conductance (real component of the admittance) for a normal, moderately severe (30% motor stalling), and very severe (60% motor stalling) delayed voltage recovery after a disturbance. The first observation is the voltage waveforms for both normal recovery and delayed recovery have oscillations due to the behavior of the other components in the system. In comparison, the conductance waveform is much better behaved for the normal recovery and delayed recovery. The oscillations in the voltage are due to the dynamic behavior of the external system (e.g., the generator exciter) and so the impact of these oscillations in the conductance are minimal as the oscillations do not impact the load behavior.

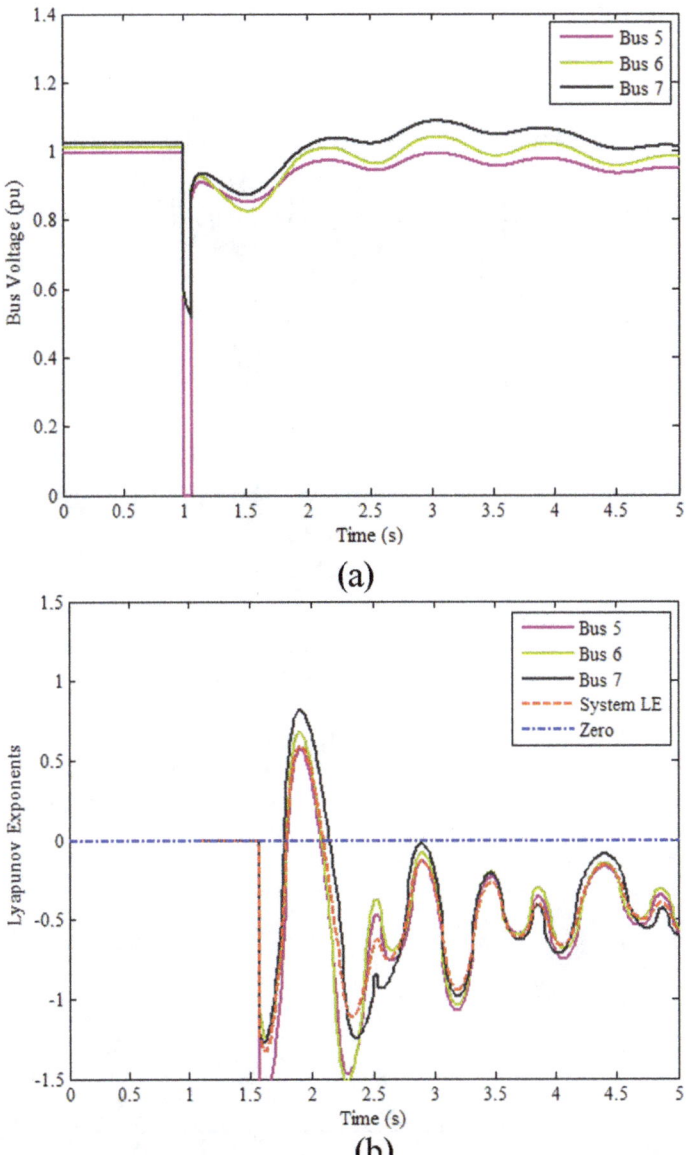

Fig. 2.34 The bus voltages and the Lyapunov Exponents for the different fault clearing times leading to stable and unstable scenarios [51]. (**a**) Bus voltage response (t_{cr}=0.05 s). (**b**) Lyapunov Exponent (t_{cr}=0.05 s). (**c**) Bus voltage response (t_{cr}=0.2 s). (**d**) Lyapunov Exponent (t_{cr}=0.2 s)

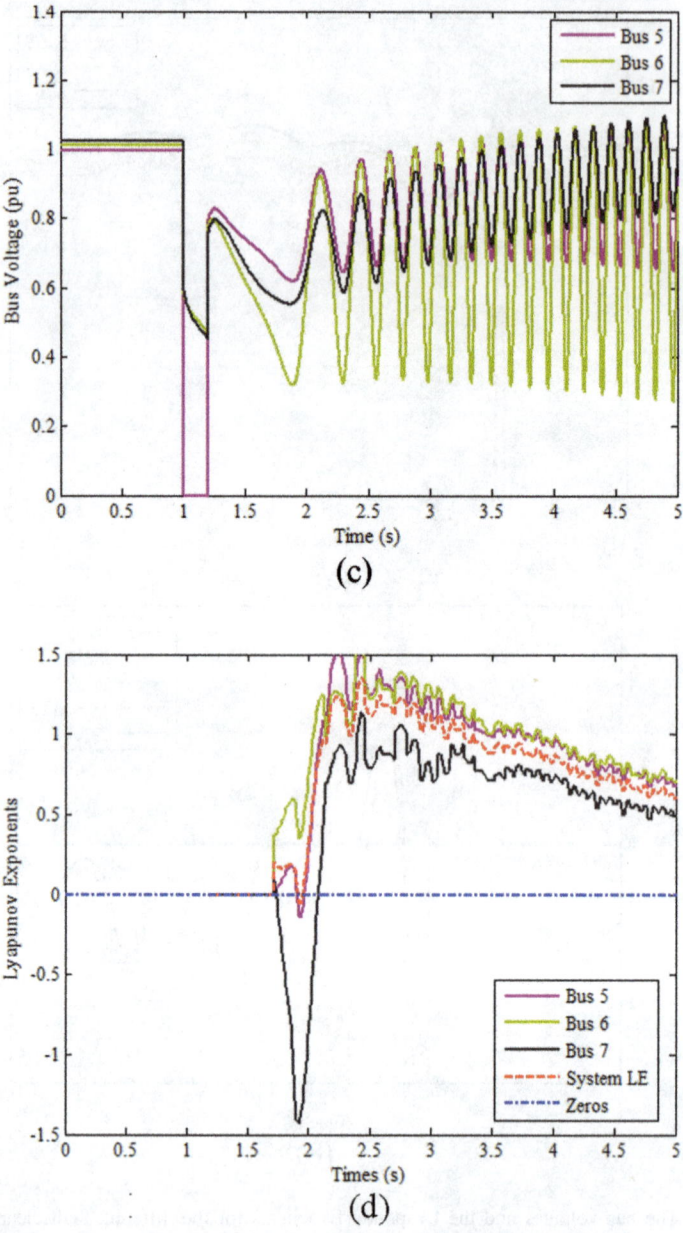

Fig. 2.34 (coninued)

The next observation is that the voltage immediately after the fault is lower for higher amount of motor stalling. Similarly, the load conductance after the fault is cleared increases as the percent of motor stalling increases. However, it is not easy to quantify the severity of the FIDVR event from the voltages as the reduction in

2 Basics of Voltage Stability Assessment

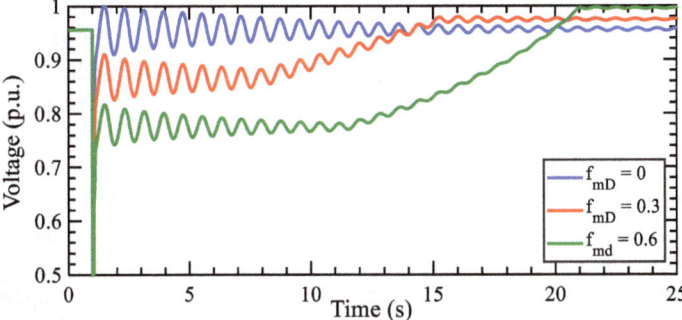

Fig. 2.35 Voltage response with various motor stalling proportion [43]

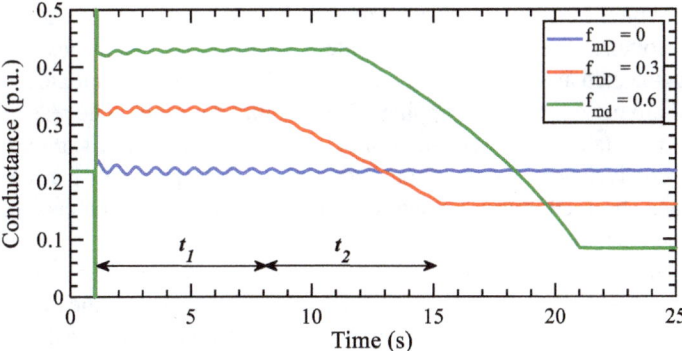

Fig. 2.36 Load conductance with various motor stalling proportion. The times t_1 and t_2 are indicated for $f_{mD} = 0.3$ [43]

voltage is not easily related to the severity and depends on the external network parameters. In contrast, the conductance makes it easy to quantify the severity of the event as the conductance increases in a nearly linear manner to the amount of motors stalled. Thus, it provides a quick way to characterize the severity of the FIDVR and enables monitoring and control schemes based on this quantification. The conductance during normal recovery quickly (<1 s) returns to the pre-contingency value. On the other hand, the conductance of the delayed voltage scenario has a sudden rise due to the stalling of the 1-ϕ IMs.

The sudden rise can be used as a reliable indicator of the onset of the FIDVR phenomenon. The same cannot be said for the voltage as a severe FIDVR on a bus will depress voltages in neighboring buses even if there is no stalling in the neighboring buses. Finally, the conductance for the delayed voltage scenario can be split into two parts—a flat region and a monotonically decreasing region. The flat region corresponds to the time to initiate the thermal tripping of 1-ϕ IM (t_1) and the region where the conductance reduces which corresponds to the time taken to complete the thermal tripping of 1-ϕ IM (t_2). It is much easier to distinguish between these phases of operation from the conductance plots compared to the voltage plots as the oscillations and other phenomena can mask the exact time of transition [43].

The load susceptance has a similar behavior as the load conductance for the FIDVR scenario. By observing various conductance (susceptance) plots for various proportions of stalled motor, two observations can be made: (1) the load conductance (susceptance) is nearly constant till the motor thermal protection triggers and (2) the slope of conductance (susceptance) due to the thermal disconnection is almost constant. Similar observations can be made for FIDVR events in the field both in distribution and transmission systems. For example, Fig. 2.37 plots the load conductance for the FIDVR event described in Fig. 2.26. The conductance plot for the event is less noisy than the voltage plot and also has a similar profile of simulated events (Fig. 2.36). The conductance has a large jump at the stalling condition and is flat till the disconnections begin. The voltage profile (Fig. 2.26) is not flat and there are several voltage sags (e.g., at around 5 s) which make it hard to quantify FIDVR just from voltage. The stair-shaped profile for conductance indicates that several loads disconnected approximately 6 s after the FIDVR event was initiated and is due to thermal protection schemes tripping off residential A/C units. The reason why the waveform is a staircase and not smooth is that the number of A/Cs are around 10–15. If the conductance of a few thousand motors (corresponding to a load of tens of MW) are observed, the individual disconnections cannot be perceived, and the resulting conductance looks smooth as in Fig. 2.36.

Based on these observations and the admittance-based model, it is shown in [43, 44] that by measuring the admittance just after the FIDVR begins, the time durations t_1 and t_2 can be estimated from the load parameters and can be used as a way to quantify the severity of the FIDVR event. This will enable the localization of the FIDVR event both in offline simulations and in online stability monitoring from PMU measurements and will enable effective assessment of FIDVR.

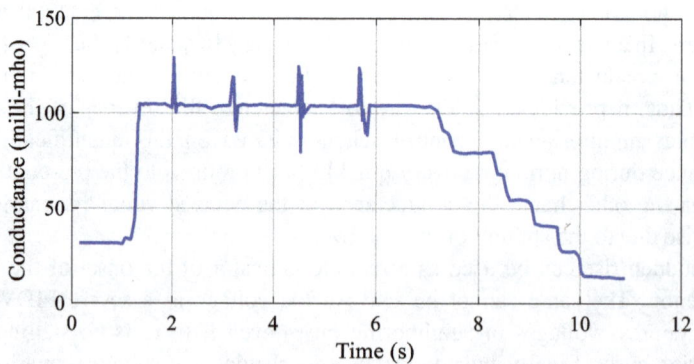

Fig. 2.37 Load conductance for the southern California FIDVR event on July 10, 2012 [43]

2.11 Effect of Distributed Energy Resources on Short-Term Voltage Stability

The significant increase in distributed energy resources (DERs)/distributed generation (DGs) is leading to the development of new performance and reliability standards. The Federal Energy Regulatory Commission (FERC) has recently announced [52] that DERs must ride through abnormal frequency and voltage events. It states that the specific ride through settings must be consistent with Good Utility Practice and any standards and guidelines applied by the transmission provider to other generating facilities on a comparable basis. It is also stated that they should have appropriate ride-through requirements comparable to large generating facilities.

The strong motivation behind these requirements is to ensure high reliability of the interconnected power system, and so, the DERs must continue to remain connected during disturbances and at the same time they cannot be connected indefinitely in the event of a fault or power system malfunction. The IEEE standard 1547 [53] provides the technical specifications for, and testing of, the interconnection and interoperability between utility electric power systems (EPSs) and DG sources. In this section, the focus is on the Voltage Ride Through (VRT) requirements pertaining to the IEEE standard 1547 as shown in Fig. 2.38.

Section R5 of the NERC standard TPL-001-4 [37] states that each transmission system planner shall have criteria for acceptable system voltage limits including voltage transients. Considering that the DGs can affect the transient voltages depending upon the ride-through capabilities, it is therefore imperative to examine

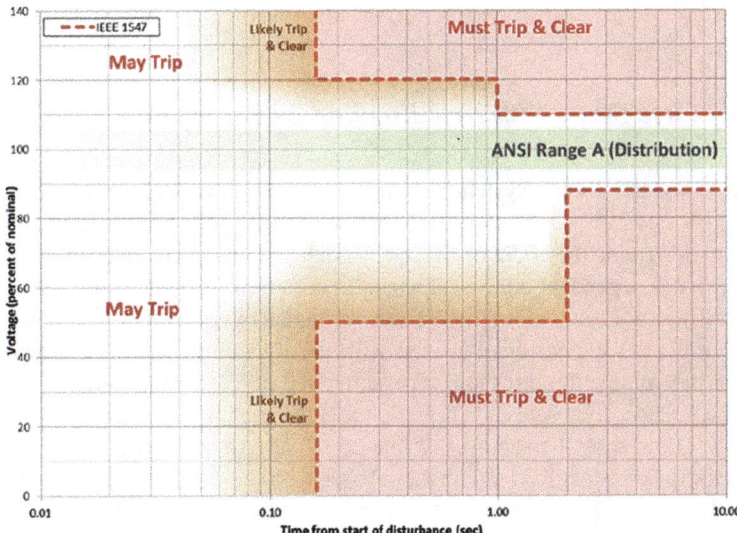

Fig. 2.38 IEEE standard 1547 voltage ride through requirement (extract from [53])

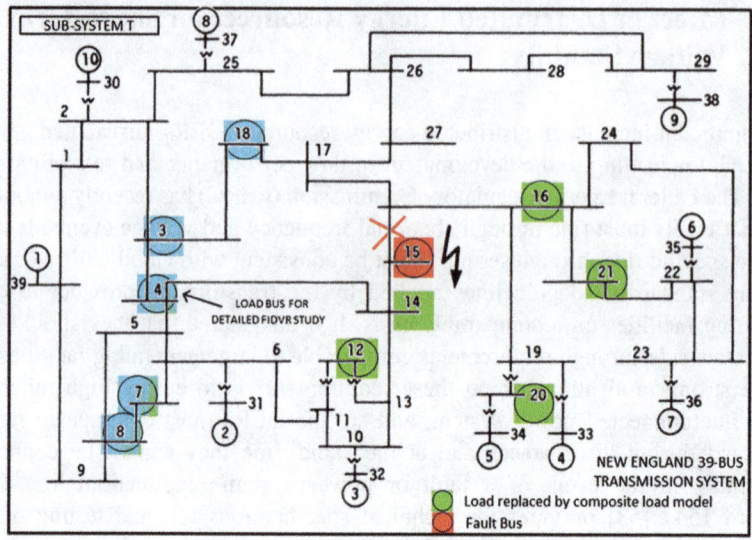

Fig. 2.39 New England 39-Bus transmission system with PQ load att 10 load buses replaced by equivalent composite load models with DG

the effect of the DG penetration on the power system dynamics. The DG present in the downstream feeders is lumped together into a single DG model that is modeled as a phasor representation of the DER inverter [54]. The power supplied by the DG is given by the fraction Fdg that specifies the DG power in terms of the power demanded by the load.

2.11.1 Case Study with New England 39-Bus System

The New England 39-bus system shown in Fig. 2.39 is considered for this study. This system has 29 load buses and 10 generator buses. The objective of this case study is to determine the effect of DGs on the delayed voltage recovery behavior. As this behavior is dependent on the voltage level at fault which determines the stalling characteristics of the induction motors located on that bus, we can selectively choose buses to replace the constant PQ load with the dynamic composite load model. Based on the voltage dip threshold criteria derived in [55], we identify those buses where the voltage goes below 0.75 pu due to a three-phase to ground fault applied at bus 15 followed by the removal of line 15–16 after a fault duration of 5 cycles, and replace the constant PQ load on these buses with the composite load model. The identified buses are 3, 4, 7, 8, 12, 14, 16, and 18. Buses 20 and 21 are also included due to their proximity to the fault. The fractions FmA, FmB, FmC, FmD, and Fel are all equal to 0.12 and the remaining power is in the static load. The fraction of

2 Basics of Voltage Stability Assessment

Table 2.4 DG VRT settings

Voltage range (% nominal)	Max clearing time (s)
<50%	0.16
50–88%	2.0
<110–120%	1.0
>120%	0.16

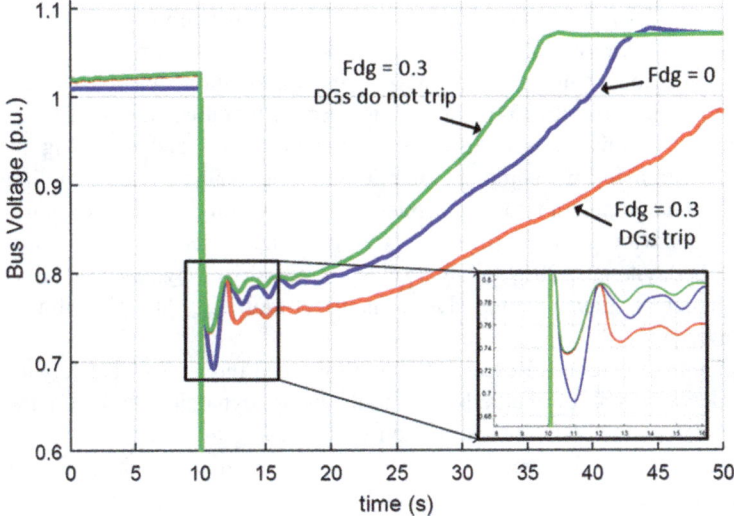

Fig. 2.40 Comparison of voltage recovery with DG tripping, not tripping against total absence of DG [54]

the DG (Fdg) for this case is 0.3. For the purpose of the study, the DG Voltage Ride Through (VRT) shown in Table 2.4 are applied.

In order to meet the Voltage Ride Through criteria, the DGs are modeled to meet the standards shown in Table 2.4. Figure 2.40 compares this case with (a) fault recovery in the absence of DG and (b) fault recovery in the presence of an always connected DG. From the plot, it can be ascertained that when DG exists, but then eventually trips, the recovery is much slower and can possibly cause a violation of the transient voltage criteria. This behavior is due to the fact that the DG, which was providing local active and reactive power, is suddenly disconnected, the voltage drops as a result of insufficient reactive and active power. In this case study, it is assumed that all the DGs will trip when their voltage levels are outside the no-trip boundary. According to IEEE standard 1547, the DGs cannot restart for 5 mins after tripping, provided that voltage and frequency have recovered to within tolerance. Therefore, in this case study, where the simulation is expected to run to approximately 1 min, the DGs are not set to restart after they trip.

2.12 Future Research Directions

As the key components that impact the short-term voltage stability of a system are induction motor loads and DERs with smart inverters that are physically present in the distribution system, representing the full distribution system dynamics instead of the aggregated composite load model has been of recent interest in utilities and in academia. As the traditional grid simulators for transmission and distribution systems have been developed and optimized over several years, applying a single tool to study the combined transmission and distribution dynamics often leads to numerical instabilities [54]. Instead, the recent literature has been focused on interfacing transmission and distribution system solvers and operating them in tandem by transferring common quantities at the boundary at each time step to perform co-simulation of Transmission-Distribution Systems. Capturing the dynamic behavior of the distribution system components is very important, an example is the August 2019 UK blackout that was driven by cascaded tripping of large number of smart inverters due to incorrect fault ride through settings in the smart inverters [56].

Academic researchers, Research Laboratories, and Industry are actively involved in the development of methods and tools using dynamic T&D co-simulation and data-driven methods for distribution system using micro-PMUs [54–56]. The industry and various utilities in association with various research laboratories are conducting research to further understand and explore the utilization of the distribution system assets to mitigate the short-term voltage instabilities by developing suitable control methods.

References

1. P. Kundur, J. Paserba, V. Ajjarapu, et al., Definition and classification of power system stability IEEE/CIGRE joint task force on stability terms and definitions. IEEE Trans. Power Syst. **19**(3), 1387–1401 (2004). https://doi.org/10.1109/TPWRS.2004.825981
2. N. Hatziargyriou, J.V. Milanovic, C. Rahmann, et al., "Definition and Classification of Power System Stability Revisited & Extended," in IEEE Transactions on Power Systems, https://doi.org/10.1109/TPWRS.2020.3041774
3. N. Hatziargyriou, J. Milanovic, C. Rahmann, et al, "Stability definitions and characterization of dynamic behavior in systems with high penetration of power electronic interfaced technologies," IEEE, Piscataway, NJ, USA, Tech. Rep. PES-TR77, 2020
4. P. Kundur, "Power System Stability and Control" textbook
5. P.W. Sauer, M.A. Pai, Power system steady-state stability and the load flow Jacobian. IEE Trans. Power Syst. **5**(4), 1374–1383 (1990)
6. V. Ajjarapu, *Computational Techniques for Voltage Stability Assessment and Control* (Springer Science Business Media, LLC, 2006)
7. S. Greene, I. Dobson, F.L. Alvarado, Sensitivity of the loading margin to voltage collapse with respect to arbitrary parameters. IEEE Trans. Power Syst. **12**(1), 262–272 (1997). https://doi.org/10.1109/59.574947
8. T.V. Cutsem, C. Vournas, *Voltage Stability of Electric Power Systems* (Text Book, Springer Science Business Media, LLC, 2008)

9. V. Vittal, Consequence and impact of electric utility industry restructuring on transient stability and small-signal stability analysis. Proc. IEEE **88**(2), 196–207 (2000). https://doi.org/10.1109/5.823998
10. "Distributed Energy Resources" NERC Report, Feb 2017. https://www.nerc.com/comm/Other/essntlrlbltysrvcstskfrcDL/Distributed_Energy_Resources_Report.pdf
11. Staff Report, "Distributed Energy Resources—Technical Considerations for the Bulk Power System", Docket No. AD18–10–000, 2018
12. A.K. Bharati, V. Ajjarapu, Investigation of relevant distribution system representation with DG for voltage stability margin assessment. IEEE Trans. Power Syst. **35**(3), 2072–2081 (2020). https://doi.org/10.1109/TPWRS.2019.2950132
13. L.R. de Araujo, D.R.R. Penido, J.L.R. Pereira, S. Carneiro, Voltage security assessment on unbalanced multiphase distribution systems. IEEE Trans. Power Syst. **30**(6), 3201–3208 (2015). https://doi.org/10.1109/TPWRS.2014.2370098
14. X.-P. Zhang, J. Ping, E. Handschin, Continuation three-phase power flow: A tool for voltage stability analysis of unbalanced three-phase power systems. IEEE Trans. Power Syst. **20**(3), 1320–1329 (2005). https://doi.org/10.1109/TPWRS.2005.851950
15. H. Sun, Q. Guo, B. Zhang, Y. Guo, Z. Li, J. Wang, Master–slave-splitting based distributed global power flow method for integrated transmission and distribution analysis. IEEE Trans. Smart Grid **6**(3), 1484–1492 (2015)
16. Z. Li, *Distributed Transmission Distribution Coordinated Energy Management Based on Generalized Master-Slave Splitting Theory* (Springer Publications) Text Book
17. K. Vu, M.M. Begovic, D. Novosel, M.M. Saha, Use of local measurements to estimate voltage-stability margin. IEEE Trans. Power Syst. **14**(3), 1029–1035 (1999)
18. S. Corsi, G.N. Taranto, A real-time voltage instability identification algorithm based on local phasor measurements. IEEE Trans. Power Syst. **23**(3), 1271–1279 (2008)
19. F. Hu, K. Sun, A. Del Rosso, E. Farantatos, N. Bhatt, Measurement-based real-time voltage stability monitoring for load areas. IEEE Trans. Power Syst. **31**(4), 2787–2798 (2016)
20. M. Glavic, D. Novosel, E. Heredia, D. Kosterev, A. Salazar, F. Habibi-Ashrafi, M. Donnelly, See it fast to keep calm: Real-time voltage control under stressed conditions. IEEE Power Energy Mag. **10**(4), 43–55 (2012)
21. P. Kessel, H. Glavitsch, Estimating the voltage stability of a power system. IEEE Trans. Power Deliv.., PWRD- **1**, 346–354 (1986)
22. Y. Wang, I.R. Pordanjani, W. Li, W. Xu, T. Chen, E. Vaahedi, J. Gurney, Voltage stability monitoring based on the concept of coupled single-port circuit. IEEE Trans. Power Syst. **26**(4), 2154–2163 (2011)
23. H.Y. Su, C.W. Liu, Estimating the voltage stability margin using PMU measurements. IEEE Trans. Power Syst. **31**(4), 3221–3229 (2016)
24. L. Ramirez and I. Dobson, "Monitoring voltage collapse margin with synchrophasors across transmission corridors with multiple lines and multiple contingencies," 2015 IEEE Power & Energy Society General Meeting, Denver, CO, 2015, pp. 1–5
25. S.S. Biswas and A.K. Srivastava, "Voltage Stability Monitoring in Power Systems, " U.S. Patent, Feb. 25, 2014
26. A. Ramapuram-Matavalam, V. Ajjarapu, Sensitivity based Thevenin index with systematic inclusion of reactive power limits. IEEE Trans. Power Syst. **33**(1), 932–942 (2018)
27. C. Canizares, A. De Souza, V. Quintana, Comparison of performance indices for detection of proximity to voltage collapse. IEEE Trans. Power Syst. **11**(3), 1441–1450 (1996)
28. V. Ajjarapu, C. Christy, The continuation power flow: A tool for steady state voltage stability analysis. IEEE Trans. Power Syst. **7**(1), 416–423 (1992)
29. A.R. Ramapuram Matavalam, A. Singhal, V. Ajjarapu, Monitoring long term voltage instability due to distribution and transmission interaction using unbalanced μPMU and PMU measurements. IEEE Trans. Smart Grid **11**(1), 873–883 (2020). https://doi.org/10.1109/TSG.2019.2917676
30. A.R. Ramapuram Matavalam, A. Singhal and V. Ajjarapu, "Identifying Long Term Voltage Stability Caused by Distribution Systems vs Transmission Systems," 2018 IEEE Power &

Energy Society General Meeting (PESGM), Portland, OR, 2018, pp. 1–5, doi: https://doi.org/10.1109/PESGM.2018.8586328
31. DOE-NERC FIDVR Conf., Sep. 29, 2009. [Online]. Available: http://www.nerc.com/files/FIDVR-Conference-Presentations-9-29-09.pdf
32. R.D. Quint, "A Look into Load Modeling: The Composite Load Model, Dynamic Load Modeling and FIDVR Workshop," https://esdr.lbl.gov/sites/all/files/6b-quint-composite-load-model-data.pdf, 2015
33. Consortium of Electric Reliability Technology Solutions, "Fault Induced Delayed Voltage Recovery (FIDVR), https://certs.lbl.gov/initiatives/fidvr
34. Modeling and Validation Work Group, "White paper on modeling and studying FIDVR events," Western Electricity Coordinating Council, Technical Report, October 20 (2011)
35. S. Robles, "2012 FIDVR Events Analysis on Valley Distribution Circuits," LBNL report by Southern California Edison, 2013
36. A. von Meier et al., Precision micro-Synchrophasors for distribution systems: A summary of applications. IEEE Trans. Smart Grid **8**, 6 (2017)
37. NERC, "Standard TPL-001-4—Transmission System Planning Performance Requirements," http://www.nerc.com/files/TPL-001-4.pdf, 2014
38. PJM Transmission Planning Department, "EXELON Transmission Planning Criteria", March 11, 2009
39. North American Transmission Forum, Transient voltage criteria reference document, 2016
40. WECC, "WECC Dynamic Composite Load Model Specifications," https://www.wecc.biz/Reliability/WECC%20Composite%20Load%20Model%20Specifications%2001-27-2015.docx, 2015
41. P.C. Krause, O. Wasynczuk, S.D. Sudhoff, P.C. Kraus, Analysis of electric machinery and drive systems. New York: IEEE Press. (2002)
42. Siemens PTI Power Technologies Inc., PSS/E 33, Program Application Guide, Vol. II, 2011
43. A.R. Ramapuram Matavalam, V. Ajjarapu, PMU-based monitoring and mitigation of delayed voltage recovery using admittances. IEEE Trans. Power Syst. **34**(6), 4451–4463 (2019)
44. A.R. Ramapuram Matavalam," Online monitoring & mitigation of voltage instability in transmission and distribution systems using synchrophasors", PhD Dissertation, Iowa State University
45. Load Model Data Tool [online] https://svn.pnl.gov/LoadTool
46. K. Zhang, H. Zhu, S. Guo, Dependency analysis and improved parameter estimation for dynamic composite load modeling. IEEE Trans. Power Syst. **32**(4), 3287–3297 (2017). https://doi.org/10.1109/TPWRS.2016.2623629
47. M.W. Tenza, and S. Ghiocel. "An analysis of the sensitivity of WECC grid planning models to assumptions regarding the composition of loads." Mitsubishi Electric Power Products (2016)
48. M. Paramasivam, "Dynamic optimization based reactive power planning for improving short-term voltage performance", PhD Dissertation, Iowa State University
49. S. Dasgupta, M. Paramasivam, U. Vaidya, V. Ajjarapu, Real-time monitoring of short-term voltage stability using PMU data. IEEE Trans. Power Syst. **28**(4), 3702–3711 (2013)
50. M.T. Rosenstein, J.J. Collins, C.J.D. Luca, D. Luca, A practical method for calculating largest Lyapunov exponents from small data sets. Physica D: Nonlin. Phenom. **65**(1–2), 117–134 (1993)
51. A. Reddy, K. Ekmen, V. Ajjarapu and U. Vaidya, "PMU based real-time short term voltage stability monitoring—Analysis and implementation on a real-time test bed," 2014 North American Power Symposium (NAPS), Pullman, WA, 2014, pp. 1–6
52. Federal Energy Regulatory Commission, "Requirements for Frequency and Voltage Ride Through Capability of Small Generating Facilities," https://www.ferc.gov/whats-new/comm-meet/2016/072116/E-11.pdf, 2016
53. IEEE Standard for Interconnection and Interoperability of Distributed Energy Resources with Associated Electric Power Systems Interfaces," in IEEE Std 1547-2018 (Revision of IEEE Std 1547-2003), vol., no., pp.1–138, 6 April 2018

54. R. Venkatraman, S.K. Khaitan, V. Ajjarapu, Dynamic co-simulation methods for combined transmission-distribution system with integration time step impact on convergence. IEEE Trans. Power Syst. **34**(2), 1171–1181 (2019)
55. Q. Huang, V. Vittal, Application of electromagnetic transient-transient stability hybrid simulation to FIDVR study. IEEE Trans. Power Syst. **31**(4), 2634–2646 (2016)
56. National Grid ESO, "Technical Report on the events of 9 August 2019", Report 6th September 2019
57. A.K. Bharati, V. Ajjarapu, "A Scalable Multi-Timescale T&D Co-Simulation Framework using HELICS", 2021 IEEE Texas Power and Energy Conference (TPEC), College Station, TX, USA, 2021.

Chapter 3
Basics of Transient Stability Assessment

U. D. Annakkage

3.1 Introduction

Power system stability in simple terms is the ability of synchronous machines connected to a power system to remain in synchronism when subjected to a disturbance. A much broader definition is found in [1] as "the property of a power system that enables it to remain in a state of operating equilibrium under normal operating conditions and to regain an acceptable state of equilibrium after being subjected to a disturbance with most system variables bounded so that practically the entire system remains intact". This definition captures all types of stability, namely, *rotor angle stability, frequency stability and voltage stability*. A more recent publication [2] defines two additional classes of stability: *resonance stability and converter-driven stability*. The focus in this chapter is the transient rotor angle stability commonly referred to as transient stability.

Let us consider the behaviour of a synchronous generator under three conditions: (a) at steady state, (b) during a disturbance and (c) after the removal of the disturbance. Let us also assume, for simplicity, that the fault is cleared without any changes to the network (no line tripping). This means that the network is unchanged from the pre-fault system.

At the steady state, the rotor of a synchronous generator rotates at the synchronous speed. The stator winding produces balanced three-phase voltages and currents at the nominal frequency. The balanced three-phase currents in the stator winding produce a magnetic field in the air gap that rotates at the synchronous speed. Since the rotor is also rotating at the synchronous speed, the field winding

---

U. D. Annakkage (✉)
Department of Electrical and Computer Engineering, University of Manitoba, Winnipeg, MB, Canada
e-mail: udaya.annakkage@umanitoba.ca

and amortisseur windings on the rotor see a constant flux linkage (relative speed is zero). The synchronous reactance of the generator is equal to the sum of mutual and leakage reactances. The circuit model of the generator is a voltage source E in series with the synchronous reactance, X_d [3–5]. Since the rotating magnetic field produced by the stator rotates at the same speed as the rotor, they maintain a constant relative angle. This angle is referred to as the load angle δ, and it is the phase angle of E. The angle δ is responsible for transferring active power from the generator to the terminal bus.

When a disturbance occurs, for example, a three-phase short circuit at the terminals, the balance between the driving torque from the turbine and the opposing electromagnetic torque is lost and the rotor starts to accelerate. The acceleration continues during the fault. This means, unlike at the steady state, the rotating magnetic field produced by the stator currents and the rotor no longer rotate at the same speed. The field windings see this relative speed as a change of flux linkages. The effective reactance of the generator reduces to its sub-synchronous reactance, and the internal-induced voltage E starts to change. This changing voltage is known as the sub-transient voltage, E''. The sub-transient voltage is a function of fluxes in the main field winding as well as the fluxes in the amortisseur windings. The circuit model is a sub-transient voltage E'' in series with its sub-transient reactance X_d''. Please refer to [4, 5] for the mathematical derivations. During the entire transient period, E'' continuously changes as the fluxes in all rotor windings change with time.

The angular acceleration of the rotating mechanical system (the rotor, shaft and turbines) depends on the net accelerating torque and the inertia of the mechanical system. For a direct short circuit at the terminals of the generator, the electromagnetic torque is zero. For a remote fault or a fault at the terminals through a fault resistance, the electromagnetic torque is not zero, but decreases to a small value. The driving torque remains unchanged as the speed-governing system cannot respond quickly to the increasing speed of the generator. The acceleration also causes the load angle of the generator to increase.

The turbine-generator system has two main control systems, namely, the speed-governing system and the excitation system control system. As mentioned earlier, the speed-governing system is typically slow to respond. However, the excitation system responds fast to bring the terminal voltage to the reference value by adjusting the dc voltage applied to the field winding. Since the terminal voltage drops to zero or almost zero during the fault, the controllers of the excitation system demand a very large dc voltage to be applied to the field winding. Such large voltages are not practical and thus the controller imposes upper and lower limits on the dc voltage. During the fault, this field voltage reaches the upper limit or the excitation system ceiling.

Let us see what happens immediately after clearing the fault with the assumption that the protection system clears the fault soon enough to avoid any overspeeds or pole slipping. The terminal voltage and the electromagnetic torque will start to build up. The electromagnetic torque depends on the terminal voltage and the load angle. The rotor angle, δ, will advance during the fault due to the acceleration of the mechanical system. As the terminal voltage recovers to the pre-fault levels, the electromagnetic torque will increase beyond the pre-fault level because the rotor angle has increased during the fault. This means the electromagnetic torque is greater than the driving torque resulting in a deceleration of the rotating mechanical system. Although the rotating system starts to decelerate, it has acquired some overspeed during the fault. The speed will start to decrease to the synchronous speed, but the rotor angle will continue to increase as long as the speed is above the synchronous speed. If the protection system clears the fault sufficiently fast, the speed of the generator will come down to synchronous speed before the rotor angle advances too far. In these conditions one can say that the generator survived the first swing or the generator is "first swing stable". The energy-based explanation for the sequence of events explained above is that the rotating mechanical system gains kinetic energy during the fault. The stability of the generator depends on its ability to transfer this extra energy to the power system.

What happens after the first swing? If there is no damping, the mechanical system will have sustained oscillations. In practice however, there is damping due to the induced currents in the amortisseur windings as well as some damping from the system. There can also be situations where the net damping is negative and the oscillations will grow in amplitude. This is a "small signal instability or oscillatory instability" situation. In order to mitigate this, the excitation system has an auxiliary controller, the "power system stabilizer" to introduce sufficient damping. The power system stabilizer (PSS) introduces damping by modulating the reference voltage of the excitation system.

The rest of this chapter is devoted to presenting mathematical modelling of the main components of a power system, simulation of transient stability, applications, limitations and new developments.

3.2 Dynamics of the Shaft System

The shaft system is subjected to three torques. The prime mover (turbine) provides the driving torque T_m. The main opposing torque, T_e, is due to the interaction of the two rotating magnetic fields, and it is responsible for the conversion of mechanical energy to electrical energy. The third torque is proportional to the angular speed. It is the damping torque. The net torque gives rise to acceleration or deceleration of the shaft system. The dynamics of the shaft system can be expressed by (3.1). The angular speed of the rotor is ω_r, and its relative speed with respect to the synchronous speed, ω_0, expressed in per unit is $\Delta\omega_r = \omega_r - \omega_0$. The angular position, δ, is relative to a reference frame rotating at the synchronous speed. The damping coefficient is D. Time, t, is in seconds and H is the inertia constant of the rotating shaft system in seconds.

$$\frac{d^2\delta}{dt^2} = \frac{\omega_0}{2H}(T_m - T_e - D\Delta\omega_r) \qquad (3.1)$$

It is convenient to express the second-order differential equation (3.1) as two first-order differential equations (3.2–3.3).

$$\frac{d\Delta\omega_r}{dt} = \frac{1}{2H}(T_m - T_e - D\Delta\omega_r) \quad (3.2)$$

$$\frac{d\delta}{dt} = \omega_0 \Delta\omega_r \quad (3.3)$$

3.3 Dynamics of the Electrical System of the Synchronous Machine

The electrical system of the synchronous machine can be modelled as a set of six coupled coils. Table 3.1 describes these coils and the notation used to identify them. Detailed derivations of the models are not presented here. The derivations can be found in many textbooks including [4, 5]. We use a notation consistent with [5].

The coils a, b and c are stationary, while the coils fd, $1d$, $1q$ and $2q$ rotate as they are on the rotor. This makes the inductances associated with this system of coils time dependent. The most common simplification used in modelling synchronous machines is to transform the quantities a, b and c into a fictitious set of coils d, q and 0. The original contributions on a, b, c to $d, q, 0$ transformation are found in [8, 9]. References [4, 5] present the use of this transformation for transient stability simulation. The differential equations governing the dynamics of the electrical system of the machine can be expressed by the four differential equations (3.4–3.7). All quantities in (3.2–3.7) except time t are in per unit, and the definition of terms are given in Tables 3.2 and 3.3.

$$\frac{d}{dt}\psi_{fd} = \omega_0\left[e_{fd} - \frac{R_{fd}}{L_{fd}}\psi_{fd} + \frac{R_{fd}}{L_{fd}}L_{ad}''\left(-i_d + \frac{\psi_{fd}}{L_{fd}} + \frac{\psi_{1d}}{L_{1d}}\right)\right] \quad (3.4)$$

Table 3.1 Description of coupled coils in a synchronous machine

Description	Notation
Phase "a" of stator	a
Phase "b" of stator	b
Phase "c" of stator	c
Field winding	fd
Direct axis amortisseur winding	$1d$
Quadrature axis amortisseur winding 1	$1q$
Quadrature axis amortisseur winding 2	$2q$

3 Basics of Transient Stability Assessment

Table 3.2 Definition of machine parameters

Description	Notation
Resistance of the field winding	R_{fd}
Resistance of the direct axis amortisseur winding	R_{1d}
Resistance of the quadrature axis amortisseur winding 1	R_{1q}
Resistance of the quadrature axis amortisseur winding 2	R_{2q}
Leakage inductance of the field winding	L_{fd}
Leakage inductance of the d-axis amortisseur winding	L_{1d}
Leakage inductance of the q-axis amortisseur winding 1	L_{1q}
Leakage inductance of the d-axis amortisseur winding 2	L_{2q}
Sub-transient mutual inductance on the d-axis	L_{ad}''
Sub-transient mutual inductance on the q-axis	L_{aq}''

Table 3.3 Definition of machine variables

Description	Notation	Type
Relative angular speed of the rotor	$\Delta\omega_r$	State variable
Relative angular position of the rotor	δ	State variable
Flux in the field winding	ϕ_{fd}	State variable
Flux in the direct axis amortisseur winding	ϕ_{1d}	State variable
Flux in the quadrature axis amortisseur winding 1	ϕ_{1q}	State variable
Flux in the quadrature axis amortisseur winding 2	ϕ_{2q}	State variable
Voltage applied to the field winding	e_{fd}	Input
Driving torque in (3.2)	T_m	Input
Stator currents	i_d and i_q	Machine to network coupling variables
Stator voltages	e_d and e_q e_d and e_q	Machine to network coupling variables

$$\frac{d}{dt}\psi_{1d} = \omega_0\left[-\frac{R_{1d}}{L_{1d}}\psi_{1d} + \frac{R_{1d}}{L_{1d}}L_{ad}''\left(-i_d + \frac{\psi_{fd}}{L_{fd}} + \frac{\psi_{1d}}{L_{1d}}\right)\right] \quad (3.5)$$

$$\frac{d}{dt}\psi_{1q} = \omega_0\left[-\frac{R_{1q}}{L_{1q}}\psi_{1q} + \frac{R_{1q}}{L_{1q}}L_{aq}''\left(-i_q + \frac{\psi_{1q}}{L_{1q}} + \frac{\psi_{2q}}{L_{2q}}\right)\right] \quad (3.6)$$

$$\frac{d}{dt}\psi_{2q} = \omega_0\left[-\frac{R_{2q}}{L_{2q}}\psi_{2q} + \frac{R_{2q}}{L_{2q}}L_{aq}''\left(-i_q + \frac{\psi_{1q}}{L_{1q}} + \frac{\psi_{2q}}{L_{2q}}\right)\right] \quad (3.7)$$

It should be noted that this model does not have differential equations for the fluxes in the d and q axes, ψ_d and ψ_q. This is because of the assumption that the transients in the magnetic fluxes in stator windings are much faster and better damped than the electromechanical dynamics we are interested in. This is the difference between the synchronous machine models used in transient stability programs and those used in electromagnetic transient (EMT) programs.

The variables that couple the synchronous machine and the network are e_d, e_q, i_d and i_q. In transient stability simulation, these variables along with the voltages and currents in the network are treated as algebraic variables. Similar to the assumption made about the transients in stator winding fluxes, it is assumed in transient stability programs that the electromagnetic transients in the network are much faster and better damped than the electromechanical transients we are interested in. The network is modelled as constant impedances computed at the nominal frequency of 60 or 50 Hz.

The d-axis and q-axis components of the sub-transient internal voltages are given by (3.8) and (3.9). The reactances X_{ad} and X_{aq} can be replaced with their saturated values, X_{ads} and X_{aqs}, respectively, for more accurate representation.

$$E_d^{''} = -X_{aq}\left(\frac{\psi_{1q}}{L_{1q}} + \frac{\psi_{2q}}{L_{2q}}\right) \tag{3.8}$$

$$E_q^{''} = X_{ad}\left(\frac{\psi_{fd}}{L_{fd}} + \frac{\psi_{1d}}{L_{1d}}\right) \tag{3.9}$$

A further assumption made in transient stability simulation is that the sub-transient saliency is negligible ($X_d^{''} = X_q^{''} = X^{''}$). With this assumption, the synchronous machine can be modelled as a voltage source of $E^{''}$ in series with the sub-transient reactance $X^{''}$ (the armature resistance R_a can be added for better accuracy). Alternatively, the Norton equivalent of a current source in parallel with the sub-transient admittance can be used.

3.3.1 Interface Between the Network and Synchronous Generators

The interface between the network and synchronous machines is modelled as follows:

1. At the beginning of each time step of integration, determine e_d, e_q, i_d and i_q from the network quantities using $e_d = E_R \sin\delta - E_I \cos\delta$, $e_q = E_I \sin\delta + E_R \cos\delta$, $i_d = I_R \sin\delta - I_I \cos\delta$ and $i_q = I_I \sin\delta + I_R \cos\delta$. The subscripts R and I represent real and imaginary parts, respectively, of the corresponding phasor.
2. The electromagnetic torque, T_e, in (3.2) is modelled as $T_e = e_d i_d + e_q i_q$ assuming system frequency does not deviate much from the nominal frequency. A more accurate equation that includes the effect of change in ω is $T_e = (e_d i_d + e_q i_q)/\omega_r$
3. Update state variables of the synchronous machine and determine $E_d^{''}$ and $E_q^{''}$ using (3.8) and (3.9).

4. Compute the sub-transient voltages in common network reference, $E_R^{''}$ and $E_I^{''}$, using $E_R^{''} = E_d^{''} \sin \delta + E_q^{''} \cos \delta$ and $E_I^{''} = E_q^{''} \sin \delta + E_d^{''} \cos \delta$.
5. Represent the generator using the sub-transient voltage $E^{''} = E_R^{''} + j E_I^{''}$ in series with $R_a + jX^{''}$, where R_a is the resistance of the armature winding.

3.3.2 Simulation of Transient Stability

There are three stages to the simulation. The first step is to solve the steady-state power flow for the pre-fault power system and compute the initial values of all state variables. The general guideline is to set all derivatives to zero and solve for steady state. Details of the initial condition calculation are found in [5]. The second step is to apply the fault and solve differential equations of the generators and the algebraic equations of the network to interface generators to the network as described above using a suitable integration time step. The last step is to remove the fault and continue solving differential and algebraic equations.

All generators in the network are represented as described above. It is more convenient to use the Norton equivalent form of the generator model for solving the network equations. Assume that all loads are modelled using constant admittances. More detailed representation of loads will be discussed in Sect. 3.4. Since the loads are constant admittances, they can be included in the network admittance matrix. Now the voltages of all nodes can be updated by solving $I = YV$ for the network. The computational burden can be reduced considerably by eliminating all the nodes where there are no generators connected.

There are two options to proceed from here. The simple approach is to start the computations for the next time step using the updated terminal voltages and currents. This approach has an inherent delay of one time step between the network solution and the solution of differential equations. A more accurate approach is to solve the network at the end of each step of the numerical solution algorithm.

3.4 Load Models

In the previous section, the simplest load model, which is the constant admittance representation, was used. Practical loads do not always behave as constant admittances. A more accurate representation of loads can be found in [11, 12]. Load models can be divided into two main groups, namely, static load models and dynamic load models.

3.4.1 Static Load Model

The constant admittance load model is a static load model, and it is a linear model in terms of voltage and current. A more generic static load model combines linear and non-linear components. The most common non-linear static load model is ZIP model which is a constant impedance (Z), constant current (I) and constant power (P) combined model shown in (3.10) and (3.11).

$$P = P_0 \left[K_{pz} \left(\frac{V}{V_0} \right)^2 + K_{pi} \left(\frac{V}{V_0} \right) + K_{pp} \right] \quad (3.10)$$

$$Q = Q_0 \left[K_{qz} \left(\frac{V}{V_0} \right)^2 + K_{qi} \left(\frac{V}{V_0} \right) + K_{qq} \right] \quad (3.11)$$

where,

$$K_{pz} + K_{pi} + K_{pp} = 1$$

and,

$$K_{qz} + K_{qi} + K_{qp} = 1$$

The other commonly used non-linear load model is the exponential load model given in (3.12) and (3.13).

$$P = P_0 \left(\frac{V}{V_0} \right)^{np} \quad (3.12)$$

$$Q = Q_0 \left(\frac{V}{V_0} \right)^{nq} \quad (3.13)$$

In the above load model, $np = nq = 0$ gives a constant power load, $np = nq = 1$ gives a constant current load, and $np = nq = 2$ gives a constant impedance load. Reference [11] recommends the use of any other value, including values outside the range 0–2, to represent the aggregate effect of different types of loads.

3.4.2 Dynamic Load Models

The dynamic characteristics of a load can be captured using a generic dynamic load model which is an exponential recovery type model or simply using an equivalent induction motor load.

3.4.2.1 Generic Dynamic Load Model [13, 14]

The generic dynamic load model captures the following properties of an aggregate load.

- A step decrease in voltage will result in a sudden decrease in load.
- A part of the change (or all of it) will be recovered later
- rate at which the recovery takes place

The dynamics are modelled using one state variable (load state variable) x_p described in (3.14) and (3.15). The block diagram of the model is shown in Fig. 3.1. The variation of load, P_d, with time is shown in Fig. 3.2

$$\frac{dx_p}{dt} = \frac{1}{T_p}(P_s(v) - P_d) \quad (3.14)$$

$$x_p = P_d - P_t(v) \quad (3.15)$$

T_p : Time Constant
P_d : Load
$P_s(v)$: steady-state load (non-linear model) $P_0 \left(\frac{v}{v_0}\right)^{nps}$
$P_t(v)$: Transient change $P_0 \left(\frac{v}{v_0}\right)^{npt}$

Reference [13] gives theoretical derivations of exponential recovery load models for the induction motor, tap-changing transformer and heating load. A model similar to the above (3.14–3.15) can be used for reactive power too.

3.4.2.2 Induction Motor Model

The dynamics of an induction motor can be captured quite accurately for transient stability simulations using a third-order model. The electric circuit model is shown in Fig. 3.3. R_s is the resistance of the stator winding, and X'_s is the transient reactance of the stator winding. The transient voltage V' depends on the flux in the rotor windings that are defined in terms of d and q axis voltage terms v'_d and

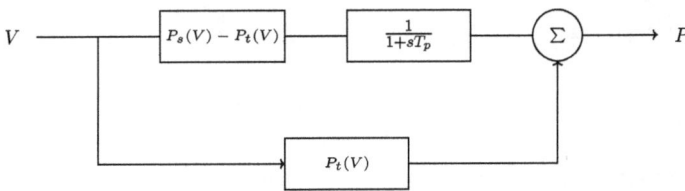

Fig. 3.1 Block diagram representation of the generic dynamic load model

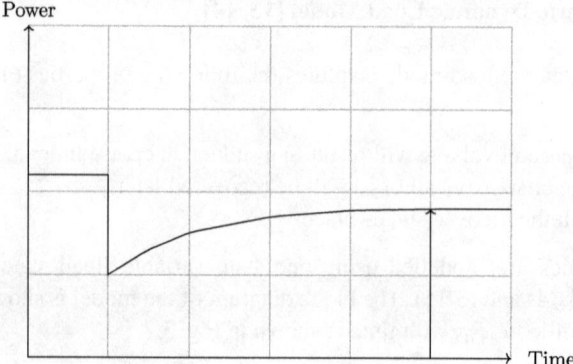

Fig. 3.2 Generic dynamic load model

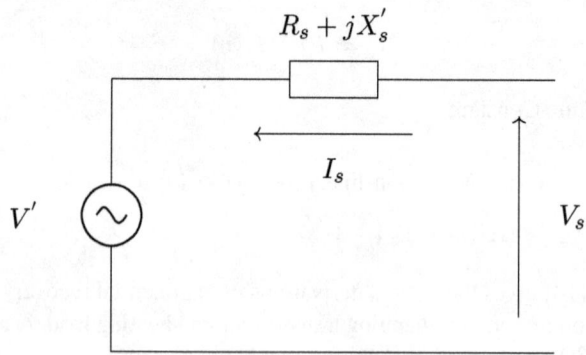

Fig. 3.3 Transient equivalent circuit

v_q'. The relevant algebraic and differential equations are given by (3.16) and (3.17). The currents $i_d = I_R \cos\theta + I_I \sin\theta$ and $i_q = -I_R \sin\theta + I_I \cos\theta$ are the d and q axis components of the stator current I_s. The angle θ is the phase angle of V_s. The currents, I_R and I_I, are the real and imaginary components of the stator current I_s. During the integration time step, V' is updated using $V' = V_R' + jV_I'$. The voltages $V_R' = v_d' \cos\theta - v_q' \sin\theta$ and $V_I' = v_d' \sin\theta + v_q' \cos\theta$ are the real and imaginary components of V'.

$$V_s = v_d + jv_q = (R_s + jX_s')(i_d + ji_q) + (v_d' + jv_q') \tag{3.16}$$

$$V_s = (R_s + jX_s')I_s + V' \tag{3.17}$$

Three differential equations, (3.18), (3.19) and (3.20), capture the dynamic behaviour of the induction motor.

$$pv_d' = -\frac{1}{T_0'}\left(v_d' + (X_s - X_s')i_q\right) + p\theta_r v_q' \tag{3.18}$$

$$pv_q' = -\frac{1}{T_0'}\left(v_q' + (X_s - X_s')i_d\right) - p\theta_r v_d' \tag{3.19}$$

$$T_e = (v'_d i_d + v'_q i_q)/\omega_s$$

$$\omega_s = 1.0 \quad \text{pu}$$

3.4.2.3 Inclusion of Load Models in a Transient Stability Program

As discussed in Sect. 3.3.2, constant admittance loads can be added to the bus admittance matrix and the load buses are eliminated. This leaves the network with only the nodes where generators are connected. When the load is represented by a non-linear static load, the Norton equivalent has to be used where the current and admittance are calculated using the node voltage at the previous time step. These nodes are not eliminated when the network is simplified.

When the load is represented by a dynamic load (generic dynamic model or induction motor), it is modelled as a current source as described above for static non-linear loads. In addition, the state variables are updated by solving differential equations in the same way as how the synchronous generator was treated in Sect. 3.3.2.

It should be noted that the above procedure introduces inaccuracies due to time step delay. One may solve the network and differential equations iteratively to reduce the errors due to time step delay at the expense of computing time. This is not necessary if a sufficiently small integration time step is used (e.g. quarter of a cycle).

3.5 Auxiliary Controllers

The models of dynamic devices presented in this chapter consist of state variables, algebraic variables and inputs. In the case of the synchronous generator, the inputs are the dc voltage applied to the field winding (E_{fd}) and the driving torque of the prime mover (T_m). These inputs to the model are the outputs of auxiliary controllers, excitation system and turbine governor system, respectively. Auxiliary controllers are always associated with a dynamic device that injects a current into the network, but they do not inject currents. Two examples of auxiliary controllers, excitation system and power system stabilizer, are discussed in Sects. 3.5.1 and 3.5.2 respectively. The complete model of a dynamic device has its current injection device model along with the models of auxiliary controllers.

3.5.1 Excitation System

The excitation system provides the dc field voltage, E_{fd}, to the field winding of the synchronous generator. There are three types of excitation systems, namely, the DC excitation systems, AC excitation systems and static excitation systems. Classification of excitation systems and standard types along with their components is found in [5] and [6]. The excitation system tries to maintain the terminal voltage of the synchronous machine at a specified voltage reference by adjusting the field voltage according to the error between the terminal voltage and its reference level (V_{ref}). In addition to this, there are protection functions. For example, there are protection functions to limit over-excitation and under-excitation. It is important to model the excitation system with limiters that introduce non-linearities accurately to capture the transient behaviour of the synchronous machine during a fault and recovery from a fault. For voltage stability simulation, it is important to model the excitation system with relevant details of protection systems.

3.5.2 Power System Stabilizer

The power system stabilizer (PSS) modulates the voltage reference of the excitation system, and it is designed to introduce a damping torque to electromechanical oscillations. The PSS plays an important role in keeping the synchronous machines synchronized with the rest of the power system when disturbances occur in the power system. Classification of PSS types and their components is found in [5] and [6].

3.6 Other Dynamic Devices

Any dynamic device connected to the network is modelled using algebraic and differential equations. The algebraic equations give a Norton equivalent as the interface between the device and the network. The current source is updated as the state variables evolve with time according to the differential equations. This approach is not different to the model of synchronous machines and induction machines described in Sects. 3.3.2 and 3.4. The input variables of the device are controlled by auxiliary controllers. They are modelled using differential equations similar to the way the excitation systems and power system stabilizers are modelled for synchronous machines.

Commercially available transient stability simulation programs have a library of dynamic models available to the user. Most programs allow the user to create their own dynamic models (user-defined models). The concept of creating a dynamic model is simple. There are two types of dynamic devices. They are (a) devices that inject a current into the network and (b) devices that serve as auxiliary controllers and modify the input variables of a dynamic device that injects a current into the network.

Devices that inject a current are modelled using a Norton equivalent. The current source is updated as the state variables evolve with time. This is identical to the way the synchronous machine, induction machine and dynamic load are modelled earlier in this chapter.

3.6.1 Modelling an HVDC Line

An HVDC system is treated as a dynamic device that injects currents into two (two terminal) or more (multiterminal) nodes of the ac power system. The modelling approaches are different for line-commutated converter (LCC) and voltage source converter (VSC) HVDC systems.

For the LCC HVDC systems, the current injected from a converter to the ac system is a non-linear function of the ac bus voltage, dc line current and the firing angle [5, 15]. A small admittance can be used for the Norton equivalent. The firing angle is determined by the HVDC controls, and therefore, it is important to consider the major control functions such as rectifier dc current/power control and inverter DC voltage/extinction angle control. Furthermore, the performance during the recovery from an AC fault such as inverter commutation failures is determined by the voltage-dependent current limiting functions. If there is a long DC transmission system, the DC line dynamics also need to be considered.

For the VSC HVDC systems, the current injections mainly determined by the d-q decoupled controllers are used. It can be assumed that the inner current controllers are fast and therefore, the current orders determined by the outer controllers such as d-axis power/DC voltage control and q-axis AC voltage/reactive power control functions can be directly injected into the AC system. Special control functions such as dynamic reactive current injection also need to be considered. The impedance of the converter transformer and a half of the phase reactor impedance are used to calculate the Norton admittance. If there is a long DC transmission system, the DC line dynamics also need to be considered [5, 15].

3.6.2 Non-linearities

It is important to identify and incorporate the non-linearities that influence the dynamic behaviour of the power system in the interested frequency bandwidth and time frame. These non-linearities include magnetic saturation of electrical machines, magnetic saturation of rotating machines in excitation systems, non-linear loads, limits in auxiliary control systems and dependent voltage and current sources in auxiliary control systems.

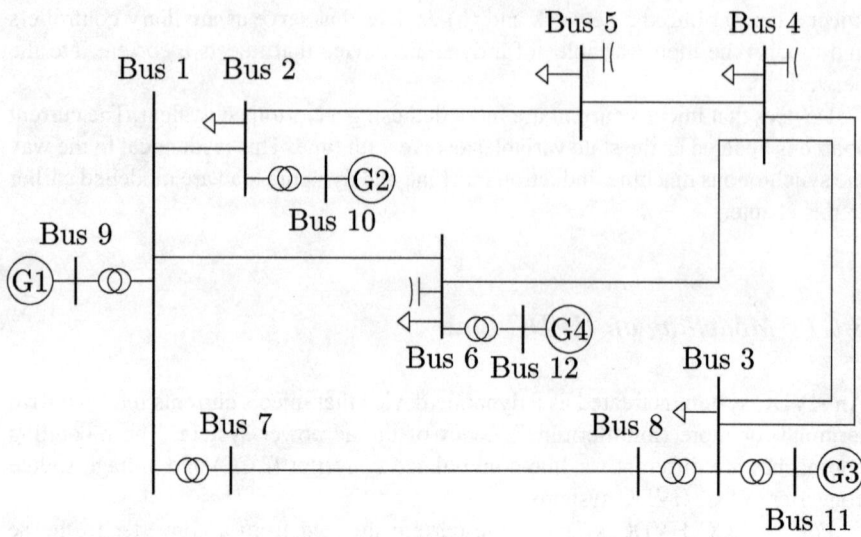

Fig. 3.4 One-line diagram of the 12-bus power system

3.7 A Case Study of Transient Stability Simulation

Transient simulations are performed at planning and operational stages by utilities, manufacturers and consultants. The size of the network can be as large as 100,000 buses with hundreds of generators. In order to present how to interpret simulation results, a small test system with 12 buses [16] shown in Fig. 3.4 is used in this section.

A three-phase fault is applied at Bus 12 for a duration of five cycles. The fault is removed without tripping any lines.

The voltage at Bus 12 is shown in Fig. 3.5. When the three-phase fault occurs at the terminals of the generator G4, the voltage of the bus drops to zero. The voltage then recovers quickly when the fault is cleared. It is important to have this quick recovery to maintain stability of the generator; the quicker the stored energy during the fault is released to the network, the better the chances will be for the generator staying synchronized to the rest of the system. This is because the electrical power transferred depends on both the voltage and the phase angle. After the recovery the voltage settles back to the pre-fault level after a few oscillations.

Figure 3.6 shows the speed of the generator G4. During the fault the generator accelerates. When the fault is cleared, it starts to decelerate due to (a) the rotor angle having advanced during the fault to a larger value during the acceleration period and (b) voltage recovering quickly as shown in Fig. 3.5. The result is to deliver a larger amount of power than the driving mechanical power as shown in Fig. 3.7. The speed settles to a constant level after a few oscillations. There is a PSS on generator G4. Without the PSS the oscillations would have lasted for a long period.

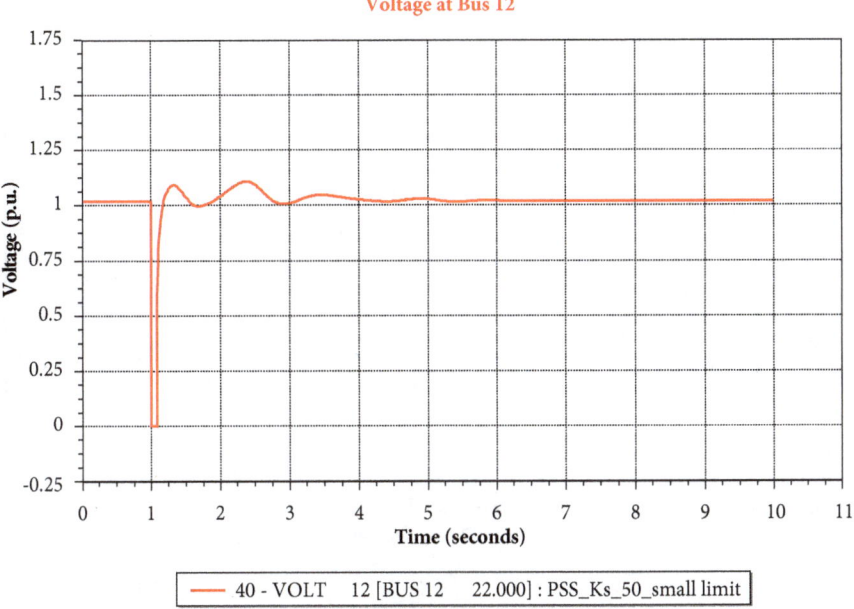

Fig. 3.5 Terminal voltage of generator G4

In this simulation the turbine and governor system was not modelled. Therefore, the speed of the generator does not return to the nominal level.

During the fault, the excitation system sees a large error in voltage and tries to rapidly increase the dc voltage applied to the field winding. However, the control system enforces a limit on the maximum value of the voltage as shown in Fig. 3.8. After clearing the fault, as the voltage recovers quickly, the error in voltage decreases and E_{fd} decreases. The field voltage E_{fd} goes below its pre-fault level when the terminal voltage is higher than the reference voltage.

The behaviour of a remote generator G3 is shown in Fig. 3.9. In this simulation, only the generator G4 has a tuned PSS. The oscillations in the speed of generator G3 can be damped out faster if a properly tuned PSS is added to that generator.

3.8 Limitations of Modelling

In conventional transient stability simulations, the network is modelled as a constant admittance matrix computed at the nominal frequency. This assumption leads to ignoring high-frequency electromagnetic transients. Typically those transients are highly damped. The purpose of a transient stability simulation is to observe electromechanical oscillations with frequencies in the range of up to about 5 Hz. Therefore, ignoring the electromagnetic transients in the network is acceptable.

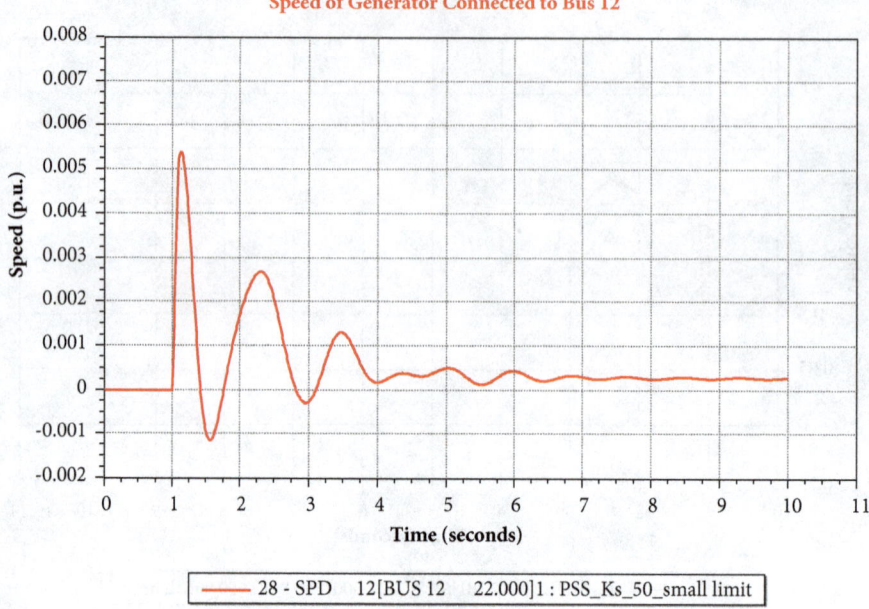

Fig. 3.6 Speed of generator G4

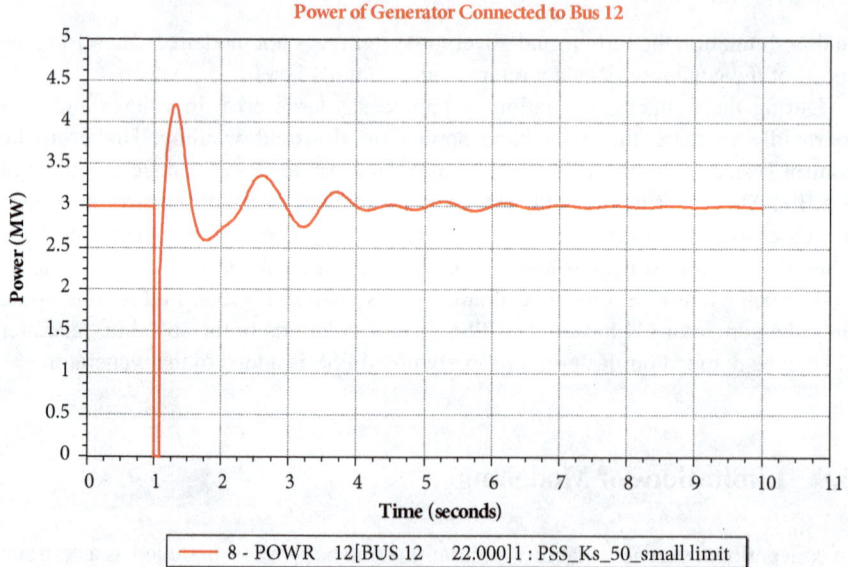

Fig. 3.7 Power output of generator G4

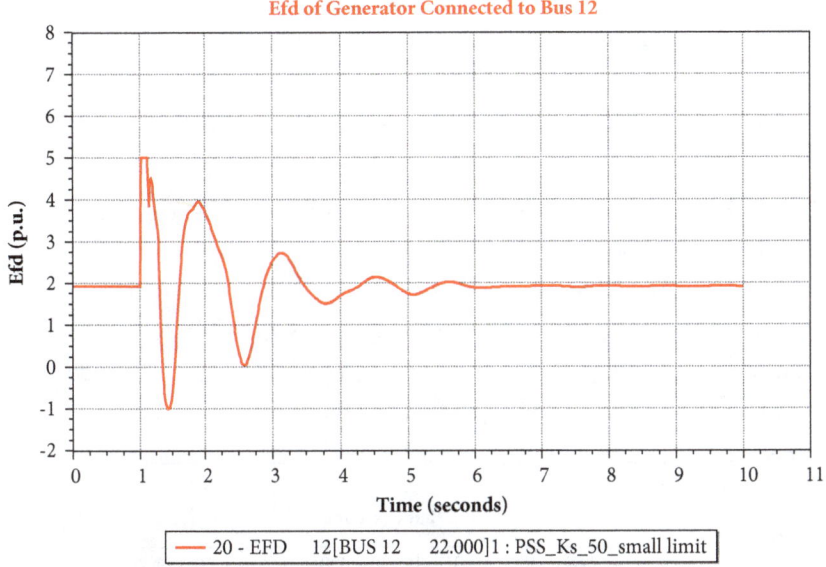

Fig. 3.8 Field voltage E_{fd} of generator G4

Similarly, the transients in the flux in stator windings are ignored in the models of rotating machines. This assumption is consistent with the above assumption of fast decaying electromagnetic network transients.

While the modelling approach presented in this chapter has well served the power engineers over several decades, there are new challenges due to the increasing presence of power electronic-based devices in the power grid. Most notable is the converter integrated renewable generation. The increased level of power transmission has also necessitated adding series capacitors to transmission lines. This has also contributed to the presence of resonance frequencies with the potential of adverse interactions with the dynamic devices. Therefore, in the modern power grid, the frequency bandwidth of dynamics of interest has extended into the sub-synchronous frequency range and in some occasions to even higher frequencies. The standard transient stability models do not capture those dynamics accurately. Although electromagnetic transient simulations can capture all these frequencies, it takes a prohibitive amount of computing time. Some new developments to address these concerns are discussed in the next section.

3.9 New Developments

In electromagnetic transient (EMT) simulations, differential equations are solved to obtain instantaneous values of voltages and currents. In transient stability (TS)

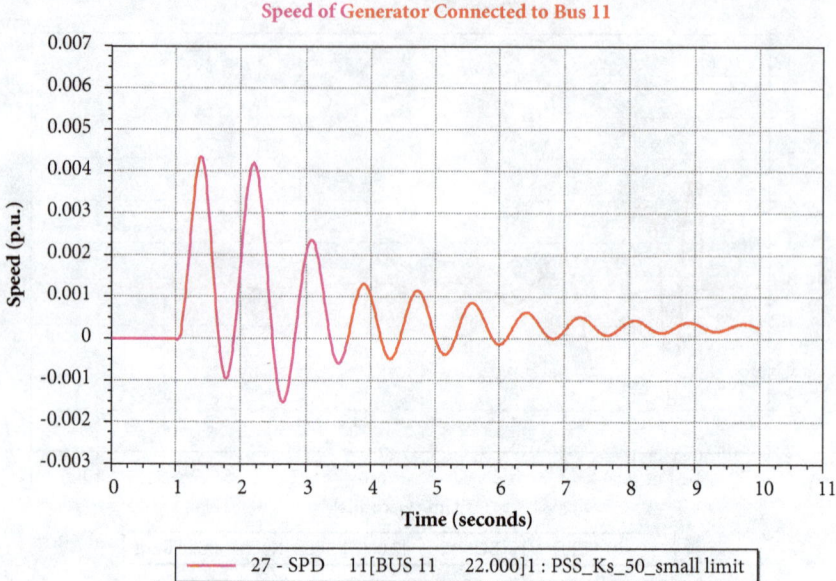

Fig. 3.9 Speed of generator G3

simulations, the phasors of voltages and currents are obtained by solving $I = YV$. This difference makes the integration time step in TS longer than in EMT and thus the computational burden of TS simulations less than EMT simulations. In TS a typical integration time step is 5–10 ms, whereas in EMT it is 10–50 µs. In TS the phasors are treated as algebraic variables.

A method that was introduced into power system simulation in [17–19], shifted frequency analysis (SFA), allows extraction of the envelope of sinusoidal quantities [17]. As its name implies, the idea is to shift the frequency by $-\omega_0$ so that the sinusoidal quantities are transformed into time-varying phasors or dynamic phasors. In this approach phasors are treated as state variables governed by differential equations. This allows the network dynamics and transients in stator windings of rotating machines to be included in the simulation. The result is to increase the frequency bandwidth of the simulation while using a larger integration time step than an EMT simulation.

In recent years co-simulation has been proposed where the portion of the network for which the investigation is performed is modelled using accurate models (EMT or DP) and the rest of the power system is modelled using TS models. Two parts of the network are interfaced during the simulation by exchanging voltages and currents at the boundary busses [20–23]. These co-simulation options are likely to be standard options available in commercial software in the near future.

Bibliography

1. IEEE/CIGRE Joint Task Force on Stability Terms and Definitions, Definitions and classification of power system stability. IEEE Trans. Power Syst. **19**, 1387–1401 (2004)
2. IEEE PES Power System Dynamic Performance Committee, Stability definitions and characterization of dynamic behavior in systems with high penetration of power electronic interfaced technologies. IEEE Power and Energy Society, TECHNICAL REPORT PES-TR77, Apr (2020)
3. J. Granger, W. Stevenson, *Power System Analysis* (McGraw Hill, New York, 1994)
4. P. Anderson, A. Fouad, *Power System Control and Stability*. IEEE Series on Power Engineering (Wiley, New York, 2002)
5. P. Kundur, *Power System Stability and Control* (McGraw-Hill International Book Company, New York, 1994)
6. Excitation systems models for power system stability studies, IEEE committee report, in *IEEE Transactions on Power Apparatus and Systems*, vol. PAS-100 (1981), pp. 494–509
7. N. Hingorani, L. Gyugyi, *Understanding FACTS* (Wiley Interscience, New York, 1999)
8. R. Park, Two-reaction theory of synchronous machines generalized method of analysis-part I. Trans. Am. Inst. Electr. Eng. **48**, 716–727 (1929)
9. R. Park, Two-reaction theory of synchronous machines generalized method of analysis-part II. Trans. Am. Inst. Electr. Eng. **52**, 352–354 (1933)
10. IEEE recommended practice for excitation system models for power system stability studies. IEEE Std 421.5-1992
11. IEEE Task Force on Load representation for Dynamic Performance, Load representation for dynamic performance analysis, in *IEEE Transactions on Power Systems*, vol. 8 (1993), pp. 472–482
12. IEEE Task Force on Load representation for Dynamic Performance, Standard load models for power flow and dynamic performance simulation, in *IEEE Transactions on Power Systems*, vol. 10 (1995), pp. 1302–1313
13. D. Hill, Nonlinear dynamic load models with recovery for voltage stability studies. IEEE Trans. Power Syst. **8**, 166–176 (1993)
14. I.A. Hiskins, J.V. Milanovic, Load modeling in studies of power system damping. IEEE Trans. Power Syst. **10**, 1781–1788 (1995)
15. C. Karawita, U.D Annakkage, Control Block Diagram Representation of an HVDC System for Sub-Synchronous Frequency Interaction Studies, *Stevenage: IET* (2010)
16. S. Jiang, U.D. Annakkage, A.M. Gole, A platform for validation of FACTS models. IEEE Trans. Power Syst. **21**, 484–491 (2006)
17. P. Zhang, J.R. Marti, H.W. Dommel, Shifted-frequency analysis for EMTP simulation of power-system dynamics. IEEE Trans. Circuits Syst. **57**(9), 2564–2574 (2010)
18. V. Venkatasubramanian, H. Schattler, J. Zaborszky, Fast time-varying phasor analysis in the balanced three-phase large electric power system. IEEE Trans. Autom. Control **40**, 1975–1982 (1995)
19. A.M. Stankovic, S.R. Sanders, T. Aydin, Dynamic phasors in modeling and analysis of unbalanced polyphase ac machines. IEEE Trans. Energy Convers. **17**, 107–113 (2002)
20. U.D. Annakkage, et al., Dynamic system equivalents: a survey of available techniques. IEEE Trans. Power Delivery **27**, 411–420 (2012)
21. D. Shu, X. Xie, V. Dinavahi, C. Zhang, X. Ye, Q. Jiang, Dynamic phasor based interface model for emt and transient stability hybrid simulations. IEEE Trans. Power Syst. **c0l**(33), 3930–3939 (2018)
22. H. Konara, U.D. Annakkage, C. Karawita, Interfacing electromagnetic transient simulation to transient stability model using a multi-port dynamic phasor buffer zone. *CIGRE Science & Engineering (CSE) Journal*, Oct (2019)
23. X. Lin, A.M. Gole, M. Yu, A wide-band multi-port system equiv-alent for real time digital power system simulators. IEEE Trans. Power Syst. **24**(1), 237–249 (2009)

Chapter 4
Implementation of the Online Transient Security Assessment Tool for RAS Real-Time Operation Monitoring

Hongming Zhang

Nomenclature

NERC	North American Electric Reliability Corporation
WECC	Western Electricity Coordinating Council
RC/BA	Reliability Coordinator/Balancing Authority
WSM	West wide System Model
TSAT	Transient Security Assessment Tool
SOL	System Operating Limit
IROL	Interconnection Reliability Operating Limit
ICCP	Inter-Control Center Communications Protocol
RAS	Remedial Action Scheme
IFRO	Interconnection Frequency Response Obligation

4.1 Introduction

As one of the graphically largest power systems in North America, WECC Region extends from Canada to Mexico and includes the provinces of Alberta and British Columbia, the northern portion of Baja California, Mexico, and all or portions of the 14 Western states between. In history, WECC system consists of many well-defined transmission corridors/interfaces called WECC Paths. A portion of WECC Paths is limited by voltage stability, transient stability, or both constraints. Also, many of WECC path operating limits and the existing SOLs and IROLs are heavily

---

H. Zhang (✉)
coreWSM Consulting LLC, Fort Collins, CO, USA

© Springer Nature Switzerland AG 2021
S. (NDR) Nuthalapati, *Use of Voltage Stability Assessment and Transient Stability Assessment Tools in Grid Operations*, Power Electronics and Power Systems, https://doi.org/10.1007/978-3-030-67482-3_4

Fig. 4.1 WECC map with WSM RAS models relevant substations

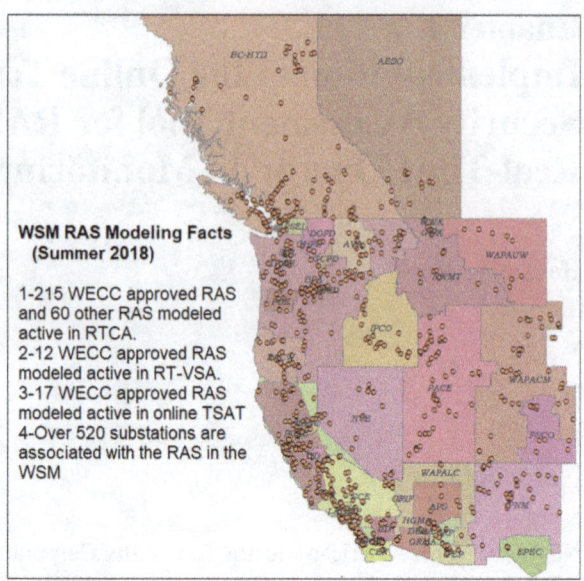

dependent on actual operations and settings of hundreds of RAS widely deployed over the entire West Interconnection, as illustrated in Fig. 4.1.

Meanwhile, many renewable and inverted based generation resources, such as solar PV and wind farm units, were continuously integrated into the WECC system over the last decade. The 2018 PEAK summer study case reported approximately 11,000 MW of solar PV generation and roughly 7000 wind power, which represents 20% of the online generation in total for certain hours. It is envisioned that the penetration rate of overall renewable generations will continuously increase and fundamentally impact WECC system operating paradigms for the next decades. The sustained declining of traditional generators with a large inertia mass requires a RC and/or a BA to monitor system frequency response sufficiency in real-time and assure for compliance of NERC's Interconnection Frequency Response Obligation (IFRO) persistently.

Since the 1996 blackouts, many original work were published in IEEE papers [1–5] for the development of efficient real-time transient stability limit computation algorithms. In the wake of September 8, 2011 blackout, WECC RC/PEAK endeavored to improve the WSM accuracy and real-time advanced network applications, such as State Estimator (SE) and Real-Time Contingency Analysis (RTCA) [6, 7]. One of the major objectives was intended for RC to obtain real-time RAS operation situational awareness by means of RTCA with inclusive RAS models. But later it is found that RTCA was unable to mimic certain transient RAS models and resulted in false alarms of violations.

In late 2014 PEAK initiated a new project with Powertech Labs to implement an online DSA Manager/TSAT tool on the WSM for the underlying operation objectives/issues:

- Unacceptable frequency response that can result in under frequency load shedding or tripping of thermal and nuclear units.
- Back up RTCA to monitor contingencies associated with a transient RAS.
- Monitoring transient or dynamic RAS operation and evaluating RAS Gen drop impact to system frequency performance and cascading outage scenarios.
- Loss of synchronism of cluster of generators and consequently their removal from the grid on a fault.
- Real-time assessment of stability limited SOL/IROL.
- Fast voltage collapse due to induction motor instability.

Since then PEAK's WSM-TSAT project has gone through three implementation phases:

A. *Phase I-Proof of Concept with a Pilot Project.*

Starting from 2014 to 2015, Peak completed online WSM-TSAT tool prototyping project in Test environment with the following key deliverables:

- Enabled EMS data transfer to TSAT Servers.
- Built master unit mapping between the WSM and WECC basecase *.dyd file.
- Migrated COI, Path 3 et al. transfer scenarios, PDCI and relevant RAS models from CAISO and BC Hydro's online TSAT tools.
- Started model validation for system events disturbances using the WSM-TSAT.

B. *Phase II-Production Implementation.*

Starting from 2016 to 2017, Peak implemented online WSM-TSAT in Production and focused on model validation against various system events [8–12]. The project deliverables were as follows:

- Continuously improving dynamic models and validating TSAT simulation results by PMU data.
- Adding dozens of RAS models in support of TSAT basecase simulation against contingencies and power transfer analysis on three scenarios: i.e., California Oregon Intertie, i.e., COI, and SDGE Import and SDGE/CFE IROLs.
- Building underfrequency load shedding (UFLS) and under-voltage load shedding (UVLS) models in online TSAT simulation.
- Implementing new software enhancements, including real-time assessment of Interconnection Frequency Response Obligation (IFRO) in accordance with the NERC BAL-003-1 standard, enhanced bus angle pair limit calculation, and enabling the use of real-time Arming ICCP data from EMS for RAS modeling.

C. *Phase III: Operational Use in Control Room.*

Since Q3–2017, Peak had focused on rolling out WSM-TSAT in the Control room for operational use. Main deliverables and targets on this phase are:

- Completed Real-time Operation Engineer (ROE) Training by 10/2017.

- Began production use by ROEs for TSAT solution monitoring and validation at 7 × 24 in 10//2017. ROE validation would be in progress through 2018.
- Finished coordination and sharing of results with TOPs by 07/2018.
- Developed operating procedures with TOPs on the use case to be rolled out by end of Q3–2018.
- Delivered Reliability Coordinator System Operator (RCSO) Training in Peak RC Q3 2018 Training Cycles.
- Started operational use by RCSOs on October 15, 2018.

The rest of the Chapter is organized as follows:

In Sect. 4.2, we first provide an overview of WSM-TSAT, and then introduce new software enhancements with emphasis on RAS modeling capability. In Sect. 4.3, we present PEAK's practical experience in validating WSM-TSAT RAS models and the impact of simulation results on two 2017 system events. In Sect. 4.4, we first present the first use case to be operational and then review other potential use cases. We conclude the chapter in Sect. 4.5 with lessons learned and future work.

4.2 An Overview of WSM-TSAT

PEAK RC used GE Grid EMS tool suite to perform SCADA monitoring and advanced network applications analysis. PEAK team developed an internal framework to support real-time data file exchange between EMS and DSA/TSAT servers.

4.2.1 WSM-TSAT System Architecture

The WSM-TSAT framework includes a few components shown in Fig. 4.2. Speaking specifically,

- Export EMS network model data (i.e., valid SE basecase raw file in PTI v30 format plus EMS equipment IDs, unit D-curves and Interface definition tables) via Peak custom EMS application called XTSAT for every 5 min.
- Run a script from EMS servers to auto move EMS input data files to corresponding TSAT servers. TSAT software auto checks if there are new SE data files coming for the next run.
- After each run, the WSM-TSAT generates a Solution Results Summary file in *.csv format. The file includes solution status, transfer limits, limiting factor/contingencies, insecure contingencies, and RAS gen drop unit IDs, etc.
- The TSAT Summary file is processed by internal Filelink task based on a user configurable mapping table. The Filelink will transfer the WSM-TSAT solution data to ICCP server first and then to EMS server.
- SCADA will issue Operator alarms as needed to ensure system situational awareness. The TSAT results are historized in PI.

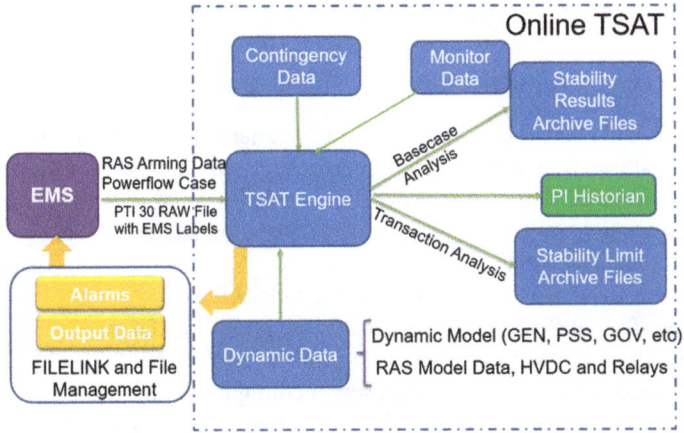

Fig. 4.2 WSM-TSAT system architecture and interface with EMS

- Archiving by automated scripts. Move 1-month old TSAT case zip files from local servers to the data servers.

4.2.2 WSM-TSAT Model Scales

The WSM represents a large-scale and full western system operational model. PEAK's TSAT runs every 7–12 min to solve for the following:

- A bus-branch model basecase with 16,000 buses and 20,000 branches (dynamic). The basecase raw file not only follows PTI v30 format and but also includes an EMS long name labeled for each network equipment.
- 3900 generating units with nearly 90% online generation capacity mapped to WECC *.dyd file.
- 23 contingencies lastly modeled active.
- Three transfer scenarios are lastly modeled, i.e., SDGE Import IROL, SDGE/CFE Import IROL, and Northwest Net Export/Oregon Net Export IROL.
- PDCI and Intermountain DC Ties are modeled.
- 17 dynamic or transient RAS are modeled for Path 26, COI, PDCI, FACRI, MATL, BC Hydro RAS, and 11 other RAS selected for SDGE Import and SDGE/CFE Import IROL scenarios.
- ~6000 UFLS and UVLS models are built in to represent in lsdt1, lsdt2, lsdt3, and lsdt9.
- User-Defined Model (UDM) is used to model the RAS and consolidate unmapped small units between EMS model and WECC *.dyd file.

4.2.3 Custom Software Enhancements

The WSM-TSAT was continuously enhanced by new features. In particular,

- Enable/disable dynamic models on a given generator or load for simulation via UI. This allows Engineer to fix common generator/load model issues easily.
- Support of CCOMP (cross compensation model).
- Display of angle separation between key substation buses or "Pseudo bus angle pairs" for transfer limits.
- Export of extended TSAT results to a PI database.
- User-interface to configure distributed swing generators for power flow solution and enabled dynamic models.
- Development of WAN computation server configuration tool.
- Compute system inertia and NERC defined frequency response measures: A-B, A-C, and C-B for the whole system and individual Balancing Authority (BA) companies against selected unit tripping contingencies.
- Add a unit MW loss threshold to allow TSAT simulation to continue when small units become unstable for unknown reasons. The feature is essential for transient cascading assessment.
- Most importantly, enhance RAS modeling capability and enable using real-time RAS Arming data from EMS ICCP measurements.

4.2.4 RAS Software Enhancement Implementation

To enable using real-time RAS Arming data from EMS in TSAT, we modified custom EMS code and made software enhancements on top of Powertech base Product. The overall flowchart of the software changes is given in Fig. 4.3.

4.2.4.1 Custom EMS Software Changes

Modify EMS application task-XTSAT to export real-time RAS Arming data from the last SE solved basecase. Modify internal File Transfer scripts to copy RAS real-time Arming data file along with both basecase Raw file, unit D-curves files to DSA/TSAT servers.

4.2.4.2 Powertech DSA Manager Changes-Part A

Create a new configuration to allow the user to supply TSAT UDM files separately from the dynamic model data files, enabling the ability to define and update these models from DSA Manager and any changes made within DSA Manager will be synchronized between primary and backup systems.

4 Implementation of the Online Transient Security Assessment Tool for RAS...

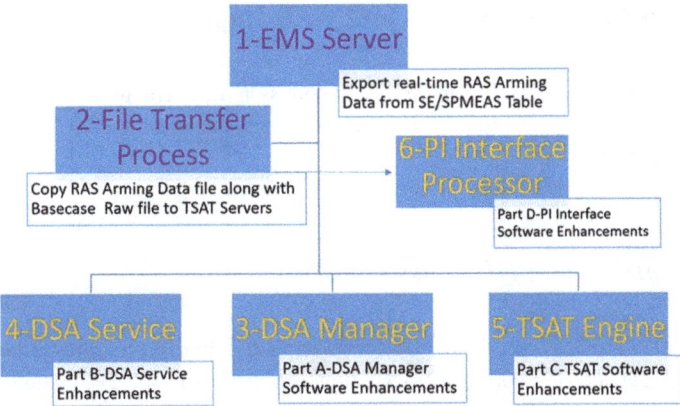

Fig. 4.3 Software enhancements for using real-time RAS Arming data

4.2.4.3 Powertech DSA Service Code Changes-Part B

(a) Parse CSV file from EMS Special Measurement (SPMEAS) files to obtain current arming values;
(b) Convert fields in the CSV file to create a unique identifiable string of RAS arming data fields to be used within the TSAT SPS layout models;
(c) Update RAS models with current arming values; and
(d) Parse outputs from TSAT to obtain and provide generation tripping values to the PI Tool.

4.2.4.4 Powertech TSAT Code Changes-Part C

(a) Create a feature within the switching progress report that indicates when a generator is tripped by stability;
(b) Generate a new report for each basecase contingency in a consolidated XML file that contains the names of RAS models, the total megawatt generation, and the load tripped by the model in the contingency.

4.2.4.5 Powertech PI Interface Code Changes-Part D

Create the following quantities output to the PI OSIsoft software tool ("PI Tool") for basecase contingencies. Export Total generation tripped by each RAS scheme; and Total generation tripped, not related to RAS scheme, and not in a contingency definition to the PI Historian.

4.3 RAS Model Validation and RAS Impact Study

WSM-TSAT simulation solution quality depends on the correctness of RAS models. From below two real events, we will demonstrate the impact of RAS models on TSAT simulation results.

4.3.1 April 2017 PDCI Loss Event

On April 06, 2017, 23:00 PST, PDCI was tripped initially on a single pole when the DC ties carried on 3000 MW. PDCI flow dropped to 1600 MW instantaneously. After 8 min the second pole was ramped down to 366 MW and then caused loss of 2482 MW generation and the frequency dip to 59.704 Hz (NADIR). A summary of the event is given in Table 4.1, where WSM-TSAT mimic PDCI loss contingency by pole blocking. Before the real event occurred, the online TSAT indicated the contingency "insecure" due to voltage collapse and angle instability, as shown in Fig. 4.4. However, the real system recovered from loss of PDCI to normal in minutes.

We reproduced the online TSAT simulation results in offline study mode. By comparing the TSAT simulated RAS actions with the real RAS operations, we confirmed PDCI gen drop RAS was not fired because the PDCI RAS was using an outdated definition of the PDCI interface. After the correct gen drop amount based on EMS ICCP data was applied manually, we re-ran the TSAT simulation and achieved creditable simulation results matching with PMU data in Fig. 4.5.

By fixing PDCI interface definition in the RAS model, we also obtained good simulation results similar to the one with the manual application of real-time RAS Arming data. However, the static RAS lookup table does not line up with the real-time RAS Arming data for gen drop action. From Fig. 4.6, the RAS actions from the lookup table remove 7 Chief Joseph units with loss of 656.17 MW, while the RAS actions per ICCP real-time RAS Arming data drop a single Grand Coulee unit with 710.57 MW output. Our TSAT simulation study on the PDCI loss event shows

Table 4.1 April 06, 2017, PDCI loss event (Data Source: BPA)

April 06, 2017, PDCI loss event started at 23:00 PPT	
Event description/cause (unknown)	1st Pole tripped, second Pole ramped down
Unit MW lost	2482 MW
BA/TOP equipment/facilities lost	Multiple BPA units
Pre-disturbance Hz/post-disturbance Hz	60.023 Hz/59.704 Hz
FTL/FRL/FAL duration	NA
Time Hz returned to normal & scheduled Hz	~8 mins
Pre-disturbance ACE	113 MW
Disturbance ACE	1493 MW

4 Implementation of the Online Transient Security Assessment Tool for RAS... 107

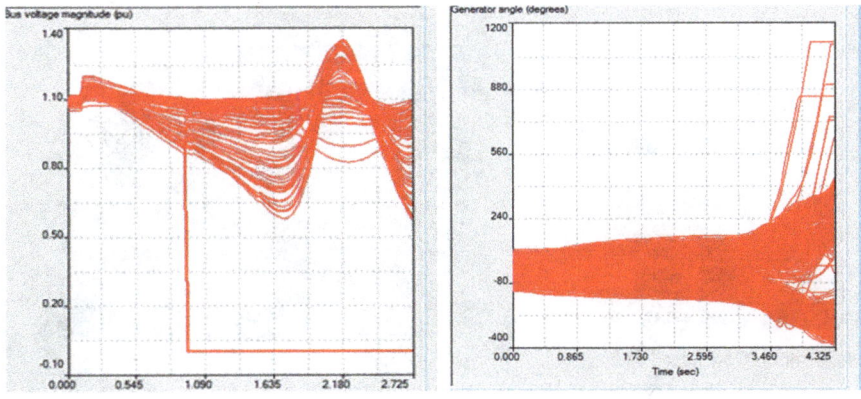

Fig. 4.4 TSAT simulation on April 06, 2017, PDCI Trip Event

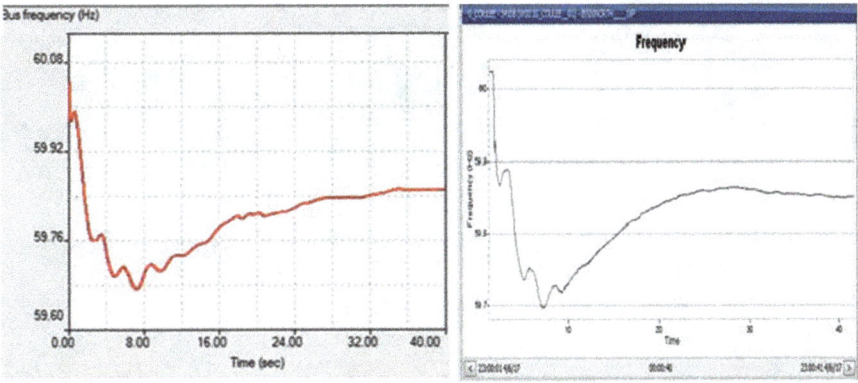

Fig. 4.5 TSAT simulation vs PMU on April 06, 2017, PDCI Trip Event

interaction of RAS models and importance of using real-time RAS Arming data from EMS ICCP measurement.

4.3.2 October 2017 US-Canada Separation Event

In mid-October 2017, Canada system was separated from US system. Figure 4.7 shows some screenshots of the playback video that highlights the event.

The first plot was the PMU recorded frequency response during the event. In the very beginning, the system's frequency is around 60 Hz where we can see green color across WECC. On that particular day, one of the two 500 kV lines between BC and Washington state, Custer-Inglewood 500 kV #1 line, was under planned outage/maintenance. Around 14:45:20 MST, we had a fault on the second 500 kV

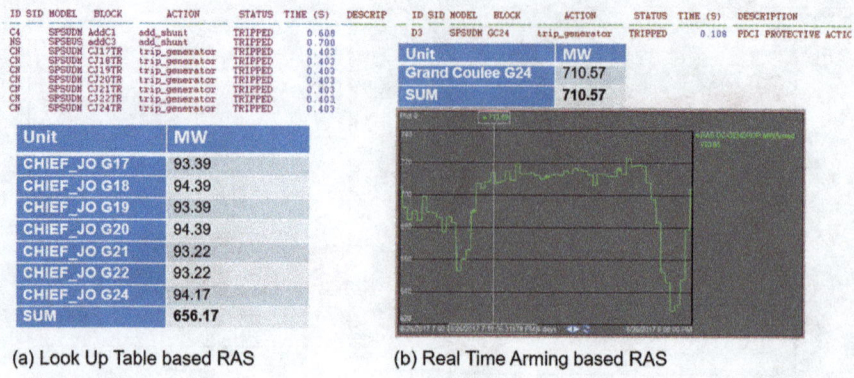

Fig. 4.6 Comparison of RAS look up table (**a**) and R-T Arming RAS (**b**)

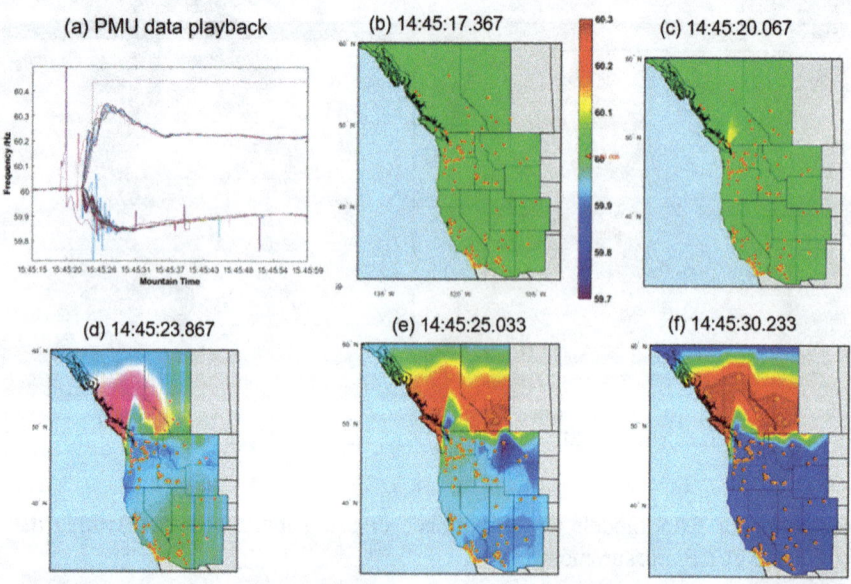

Fig. 4.7 US-Canada separation event playback: (**a**) PMU data playback; (**b**), (**c**), (**d**), (**e**) and (**f**)- the playback snapshots at 14:45:17.367, 14:45:20:067, (**d**) 14:45:23:867, (**e**) 14:45;25.033 (**f**) 14:45;30.233, respectively

line and zone 1 relay open the remaining 500 KV line which before the event had around 1500 MW flow on it. The yellow dot in figure (b) indicates where the fault was. Following the line tripping, the frequency of the US and Canada system starts to deviate. At the same time, BC hydro's RAS scheme tripped around 600 MW units in their footprint and opened their 230 KV tie line with the US in east Washington. Fig. (c) was the plot after these RAS actions.

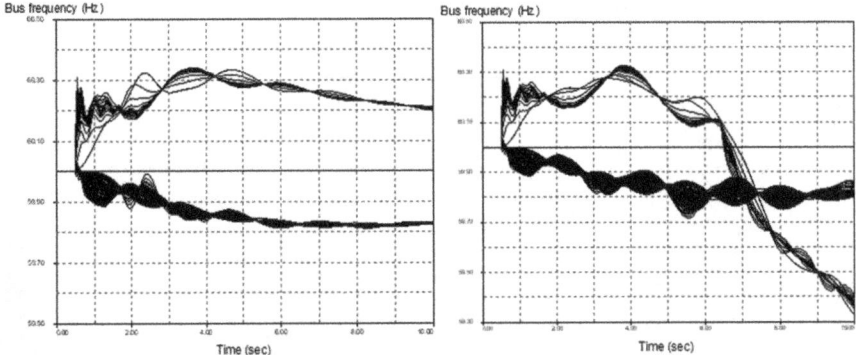

Fig. 4.8 TSAT area frequency response summary: (**a**) Simulated frequency with the OOS modeled vs (**b**) Simulated frequency without OOS modeled

It is observed that the frequency in BC was high around 60.2 and the frequency in Northwestern US started to decrease. At this moment the US and Canada system was still connected through a 230 KV line between Alberta and Montana. BC Hydro's controller could not open that line. There was a local OOS relay on that 230 KV line. Two seconds later, that line was tripped which was shown in (d). Five more seconds later, the system settled down to a new stable operating point with US and Canada separated. In this event, many actions happened during a 10 s window and the system eventually survived.

During this event, PEAK RTCA showed unsolved for this contingency due to missing of OOS relay. But WSM-TSAT successfully predicted the separation thanks to correct modeling for AESO MATL RAS. As a result, online WSM-TSAT showed this contingency as secure and correctly predicted all the actions leading to the system separation. The underlying Fig. 4.8 shows the frequency plots from two TSAT simulation scenarios: the left figure (a) is the TSAT simulation results with OOS modeled correctly; the right figure (b) showed the TSAT simulation by removing OOS. One can see the system will not survive and many units in Canada tripped.

The simulation results shown in Fig. 4.9 were retrieved from one pre-disturbance TSAT archive case saved in Primary production server.

By importing PMU data on the event from GE PhasorPoint or Phasor PI, Engineer can perform system model validation against the event with a limited workload. From the difference between TSAT simulated post-disturbance frequency and PMU frequency signal, one can see the modeling gap in the Canada system. PEAK continued ongoing efforts with BC Hydro and AESO to improve their models represented in the WSM and WECC planning *.dyd file.

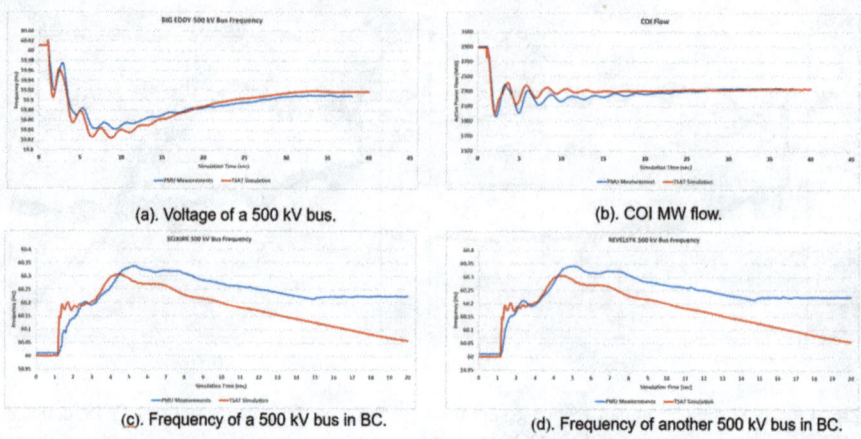

Fig. 4.9 TSAT simulation vs PMU data on the system separation: (**a**) Voltage of a 500 kV bus, (**b**) COI MW flow, (**c**) Frequency of a 500 kV bus in BC and (**d**) Frequency of another 500 kV bus in BC

4.4 Development of Use Cases

PEAK successfully rolled out its first online TSAT Use Case-Colstrip ATR RAS monitoring for operation decision on October 15, 2018. Development of other use cases are in good shape.

4.4.1 Colstrip ATR RAS Monitoring

NWMT Colstrip Generation Station has a Remedial Action Scheme (RAS) called the Acceleration Trend Relay (ATR). The device consists of a computer-based relay which monitors the real-time speed, acceleration, and angle of the four Colstrip units. It uses this information to assess the dynamic performance of the system and is able to detect unstable events in progress. The ATR will trip generators as necessary to restore the system to a stable operating condition when instability occurs.

The dynamic nature of the ATR precludes modeling its actions in RTCA thus presenting potentially false N-1 violations for contingencies related to the 500 kV transmission corridor elements between Colstrip in Montana and BPA system in Washington. NWMT and PEAK engineers have coordinated process studies that demonstrate WSM-TSAT is very accurate in simulating ATR RAS actions. Detailed model validation for Colstrip ATR RAS using WSM-TSAT was presented in [10–12]. Currently, TSAT is monitoring nine 500 kV corridor contingencies and reports which Colstrip unit(s) is tripped for each one. RCSO now have valid information to complete a study and verify ATR RAS actions mitigate the RTCA violations. Alternatively, WSM-TSAT can verify when no unit(s) at Colstrip are tripped and another mitigation is needed in real-time.

An operating procedure for monitoring ATR RAS by WSM-TSAT has been developed. It consists of several steps:

(a) Identify RTCA SOL exceedances on the intended contingencies and verify the exceedances in Study mode.
(b) Review WSM-TSAT ATR Monitoring PI display to check if any Colstrip unit(s) is tripped per the last simulation run. Note the same PI display is made available to NWMT and relevant entities via a Web service for sharing real-time data.
(c) Simulate Colstrip unit(s) actions by EMS Study power flow.
(d) Verify unit(s) tripping mitigate post-contingency exceedance. If unit actions do not mitigate exceedance, then proceed to the next step.
(e) Initiate a call to affected TOP(s) to discuss the RTCA results and TSAT solution indicating Colstrip ATR actions either insufficient to mitigate exceedance or ATR not activated. Generation reduction or other acceptable pre-contingency mitigation is necessary as described.

Till date, there was no exception case found. All RTCA exceedances on these nine contingencies were mitigated successfully per engineer testing. Initial operation training on Colstrip ATR RAS monitoring has been delivered to RSCO and control room engineers. A computer-based training (CBT) module on the subject will be delivered before October 2018.

In addition to ATR RAS monitoring, WSM-TSAT can back up RTCA to monitor a handful of [N-2] contingencies that are falsely unsolved due to RTCA's inability to model out of step relays, frequency rate of change and time delay protective actions, etc.

4.4.2 IFRO Measure Calculation (BIAS: MW/0.1 Hz)

WSM-TSAT has a new feature to calculate system inertia and IFRO Measure A-C et al. on WECC system or individual company level. Figure 4.10 shows a snapshot of the Area Frequency Response Summary table calculated against Palo Verde unit tripping contingency. PI trend on frequency BIAS of WECC, BPA, and SCE systems between 2/20/2018 and 9/5/2018 are given in Fig. 4.11. It is interesting to see the patterns of WECC system and two BAs change noticeably from Spring to Summer. Real-time calculation of system inertia and frequency response measures will be essential for monitoring the impact of high penetration of renewable generation in the future.

Area Frequency Response Summary - APS0U005 (PALO VERDE UNIT 1)

No.	Name	Inertia	A-B	A-C	C-B
1	AESO	55652.16	102.177	83.025	1.231
2	APS	32395.16	26.730	36.635	0.730
3	AVA	7949.87	1.970	2.217	0.888
4	AVRN	0.00	7.042	6.761	1.042
5	BANC	10800.65	10.409	12.213	0.852
6	BC-HYD	53549.15	56.376	48.623	1.160
7	BPA	63144.30	111.720	98.500	1.134

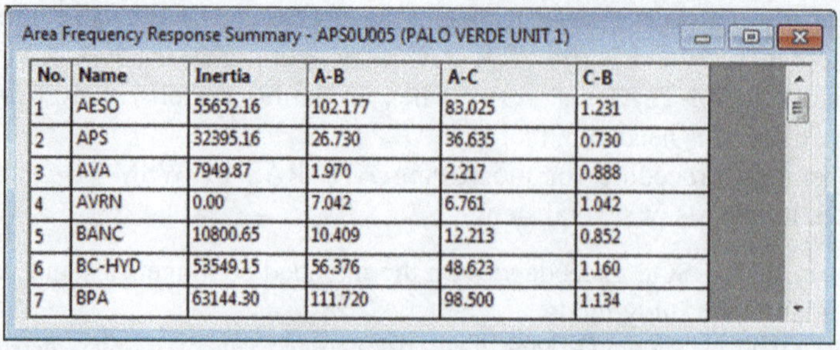

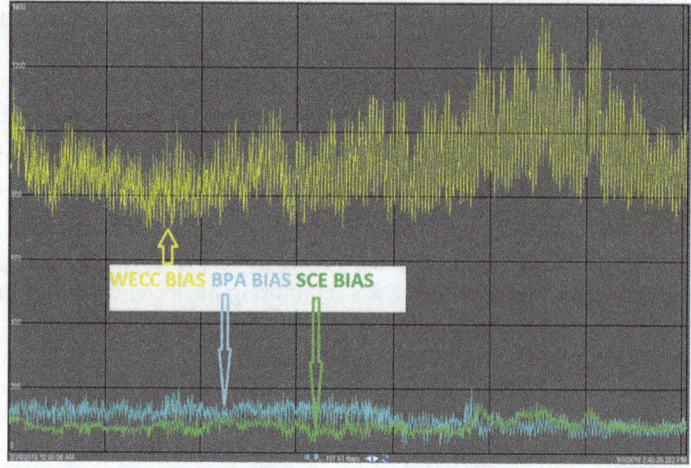

Fig. 4.10 WSM-TSAT calculated area frequency response summary

Fig. 4.11 Calculated frequency BIAS for the system and areas

4.4.3 Backup RT-VSA for Transfer Capability Analysis

Presently PEAK is running V&R real-time voltage stability analysis tool (RT-VSA) importing the WSM SE export case to perform real-time assessment of the existing IROLs: SDGE Summer Import, SDGE/CFE Import, Northwest Washington Load Import, and Oregon Net Export.

4.4.3.1 SDGE/CFE Import IROL Calculation

CAISO Bigwood RT-VSA tool also calculates the VSA limits for both SDGE Import ad SDGE/CFE Import IROL scenarios. PEAK built two IROL transfer scenarios in WSM-TSAT by the same definition as the RT-VSA to validate the IROL value when two RT-VSA tools diverge on a VSA limit drop or low margin case.

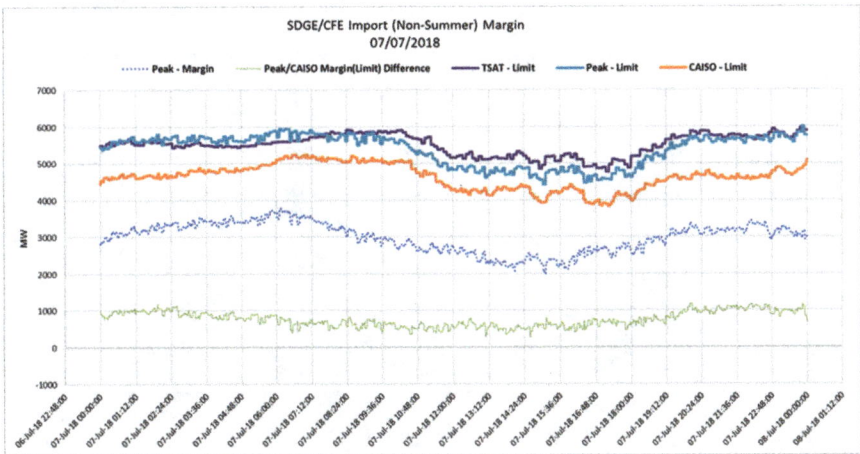

Fig. 4.12 SDGE/CFE import IROL calculations

From Fig. 4.12, one can see a gap between two RT-VSA tools due to different software and network models. In this case, WSM-TSAT and V&R RT-VSA tools solve the SDGE/CFE IROL closely compared to CAISO RT-VSA results. If either of the IROLs falls into a fast voltage collapse due to stalling of massive air conditioning and industrial motors, WSM-TSAT has the potential for detecting the risk than both RT-VSA tools.

4.4.3.2 Oregon Net Export IROL Calculation

The Oregon Net Export IROL consists of three WECC Paths including COI (California and Oregon Intertie) that dominates the IROL. This IROL is normally voltage stability limited, but under certain 500 kV line outages, the limiting factor will be angle instability. From Fig. 4.13, WSM-TSAT solves the VSA limit for the IROL between PEAK RT-VSA (highest limit) and BPA offline study VSA tool (lowest limit with a flat line for most time). In this case, WSM-TSAT will back up the RT-VSA tool in case of angle instability issues occur.

In addition, WSM-TSAT online solution archive cases were commonly used for various system event validation (offline), such as MOD-033/MOD-026/027, PV Solar momentary cessations, and oscillation mode baselining.

4.5 Conclusion

PEAK has successfully implemented WSM-TSAT in RC Control room in collaboration with Vendor and Entities.

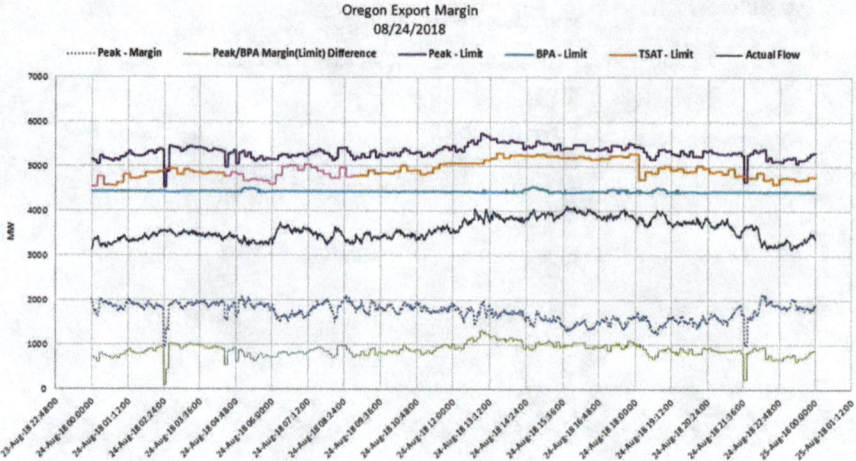

Fig. 4.13 Oregon Net Export IROL calculations

Development of WSM-TSAT involved tons of efforts on both software enhancements, system integration, model improvements, and validation against system events. Till date, WSM-TSAT remains the only online TSAT tool running with a full western system operational model and a dozen of accurate RAS model representations.

After multi-year continuous efforts on use cases development and model validation, the tool has been cut off in Control rooms for near real-time monitoring on Colstrip ATR RAS and the IROLs. The online results of Colstrip ATR RAS monitoring are being shared with NWMT and other Entities via Peak Web Portal: www.peakrc.org to improve operation situational awareness across among stakeholders.

Practical experience and lessons learned from the WSM-TSAT implementation project are summarized below:

- WSM-TSAT allows us to maximize model quality and to minimize uncertainties in order to increase transmission capacity utilization while maintaining reliability.
- Using real-time RAS Arming ICCP data for RAS modeling in TSAT is essential for accurate TSAT simulation results.
- Calculation of system inertia and frequency BIAS in real-time provides an efficient way to monitor system frequency response capability.
- Setting of Unit Base Load Flags affects TSAT simulation performance of governor responses Secondary control loop.
- Unit status availability and model accuracy matters.
- Lower frequency dips in simulation might be caused by excess governor response from some units with incorrect Pmax or base load flag settings.

- Modeling errors on wind farms and PV Solar plants are the ones usually causing TSAT solution issues.
- WSM-TSAT can backup RT-VSA for real-time assessment of the IROLs, particularly when the limiting factor is fast voltage collapse or angle instability.

Recently PEAK worked with Powertech to make TSAT software enhancements in several areas:

- Enable parallel execution of basecase and transfer analysis phases in DSA Manager. This feature is desired so that the results of basecase analysis will be submitted to PI as soon as basecase analysis is finished.
- Improve PV modeling capability in the software for more realistic simulation of PV Solar momentary cessation.
- Enhance ePMU simulation feature to allow large scale adjustment of loads and generation MW for PMU training and modal analysis baselining.
- Improve powerflow robustness against low bus voltages in basecase solution.
- Implement accurate SVC models in replacement of generator equivalent SVCs.

Those software enhancements and modeling improvements were cut off in production in early 2019. Most of the recent enhancements have been migrated in PowerTech TSAT base product release for more customers to adopt.

Acknowledgments The author gratefully acknowledges the support of PEAK Engineering and IT Departments, as well as PowerTech Labs, specifically Frederic Howell, Lei Wang, and Xi Lin. I personally thank my former colleagues Dr. Slaven Kincic, Dr. Haoyu Yuan, Mr. Ronald Evjen, Dr. Luoyang Fang, Dr. Xiaoyuan Fan, Dr. May Mahmoudi, and Dr. Yidan Lu et al. for their contributions to the work described in this chapter. The author sincerely appreciates PEAK's utility partners-CAISO, BPA, NWMT, and BC Hydro for providing their modeling data to PEAK and validating the simulation results over the past years.

References

1. D.N. Kosterev, C.W. Taylor, W.A. Mittelstadt, Model validation for the August 10, 1996 WSCC system outage. IEEE Trans. Power Syst. **14**(3), 967–979 (1999)
2. L. Wang, M. Klein, S.Y.P. Kundur, Dynamic reduction of large power systems for stability studies. IEEE Trans. Power Syst. **12**(2), 889–895 (1997)
3. H.-D. Chiang, C.-S. Wang, H. Li, Development of BCU classifiers for on-line dynamic contingency screening of electric power systems. IEEE Trans. Power Syst. **14**(2), 660–666 (1999)
4. C.Y. Chung, L. Wang, F. Howell, P. Kundur, Generation rescheduling methods to improve power transfer capability constrained by small-signal stability. IEEE Trans. Power Syst. **19**(1), 524–531 (2004)
5. R. Diao, V. Vittal, N. Logic, Design of a real-time security assessment tool for situational awareness enhancement in modern power systems. IEEE Trans. Power Syst. **25**(2), 957–964 (2010)
6. H. Zhang, B. Wangen, "Implementation of a full Western Bulk System operational model for reliability monitoring," in *Proceedings of IEEE PES General Meeting*, July 22–26, 2012

7. B. Wangen, H. Zhang, "Monitoring for post-contingency System Operating Limit exceedance in the Western Interconnection," in *Proceedings of IEEE PES General Meeting*, July 22–26, 2012
8. S. Kincic, D. Davies, D. Kosterev, H. Zhang, B. Thomas, M. Vaiman, J. Weber and R. Ramanathan, "Bridging the Gap in Between operation and Planning in WECC" PES-GM2016, Boston, MA
9. B. Thomas, S. Kincic, D. Davies, H. Zhang, and J. Sanchez-Gasca, "A New Framework to Facilitate the Use of Node-Breaker Operations Model for System Model Validation in WECC" PES-GM2016, Boston, MA
10. Y. Lu, S. Kincic, H. Zhang and K. Tomsovic, "Validation of Real-Time System Model in Western Interconnection", PES GM 2017 Chicago, IL
11. M. Mahmoudi, S. Kincic, H. Zhang and K. Tomsovic "Implementation and Testing of Remedial Action Schemes for Real-Time Transient Stability Studies", PES GM 2017 Chicago, IL
12. M. Mahmoudi, S. Kincic, H. Zhang and K. Tomsovic "Model Enhancement for Real Time Transient Stability Assessment in Western Interconnection", IEEE PES T&D Conference, April 2018, Denver, CO

Chapter 5
Implementing the Real-Time Voltage Stability Analysis Tool for the Western IROL Real-Time Assessment

Hongming Zhang

Nomenclature

WECC	Western Electricity Coordinating Council
RC/BA	Reliability Coordinator/Balancing Authority
WSM	West wide System Model
SOL	System Operating Limit
IROL	Interconnection Reliability Operating Limit
ICCP	Inter-Control Center Communication Protocal
PEAK	Peak Reliability
RAS	Remedial Action Scheme

5.1 Introduction

WECC system is graphically one of the largest power systems in North America. WECC system experienced two major blackouts on July 2, 1996 and August 10, 1996 [1–3], respectively. To enable having wide-area view monitoring capability, WECC merged three regional operational models (i.e., Northwest, California and Mountains and South Desert Areas) into a single West wide System Model (WSM) and cutover the WSM as well as the EMS tool suite in production as of January 1, 2009 [4]. Since then, WECC played a centralized Reliability Coordinator (RC) role for the entire Western Interconnection until February 2014, when the RC function was split from WECC to form a new operating Entity-Peak Reliability

---

H. Zhang (✉)
coreWSM Consulting LLC, Fort Collins, CO, USA

© Springer Nature Switzerland AG 2021
S. (NDR) Nuthalapati, *Use of Voltage Stability Assessment and Transient Stability Assessment Tools in Grid Operations*, Power Electronics and Power Systems,
https://doi.org/10.1007/978-3-030-67482-3_5

("PEAK"). WECC RC/PEAK used to solely rely on State Estimator (SE) and Real-Time Contingency Analysis (RTCA) applications to monitor system operating limits (SOLs) exceedance in real-time and post-contingency conditions [5].

However, the Pacific Southwest blackout on September 8, 2011, attested the importance of performing real-time assessment of the IROLs [6]. Most of the identified IROLs in WECC footprint are subject to voltage stability constraints. Numerous planning voltage stability studies were performed by WECC using different offline analysis tools [7], but there was no real-time voltage stability analysis tool available in WECC RC Control room. Since 2008, there were a few new attempts in the industry towards online or real-time voltage stability analysis implementation for control room use [8–11]. Lately, some original work was reported to perform voltage stability analysis under high penetration of wind and PV solar generations [12–14].

WECC RC initiated a new project with V&R Energy in 2012 to implement real-time voltage stability analysis tool on top of V&R POM (Physical and Operational Margin) core engine. Back then, the POM tool users were limited to planning engineers only working in offline and laptop type environment. To upgrade offline POM engine to real-time voltage stability analysis (called WECC or PEAK ROSE RT-VSA) tool, there were a number of challenging issues to be resolved.

For example,

- Integrate V&R software with GE (formerly Alstom) EMS system to enable retrieving SE real-time snapshot files and sending VSA solution results to EMS/SCADA flawlessly.
- Import a full node-breaker model export file from real-time SE solution and interpret the system topology and network components and network schedules the same as EMS does.
- Support parallel computing and fast power flow solution to ensure solving multiple transfer analysis scenarios in 5 mins.
- Be able to model various complicated Remedial Action Scheme (RAS) including Nomogram tables accurately.
- User-friendly visualization and alarming features for operation awareness.
- Archiving and file management for CIP compliances.

The rest of the chapter is organized as follows:

Section 5.1 provides an overview of V&R PEAK ROSE RT-VSA Tool, including system architecture integrating the RT-VSA tool with GE EMS platform, WSM key facts, and PEAK ROSE RT-VSA tool project milestones.

Section 5.2 presents custom software enhancements implemented by V&R to enable PEAK using the RT-VSA tool for assessment and monitoring of multiple IROLs in near real-time window.

Section 5.3 introduces Peak's RT-VSA tool validation work and simulation results for each IROL identified by Western utilities over the last 4 years. Main findings and resolutions are given in this section.

We conclude the chapter in Sect. 5.4 with lessons learned and future work.

5.2 An Overview of V&R PEAK ROSE RT-VSA Tool

PEAK control room relies on GE (formerly Alstom/Areva) EMS tool suite to perform SCADA monitoring and advanced network applications analysis, while V&R Energy base product has no interface with external EMS system. To integrate two vendor products for an integrated control room solution, PEAK team developed a custom framework to support real-time data file exchange between EMS and V&R servers.

5.2.1 System Architecture of Integrating with EMS

The framework includes a few components shown in Fig. 5.1. It consists of the following steps:

- Run custom scripts installed onto EMS servers to automatically retrieve the last solved SE solution snapshot into a WSM export file and special measurement output file in *.csv format. The SE files from primary EMS are sent over to all four VSA servers.
- The scripting process is scheduled to send the new SE files to V&R servers for every 4.5 min.
- V&R RT-VSA software automatically checks if there are new SE data files coming in, and then kick off the next analysis run if the last VSA solution completes.
- After each new run, the RT-VSA tool generates solution output files, e.g., EMS alarms, VSA solution summary files, P-V curves, V-Q curves, and different log files like RAS and iteration logs, etc.

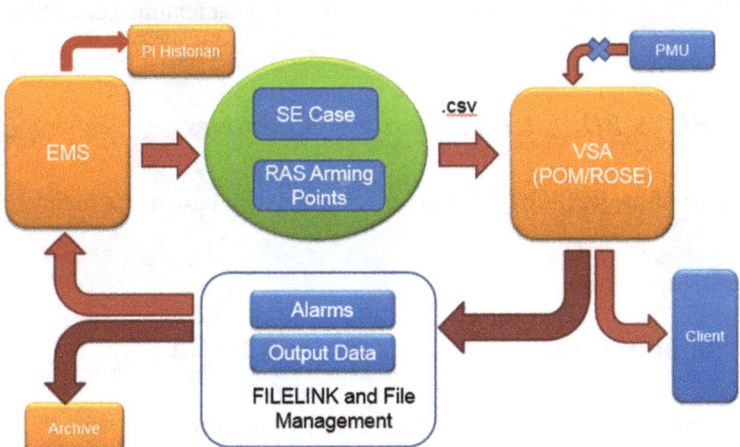

Fig. 5.1 RT-VSA tool architecture and interface with EMS

- EMS Alarm file includes all critical VSA solution results, e.g., limits, critical contingencies, weakest buses, and solution failure code. The file is read into the GE's Filelink application for data interpretation per a user-defined mapping table and data transfer to ICCP servers for real-time update on SCADA VSA points.
- SCADA has custom operator alarms and IROL monitoring displays for RC situational awareness.
- The RT-VSA results in SCADA are historized in PI.
- The scripts are created to auto move 3 days older VSA output files from the VSA servers to the data servers.

5.2.2 PEAK WSM Facts

The WSM represents a large scale and full Western system operational model as follows:

- 16,000 buses/20,000 branches (dynamic) after topology processing (or over 113,000 nodes);
- 8950 substations and 3835 units;
- 14,084 lines, 5683 transformers including 45 phase shifters;
- 11,000 individual loads within 38 areas/companies;
- 177,685 SCADA points in analog and status type;
- 8091 contingencies modeled active;
- 515 RAS models being screened by all contingencies in RTCA.

PEAK ROSE RT-VSA tool imports a full topology model SE solution for every 4–5 min. It requires to solve all required IROLs transfer scenarios in 5 min on average. Fast power flow solution and competitive RAS modeling ability is the main business driver for PEAK to select V&R POM engine from several Vendor products to build RC's RT-VSA tool for IROLs assessment in near real-time conditions.

5.2.3 PEAK ROSE RT-VSA Project Milestones

A sequence of the ROSE RT-VSA project deliverable milestones is summarized in Fig. 5.2.

5.3 Major Software Enhancements in RT-VSA

Since its initial deployment at PEAK in 2012 [15, 16] and up to now, PEAK RT-VSA has undergone many software enhancements through the past years.

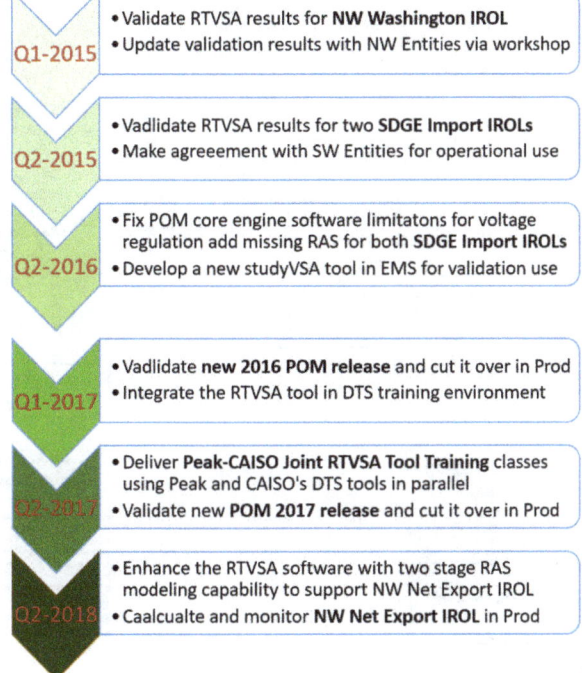

Fig. 5.2 PEAK RT-VSA tool implementation project milestones

5.3.1 Key Software Improvements

2013–2014: Updated shunt device switching logic during ROSE analysis to match EMS switching logic. This was an extensive development which had a significant impact on IROL computations. We will compare calculated IROL results with and with shunt switching and discuss the impact of enabling shunt switch in Sect. 5.4.

2015: (1) Added reverse transfer analysis and load shed calculation and associated alarms, when ROSE identifies a contingency or multiple contingencies that fail to solve at base case conditions. (2) Added a capability to stop stressing the system after the first "unhealthy" step is identified. All contingencies at the last "unhealthy" step are applied, and all contingencies causing a violation are identified. These are limiting CTG(s). (3) Added options for Negative Qload Scaling. (4) Implemented economic power factor due to considerations of economic operation and/or maintain regulation reserves.

2016: (1) Added multi-threading calculations. (2) Added computation of sensitivities dV/dQ. (3) Added PEAK network equipment labels, which are unique way of defining and addressing the network equipment.

2017: (1) Enhanced topology processing. (2) Added saving the last "healthy" (e.g., solved) step as a power flow case.

2018: (1) Added flexibility and new pre-built functionality for RAS modeling. (2) Implemented Stage 2 RAS.

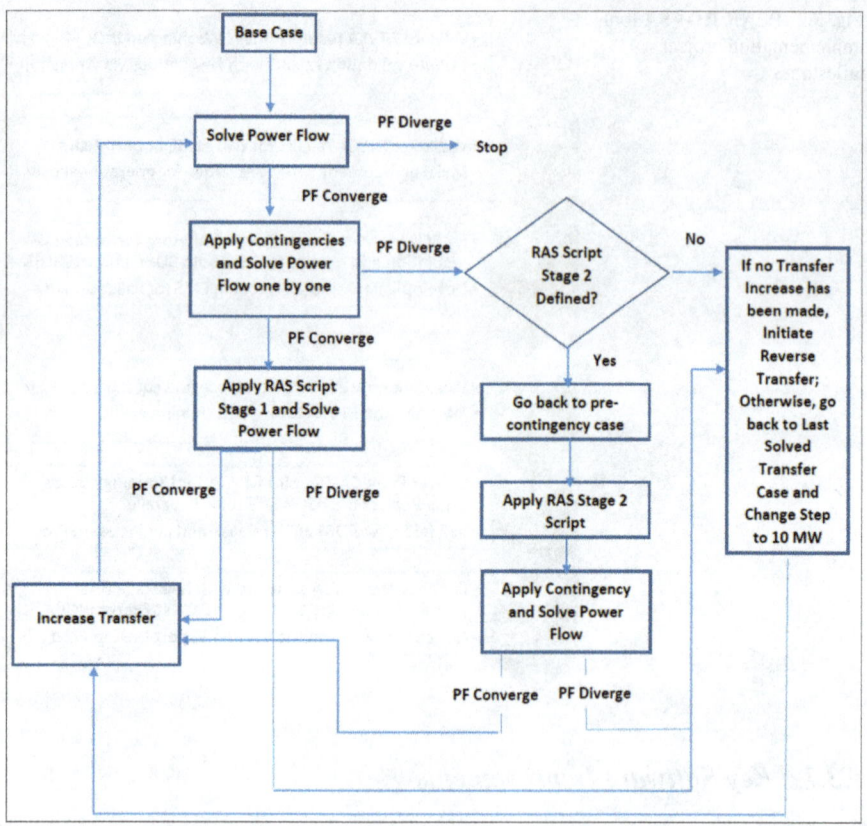

Fig. 5.3 Two-stage RAS logic diagram

5.3.2 Two Stage RAS Logic Implementation

Two stage RAS logic is illustrated in Fig. 5.3. A key point is to ensure the required RAS protection actions can be applied correctly in three steps: (1) apply RAS stage 1 (full RAS model) if post-contingency power flow solves and RAS trigger conditions are met; otherwise, go to (2) restore to the basecase at the last healthy stress level, and then apply contingency and RAS stage 2 (partial RAS model) if Stage 2 RAS does not make the power flow solve, then (3) stop and report the last solved IROL value. The new RAS logic resolves the most RAS backfire issues and produces a more creditable IROL assessment than BPA offline study SOL results.

5.3.3 Other RT-VSA Tool Features

The RT-VSA tool was enhanced with new features on top of the standard POM product:

- Applies critical contingencies and relevant RAS models by user-defined VB scripts.
- Performs both PV and QV analysis.
- Provides a library of sub-function scripts for debugging and RAS scripting.
- Supports batch mode run by custom scripts for a massive case study.
- Transfer RT-VSA output results to EMS for alarming and visualization in EMS/SCADA.

5.4 Tool Validation on Calculation of the IROLs

Since late 2014, PEAK Network Applications engineers have conducted extensive RT-VSA tool validation and comparison with other vendor software and CAISO's Bigwood RT-VSA tool. The whole RT-VSA tool validation process started with Northwest Washington Load Area IROL, then moved on to SDGE Summer Import and SGGE/CFE or SDGE/CEN Import IROLs, and lately worked on Northwest Net Export or Oregon Net Import IROL. Each IROL solution validation identified different modeling and software issues and resulted in improvements on VSA modeling and the newer version of POM engine. All RT-VSA result validation projects began with consolidating general assumptions for the intended IROL scenario to ensure all tools for comparison apply the same or similar assumptions. Below are examples of general assumptions adopted:

5.4.1 VSA Basic Assumptions

Overall
- RAS need to be modeled upon the operating procedure as per the software capabilities.
- Instability indicated by power flow divergence could come from numerical problems other than actual voltage collapse.
- Generator Automatic Voltage Regulators (AVR) (remote regulating), dynamically controlled variable susceptance devices, reactive shunts, phase shifters, and LTCs shall be allowed to move in the basecase (pre-contingency).
- Only generator AVRs (terminal regulating) and dynamically controlled variable susceptance devices will be able to move post-contingency (all others are locked).

Generator (Source)
- Only in-service units defined in the Source subsystem and enabled with AGC flag are scaled upon unit's Pmax during load stressing or transfer analysis.
- Multiple units are able to regulate the same bus voltage collectively within the range of Qmax and Qmin with respect to the unit D-curves.
- Line Drop Compensation (LDC)/Reactive Current Compensation (RCC) will not be modeled due to program limitations.

Load (Sink)
- All load will be modeled as constant MVA through the voltage band.
- All loads in the "Sink" subsystem will be scaled except dynamic loads/auxiliary loads et al. non-conforming loads.

5.4.2 Northwest Washington Load Area IROL Validation

PEAK performed rigorous VSA solution validation against the IROL between 10/2014 and 04/2015. We adopted three commercial VSA software products for benchmarking study: PowerTech VSAT, PowerWorld, and V&R POM/ROSE. Overall, we compared the VSA calculation results over several days for three scenarios: (a) Basecase Interface Flows in Fig. 5.4; (b) Interface Margin w/o Outages in Fig. 5.4; and (c) Interface Margins with Outages in Fig. 5.5. Note that on those plots of each figure, VR/PW/VSAT means V&R, PowerWorld, and PowerTech tool results, respectively.

Remarks: Figs. 5.4 and 5.5 give 48 h of VSA solution data under normal system conditions, the margins happened to be very high which did not necessarily provide the best comparison of results between the three different programs. For that reason, only one low Margin calculation will be shown as it was the highest loading seen in the Pacific NW during the program testing. This occurred on a Monday morning load pickup and continued through Tuesday evening (12/29 00:00–12/30 23:59).

To stress the system, we took several 500 kV outages on WECC Path 4 (West of Cascades North) and reran the VSA study. As shown in Fig. 5.6, these outages

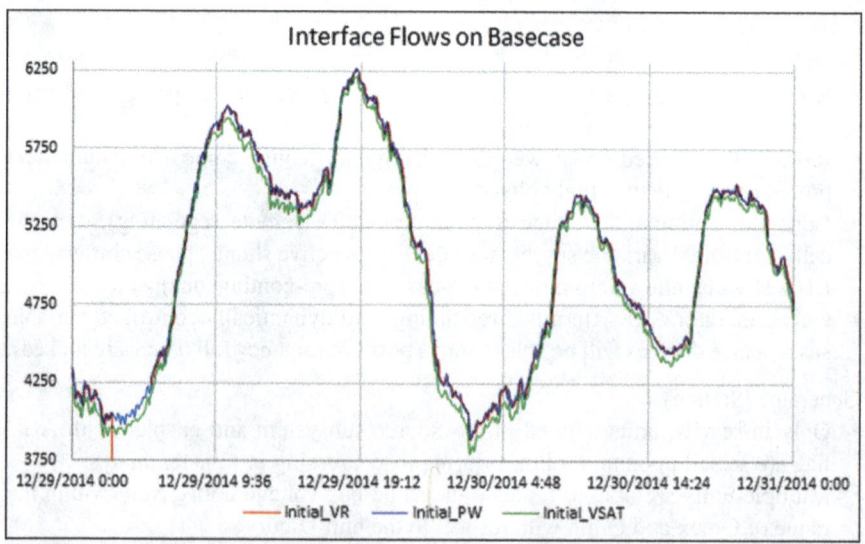

Fig. 5.4 Calculated basecase interface flow (12/29/14–12/30/14)

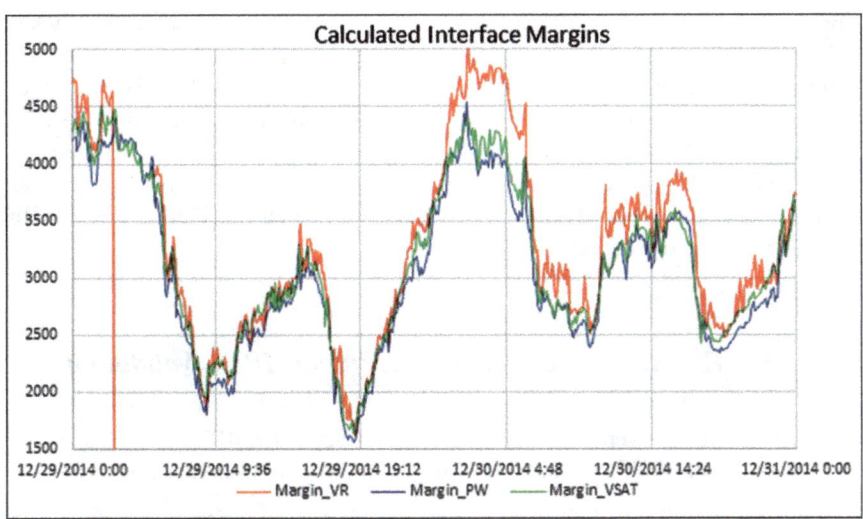

Fig. 5.5 Calculated interface margins w/o outages

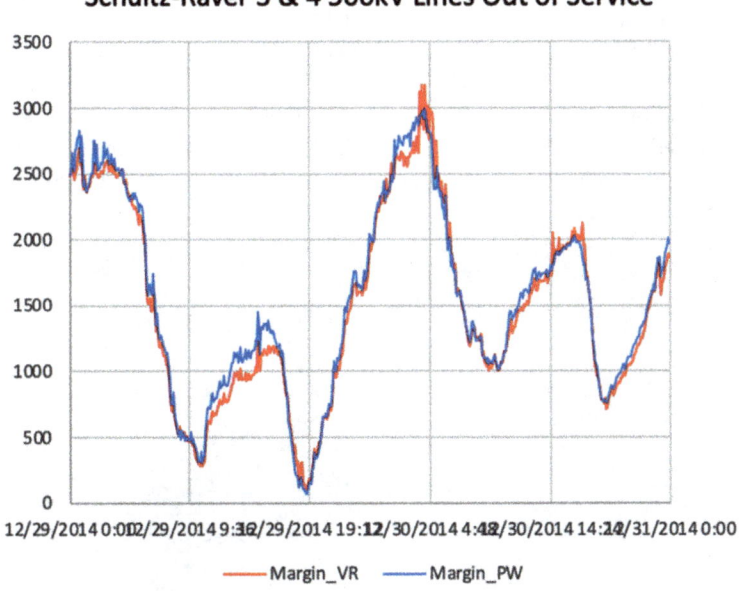

Fig. 5.6 Calculated interface margins with outages

drop the limit considerably but also cause the previously seen divergences in V&R tool to become minimal. Note that the VSAT tool was excluded for outage case comparison. It has been shown that for the given data sets, V&R ROSE VSA tool has proven to be comparable with other software vendors for the output results. To mitigate VSA solution oscillations in high margin cases, PEAK limited the max transfer level without overlooking actual operation risk. From the continuous validation test results, the ROSE RT-VSA tool has demonstrated adequacy in both its performance and reliability.

5.4.3 SDGE Import and SDGE/CFE Import IROL Validation

Both SDGE IROLs were initially validated between 01/2015 and 07/2015 and revalidated between 10/2015 and 05/2016. Key findings are summarized below:

- There are 7500KV N-1 Contingencies and 12 RAS associated with the South West IROLs. Missing or inappropriate modeling of the RAS could impact the IROL calculation significantly. Figure 5.7 shows impact of RAS model settings on the SDGE/CFE Import IROL calculations on a system event that led to unnecessary load shedding: (1) Prod_Margin is PEAK production RT-VSA results without CFE and IV RAS modeled; (2) IV_Margin is Peak offline VSA study results with IV RAS modeled; (3) CFE_IV_Margin is PEAK offline VSA study with CFE and IV RAS modeled; (4) CISO_Margin is CAISO production Bigwood VSA tool results (no CFE RAS modeled). Per coordinated ad-hoc event analysis among PEAK, CAISO, SDGE, and CFE, both CFE and IV RAS shall be modeled in the RT-VSA tool, but CFE RAS will not be fired to reduce the IROL. By validation, IV_Margin produces the best estimate of the SDGE/CFE IROL on the event.

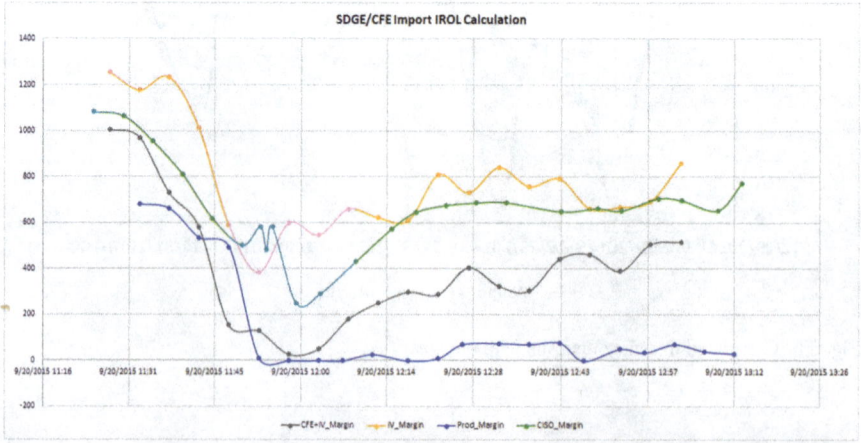

Fig. 5.7 SDGE/CFE import IROLs with different RAS settings

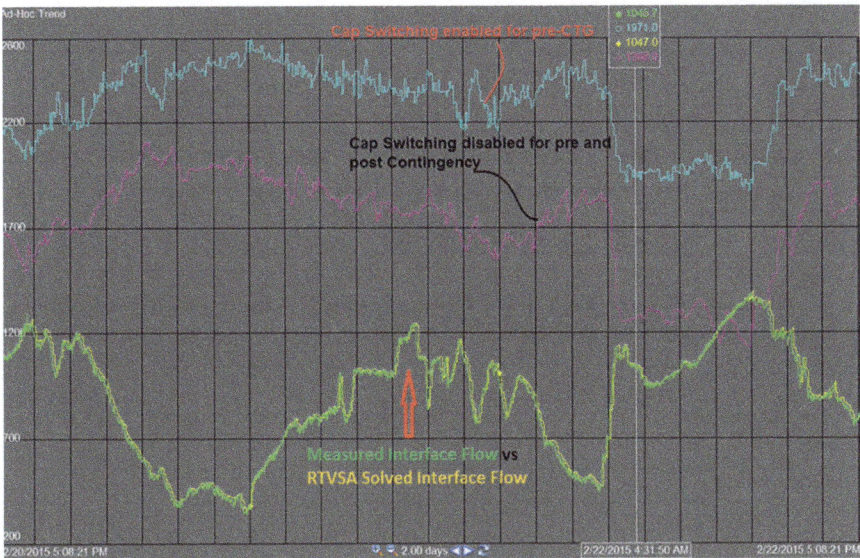

Fig. 5.8 Effect of auto shunt switching in pre-contingency on the IROL

- Enabling auto shunt Cap switching in pre-contingency increases the margin by 600–700 MW compared to the one without shunt switching. Enabling Cap switching also caused more numerical oscillation in VSA solution. The impact of auto shunt switching is shown in Fig. 5.8.
- The ROSE VSA tool increases P/Q load in proportion to solved base values, regardless of the negative Q load estimated in basecase. There are many Distributed Generators (DG) installed with capacitor banks. As the WSM does not model sub-100 kV typically, those DG resources are simply represented by equivalent loads. As a result, those reactive loads could be estimated either positive or negative. Stressing reactive loads makes the IROL solve noticeable higher mistakenly.
- Stressing step size has big impact on the VSA solution quality and performance, e.g., use of a smaller step size may produce a more accurate solution, but yet takes longer solving time. For instance, 10 MW step size provides more accurate results, while 100 MW step size can solve all IROL scenarios in 5 min. As a trade-off, a step size of 25 MW is chosen by default.

In light of the findings from validating both IROLs, PEAK worked with V&R to make corresponding software enhancements and reviewed/updated the RAS modeling scripts in collaboration with the entities.

- Improve auto shunt switching logic to be more robust.
- Add three options for reactive load stressing: (1) per basecase P/Q ratio; (2) per user-defined constant power factor (PF); (3) freeze negative reactive load while increase active power load.

- Add new logic to perform reverse transfer analysis in case the RT-VSA could not solve SE case for a contingency.
- Enhance output log files to include:
  - Limiting Contingency ID.
  - Violated Element.
  - Name, kV level, and VAR margin of the weakest bus.
- Add new logic to calculate dV/dQ sensitivity for voltage collapse point verification.
- PEAK engineer developed a StudyVSA tool in EMS study mode for control room engineer to validate correctness of RT-VSA results in case a zero or low IROL margin/limit dropout.

The above changes in software and IROL scenario definitions and RAS scripts significantly improved both SDGE Import and SDGE/CFE Import IROL solution quality.

5.4.4 Northwest Net Export IROL Validation

This IROL consists of three WECC paths that may compete with each other, e.g., Path 66 (COI), Path 14, and Path 75. As shown in Fig. 5.9, the IROL has a large

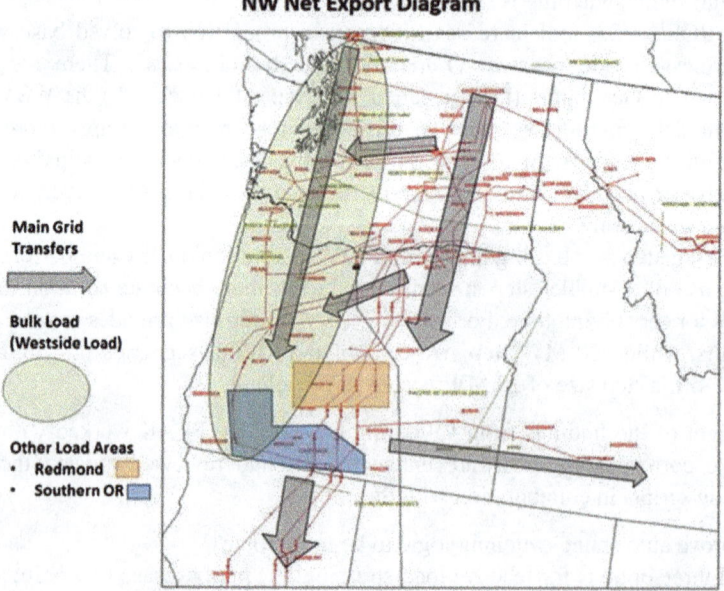

Fig. 5.9 Northwest net export conceptual diagram

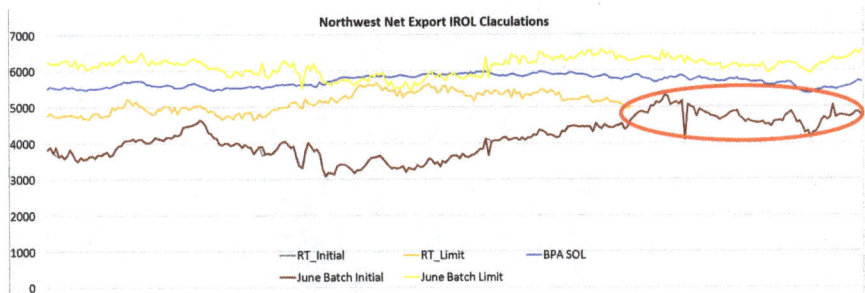

Fig. 5.10 Northwest net export IROL calculations

impact on power transfer capability among multiple areas. PEAK started to develop the IROL scenario including RAS models in the RT-VSA tool and validate it against BPA study tool results in October 2017. This validation encountered new issues.

Main challenge of calculating this IROL in real-time is modeling of two important RAS in the RT-VSA tool: PDCI outage protection scheme and FACRI. Both RAS will backfire if post-contingency power flow does not converge. RAS backfire could cause low or even zero IROL margin.

In Fig. 5.10, RT_Initial and RT_Limit are Peak RT-VSA tool calculated initial interface flow and IROL limit, where the RT-VSA could not solve the IROL for hours because RAS was not modeled. BPA SOL is BPA offline study tool that solved VSA limits using the Nomogram, e.g., RAS was modeled. June Batch Initial and June Batch Limit are the results solved by the new RT-VSA software, with new RAS added to RT-VSA. To compare three VSA study datasets, we can see the new software calculates more reliable IROLs, due to the addition of two stage RAS logic.

5.4.5 Benchmarking PEAK ROSE with CAISO's Bigwood Tool

Since ROSE RT-VSA tool cut off in PEAK Production for operational use in mid-2015, PEAK continuously worked with V&R team to improve the RT-VSA software capabilities. Meanwhile, PEAK collaborated with CAISO, SDGE, and CFE et al. entities closely to review/validate the calculated IROLs by PEAK V&R and CAISO Bigwood RT-VSA tools:

- Host bi-weekly or monthly conference call to review the real-time VSA results.
- Write and update study assumptions documents to align the study assumptions between two tools.
- Exchange online RT-VSA tool results via ICCP between PEAK and CAISO/SDGE.
- Have Seasonal study discussions.

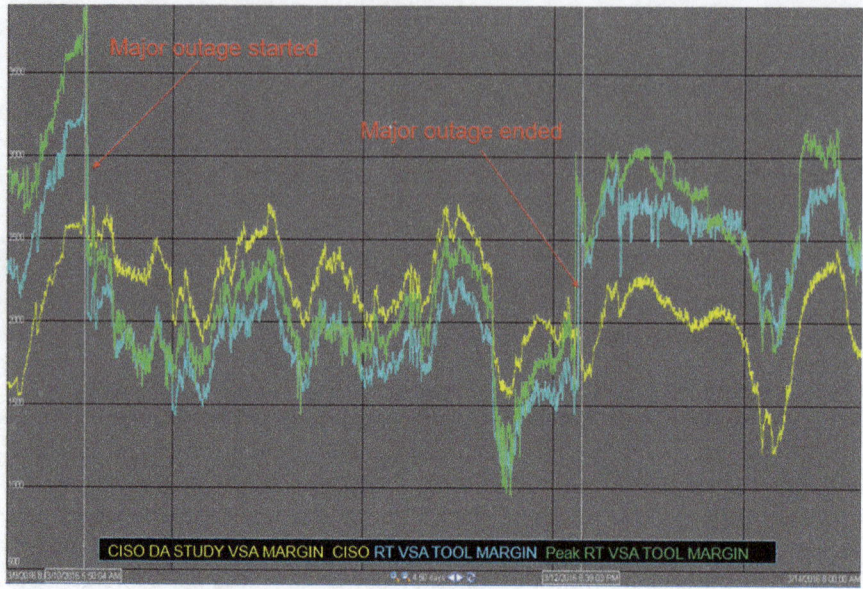

Fig. 5.11 VSA margins between PEAK and CAISOs tools

- Provide feedback and suggestions from entities to V&R for software improvement guidance via V&R monthly POM users conference calls.
- Provide RT-VSA tool update in V&R User Group meeting, and WECC Joint Synchrophasor Information Subcommittee (JSIS)/PEAK SMART meetings.

Thanks to collaboration, we are able to observe two real-time tools and one CAISO day ahead study tool to solve the IROL values closely on stressed system conditions in March 2016. This can be seen from Fig. 5.11, where PEAK's RT-VSA results lined up with CAISO's RT-VSA and Day ahead Study results closely during outage occurrence times, but the gaps among the three tools increased after the outage was removed.

In Fall 2017, PEAK and CAISO's RT-VSA tool results were getting closer on low margin study cases, regardless of major outages occurred or not. Fig. 5.12 attests two RT-VSA tool results look much closer in early October than September.

5.5 Conclusion

Peak Reliability made a long haul to implement V&R ROSE RT-VSA tool in RC Control room through close collaboration with Vendor and Entities. After 4-year continuous validation and improvement, the tool is proven adequate for providing near real-time assessment on four IROLs effective as of today. PEAK's RT-VSA

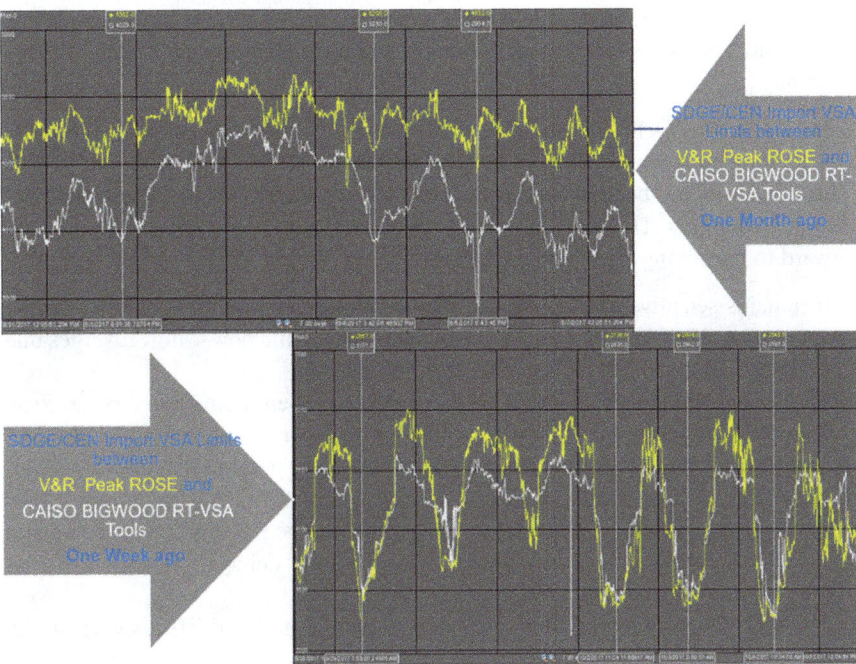

Fig. 5.12 VSA margins between PEAK and CAISO's RT-VSA tools

solution results are updated every 5 min and are shared with CAISO, BPA, SDGE, and CFE via ICCP to improve operation situational awareness across the regions. PEAK sends WSMExport cases to SDGE every 5 min, and SGDE also runs ROSE RT-VSA every 5 min for SDGE Import and SDGE/CEN IROLs in their control center. There were a lot of good practice and lessons learned from V&R PEAK ROSE RT-VSA implementation projects, such as:

- Cross validation between PEAK's RT-VSA tool and CAISO's is a critical and effective way to solve "real" SDGE Import and SDGE/CEN IROLs.
- Cross validation of PEAK's RT-VSA and BPA's Study VSA results is essential for calculating and monitoring NW Washington and NW Net Export IROLs.
- Bi-weekly conference calls on RT-VSA tool setting and VSA results review is the key to build transparency and mutual trust among all stakeholders.
- Trustworthy collaboration between PEAK and V&R Energy enables major RT-VSA solution quality issues solved productively.
- Batch mode study process developed by the PEAK team is proven very useful in testing new software patches, VSA scenario and RAS modeling script changes.
- It is important to develop practical RT-VSA tool trouble-shooting training modules from real system events. The training helps RC and Control room engineers in building their skills and confidence in the RT-VSA tool.

- Effective coordination and clear communication on RT-VSA software and IROL scenario and RAS setting changes will minimize unnecessary human errors greatly.

Presently the V&R PEAK ROSE RT-VSA tool is the only one real-time tool solving four effective IROLs using a full western system operational model. The value of the tool has been widely recognized and applauded by internal and external customers/partners. However, no tool is perfect. There are a few areas we look forward to improving with V&R's support:

- It remains sensitive to massive unit Var regulation control and shunt switching operations. The VSA limit could drop incorrectly while power flow diverges due to numerical instability upon massive switch changes.
- The tool definitely provides better VSA results when a smaller step size (say 10 MW) is applied. But it causes the RT-VSA does not compute all IROLs in 5 min. Dynamic step size searching on the last healthy point needs to be improved to balance the solution accuracy and solving time.
- It is not very convenient for users to debug problematic RT-VSA cases and invalid basecase solution issues. User-friendly error logs and debugging means are more than welcome to add on.
- The RT-VSA tool might encounter an issue while multiple islands exist in the basecase.
- The IROL computation iteration will stop when a local voltage collapse issue occurs for some reason. There is no "SMART" logic to distinguish IROL oriented voltage collapse or a local voltage instability issue.
- The RT-VSA tool can be modified to be used for TTC and ATC calculation, as well as real system reactive sufficiency assessment in the future. PEAK explores an opportunity to expand the RT-VSA use cases in this regard.

Acknowledgments The author gratefully acknowledges the contributions from the Vendor partner-Marianna Vaiman, and many who worked in the Peak Network Applications Team on the RT-VSA tool implementation projects, specifically, James O'Brien, Ran Xu, Madhukar Gaddam, May Mahmoudi, Jiawei 'Alex' Ning, and Hari Ramana. I would thank Saad Malik and Matthew Veghte for their operational input in development of various IROL scenarios and validation of the tool results. Peak IT team, specially, Lon Kepler, Murat Uludogan, Peter Tang, and Steve Pharo, provided essential support for online and offline ROSE RT-VSA software installation and integration. I also appreciate many utility partners who sponsored and participated in West Interconnection Synchrophasor Program (WISP) and Peak Reliability Synchrophasor Program (PRSP) since 2011, respectively.

References

1. NERC Report, "Review of Selected 1996 Electric System Disturbances in North America", August 2002. [Online]. https://www.nerc.com/pa/rrm/ea/System%20Disturbance%20Reports%20DL/1996SystemDisturbance.pdf

2. C.W. Taylor, D.C. Erickson, Recording and analyzing the July 2 cascading outage. IEEE Comput. Appl. Power **10**(1), 26–30 (1997)
3. D.N. Kosterev, C.W. Taylor, W.A. Mittelstadt, Model validation for the August 10, 1996 WSCC system outage. IEEE Trans. Power Syst. **14**(3), 967–979 (1999)
4. H. Zhang, B. Wangen, "Implementation of a full Western Bulk System operational model for reliability monitoring," in *Proceedings of IEEE PES General Meeting*, July 22–26, 2012
5. B. Wangen, H. Zhang, "Monitoring for post-contingency System Operating Limit exceedance in the Western Interconnection," in *Proceedings of IEEE PES General Meeting*, July 22–26, 2012
6. Federal Energy Regulatory Commission, "Arizona-Southern California Outages on September 8, 2011: Causes and Recommendations," FERC and NERC Staff, Apr 2012. [Online]. https://www.ferc.gov/legal/staff-reports/04-27-2012-ferc-nerc-report.pdf
7. C.W. Taylor, *Power System Voltage Stability* (McGraw-Hill, New York, NY, 1994)
8. S. Maslennikov, E. Litvinov, M. Vaiman, M. Vaiman, "Implementation of ROSE for on-line voltage stability analysis at ISO New England, " in *Proceedings of IEEE PES General Meeting*, July 27–31, 2014
9. S. Corsi, G.N. Taranto, A real-time voltage instability identification algorithm based on local phasor measurements. IEEE Trans. Power Syst. **23**(3), 1271–1279 (2008)
10. S.M. Abdelkader, D.J. Morrow, Online tracking of Th'evenin equivalent parameters suing PMU measurements. IEEE Trans. Power Syst. **27**(2), 975–983 (2012)
11. S.M. Burchett, D. Douglas, S.G. Ghiocel, M.W.A. Liehr, J.H. Chow, et al., An optimal Th'evenin equivalent estimation method and its application to the voltage stability analysis of a wind hub. IEEE Trans. Power Syst. **33**(4), 3644–1804 (2017)
12. E. Vittal, M. O'Malley, Andrew, A stead-state voltage stability analysis of power systems with high penetrations of wind. IEEE Trans. Power Syst. **25**(1), 433–442 (2010)
13. K. Kawabe, K. Tanaka, Analytical method for short-term voltage stability using the stability boundary in the P-V plane. IEEE Trans. Power Syst. **29**(6), 3041–3047 (2014)
14. K. Kawabe, Y. Ota, A. Yokoyama, K. Tanaka, Novel dynamic voltage support capability of photovoltaic systems for Improvement of short-term voltage stability in power systems. IEEE Trans. Power Syst. **32**(3), 1796–1804 (2017)
15. S. Malik, M.Y. Vaiman, M.M. Vaiman, "Implementation of ROSE for real-time voltage stability analysis at WECC RC", *2014 IEEE PES T&D Conference and Exposition*, 14TD0175, https://doi.org/10.1109/TDC.2014.6863542
16. S. Malik, M.Y. Vaiman, M.M. Vaiman, "Implementation of RAS actions for real-time voltage stability analysis (VSA) at peak reliability", *2014 PAC World Americas Conference*, OP062

Chapter 6
Voltage and Transient Security Assessment in ERCOT Operations

Sidharth Rajagopalan, Jose Conto, Yang Zhang, and Sarma (NDR) Nuthalapati

6.1 Introduction

The Electric Reliability Council of Texas Inc. (ERCOT) serves as the independent system operator (ISO) for most of the state of Texas. The ISO manages the delivery of electric power to over 26 million customers in Texas, representing about 90% of the state's electric load. The ERCOT electric grid consists of around 46,500 miles of transmission and more than 680 generating units. The total installed generation capacity for peak demand in ERCOT is over 82,000 MW and its recorded peak load is 74,820 MW. As of June 2020, there are almost 25,000 MW of wind generation capacity and 3095 MW of utility-scale solar capacity installed and operating in the ERCOT market, and the total wind generation reached a record 21,375 MW. ERCOT also set a wind penetration record of 59.3% in May 2020 [1].

Most of the generated power in ERCOT is consumed in industrial and urban load centers in East, Central, and South Texas, while a large portion of the wind and solar energy is installed in the north-west part of the state because of abundant wind and solar resources there. This separation between resources and load centers has created some unique challenges in the integration of renewable resources, and in meeting the growing demand for power while also effectively ensuring the reliability of the grid [2–5].

S. Rajagopalan (✉) · J. Conto · Y. Zhang
Electric Reliability Council of Texas, Inc. (ERCOT), Austin, TX, USA

S. Nuthalapati
Department of Electrical and Cumputer Engineering at Texas, A&M University, College Station, TX, USA

System Operations Engineering, Dominion Energy, Richmond, VA, USA
e-mail: ndrsarma@ieee.org

© Springer Nature Switzerland AG 2021
S. (NDR) Nuthalapati, *Use of Voltage Stability Assessment and Transient Stability Assessment Tools in Grid Operations*, Power Electronics and Power Systems,
https://doi.org/10.1007/978-3-030-67482-3_6

The transmission of power from increasingly remote generation to ever-growing load centers imposes constraints on daily operations driven by the need to maintain stability. In order to properly account for these constraints, ERCOT has defined several Generic Transmission Constraints (GTCs), each having associated Generic Transmission Limits (GTLs). The GTCs are transmission flow constraints imposed on collections of circuits that are gathered together as interfaces in the ERCOT Energy Management System (EMS). The flows on these interfaces are monitored in order to ensure that they do not exceed the corresponding GTLs.

Traditionally, such constraints have been managed by planning/ad hoc studies ("off-line" assessments) which calculated the limits. But with the growth of remotely sited renewable generation and the increase in system load, the transmission system is being operated closer to the limits than ever before. This has caused an increase in the number and the complexity of the GTCs and managing the GTCs, using the current off-line study process, is becoming more challenging. ERCOT has implemented tools in its Control Centers that can calculate GTLs for the defined GTCs as part of the periodic sequence of actions (state estimation, contingency analysis, etc.) that are carried out within its EMS. These tools are capable of assessing voltage and transient stability of the grid in sub-hour intervals and greatly improve the operators' ability to assess and maintain the security of the ERCOT system.

6.2 Voltage Stability

Voltage stability is a phenomenon concerning the eventual collapse of voltage as loading and/or power transfers are increased on the power system [6]. The techniques to perform voltage stability analysis fall into two categories: static and dynamic. This section covers ERCOTs implementation of a static voltage stability assessment framework in the Control Room.

Static voltage stability assessment is performed using power flow solutions. It can be used to identify the susceptible regions of the power system in terms of reactive power deficiency and determine the critical contingencies and voltage stability margins for various power transfers within the power system. Power–Voltage (PV) curves are the widely accepted measure of a network's vulnerability to voltage instability or collapse. PV curves directly reveal the margin to instability in terms of the relevant and measurable quantities (MW load, generation, or transfer increase) for system operators and planners.

6.2.1 Voltage Stability Assessment in ERCOT Operations

Voltage stability assessments in ERCOT were conducted primarily in the Planning horizon, with the limits (determined for a single snapshot) being applied in daily operations as is. Since the limits were determined ahead of time, without knowledge of the operating conditions under which they would be applied, they were conservative and caused over-curtailment of generation, when they were applied to security-constrained dispatch.

To address this issue, ERCOT Operations implemented a dynamic security assessment interface in the EMS, which would provide current operating conditions and contingency definitions to a voltage stability assessment tool (VSAT, Powertech Labs) [6]. The tool uses the provided information to conduct detailed PV transfer studies, yielding interface flow limits for voltage instability/collapse (divergent power flow solution).

The tools were first used to assess the limits for the North–Houston (N–H) interface, which is the major import path for Houston. They have since been successfully extended to monitor and calculate interface flow limits for multiple GTCs including the import into the Lower Rio Grande Valley area, and generation export from East Texas, West Texas, and the Texas Panhandle.

In recent years, large-scale integration of wind/solar generation in remote areas of the ERCOT grid has given rise to a bevy of voltage stability issues. The power grid in these remote areas are usually weak systems characterized by a low short circuit ratio, which is considered the primary cause of some voltage stability issues including voltage collapse, oscillation, and temporary overvoltage. ERCOT has conducted system strength and voltage stability assessments for these remote regions of the grid that interconnect mainly wind and solar generation [4, 5]. The reports indicate several problems with weak systems, including over-voltages, low-frequency resonances, and control system instability. This led to the incorporation of a system strength estimation into the voltage stability assessment in daily operations in ERCOT.

6.2.2 System Strength

It is well-known that the performance of the various elements of the power system depends on the "strength" of the system at the point where the elements are connected. "Strength" or system strength is a metric which reflects the sensitivity of the system state to disturbances. Conventionally, short circuit ratio (SCR) is used as an index of the system strength. SCR is defined as the ratio of the short circuit capacity at the bus where the device is located, to the MW rating of the device. However, this definition ignores the impact of interactions between wind and/or solar generation sites, that could have them oscillating together like a single large unit. In such scenarios, SCR would offer an overly optimistic estimation of system strength. In order to address this issue, ERCOT proposed the concept of Weighted Short Circuit Ratio (WSCR) defined by [3]:

$$\begin{aligned}
\text{WSCR} &= \frac{\text{Weighted } S_{\text{SCMVA}}}{\sum_i^N P_{\text{RMW}_i}} \\
&= \frac{\left(\sum_i^N S_{\text{SCMVA}_i} * P_{\text{RMW}_i}\right) / \sum_i^N P_{\text{RMW}_i}}{\sum_i^N P_{\text{RMW}_i}} \\
&= \frac{\sum_i^N S_{\text{SCMVA}_i} * P_{\text{RMW}_i}}{\left(\sum_i^N P_{\text{RMW}_i}\right)^2}
\end{aligned} \qquad (1.1)$$

6.2.2.1 The Texas Panhandle

The Panhandle region of the ERCOT grid, as shown in Fig. 6.1, is a prime location for wind generation development due to the favorable wind regime. As of June 2020, there is more than 5.2 GW wind generation capacity, including operational and committed new generation projects, in the Panhandle region. According to a study by ERCOT [7], the ERCOT Panhandle system is identified to require sufficient system strength and voltage support for reliable wind and solar generation operation and stable long-distance power transfer from Panhandle to the load centers.

The detailed "off-line" dynamic studies [4, 5] identified a constraint on the WSCR for the Texas Panhandle of no less than 1.5 to avoid voltage stability

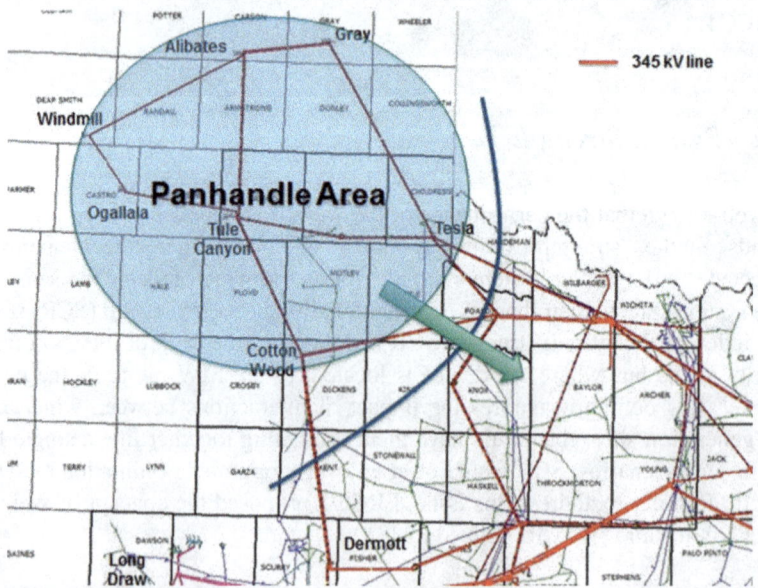

Fig. 6.1 ERCOT transmission in the Texas panhandle

problems. In 2015, Powertech Labs developed the WSCR calculation tool as a standard module embedded in VSAT [8] for ERCOT. In ERCOT Operations, the WSCR of the Panhandle system is calculated every 10 min using the tool in VSAT and the current system model available from the EMS. The limit of 1.5 on WSCR is maintained by curtailing wind and solar generation when necessary. Among all the stability limits (voltage collapse and WSCR) calculated by VSAT, the minimum is set as the Panhandle Export limit, provided it is more restrictive than any reported thermal constraints.

Since the calculation in ERCOT Operations is repeated every 10 min, two aspects of the calculation were modified to better account for the changing conditions in real-time:

(a) The MW rating of each wind and solar plant in Eq. (1) is replaced by the MW output at the time of the calculation.
(b) No contingency analysis is performed and the WSCR is evaluated in a pre-contingency state. The calculation is repeated every 10 min, any change in topology will be accounted for in the calculation that immediately follows it.

The implementation of the WSCR tool in real-time operations helps ERCOT Operators maintain stability in the Panhandle region. It provides specific information about limiting dispatch and interface flows allowing operators to determine the degree of curtailment to employ. Calculating WSCR using the real-time model, also ensures that actual system conditions are being used to determine operating limits. This helps in maximizing the output from the wind and solar plants in the Panhandle.

6.2.2.2 Applicability to Other Regions in ERCOT

The WSCR metric assumes that all the generation in the region of interest are concentrated into a coherent export area. This allows the interacting generators to be gathered into clusters for the WSCR calculation. Given the unique conditions in the Texas Panhandle, this metric is well suited to the task of determining operating limits for the region. Defining the WSCR clusters around the Panhandle resources is an obvious choice and the calculated WSCR threshold corresponds well with stability limits. Since calculating a WSCR is easier than running dynamic simulations and/or EMTP simulations, there is a clear benefit in utilizing the WSCR metric to determine operating limits for the Panhandle region.

However, the metric is not so well suited to other regions in ERCOT—primarily because of the lack of a coherent export area and that it does not account for the impact of local load. ERCOT has investigated the use of WSCR in South Texas [5], which also has a high penetration of wind generation. But the dispersed nature of the generation sites as well as the presence of local load, make it difficult to apply WSCR to this region.

6.3 Transient Stability

Transient stability issues in ERCOT Operations were largely dealt with through what are termed "static" limits—limits on transmission line flows calculated through planning/ad hoc studies. The most significant of these stability issues was the West–North (W–N) Transfer interface.

6.3.1 The W–N Interface

Before the completion of the Competitive Renewable Energy Zones (CREZ) project, the W–N interface consisted of six 345 kV transmission lines, that moved power generated in West Texas into the Dallas/Fort Worth area. Studies found that due to the limited capacity on the interface, under conditions of outage of one or more of the included circuits, inter-area oscillations in the 0.6–0.7 Hz range could occur between generators in West Texas and those elsewhere in ERCOT. As wind-generator installations increased in West Texas, this constraint became heavily restrictive—a situation exacerbated by the static nature of the limits. Since they were calculated in "off-line" studies that did not account for current system conditions, the limits were by necessity conservative, and this caused the generation in West Texas to be curtailed.

In order to address this issue, ERCOT Operations undertook the task of implementing a transient stability assessment tool (TSAT, Powertech Labs) [9] in the Control Room. These tools would use the current operating conditions in order

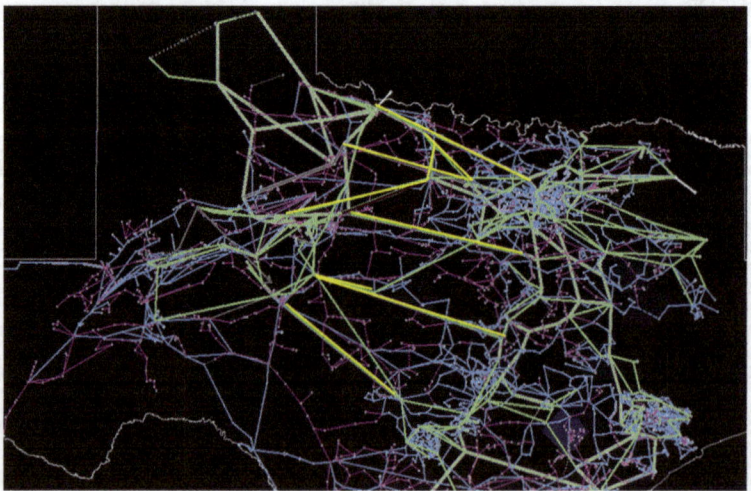

Fig. 6.2 The W–N interface after the CREZ project

to conduct the transient stability assessments that heretofore were conducted in detailed planning studies. This would ensure that the limits being applied to the W-N constraint were based on the operating conditions and therefore more apropos to the task of managing system security in daily operations.

With the completion of the CREZ project, the circuits on the W–N interface increased from 6 to 16 (Fig. 6.2), which largely resolved the issue the W–N Interface had been instated to address. As such, after extensive studies, the W–N Interface was retired from service and the transient security assessment in the Control Room was ended.

6.3.2 Transient Event Metrics

With the increase in number and complexity of the GTCs being employed in daily operations, ERCOT has renewed the effort to assess transient stability in an automated fashion in the Control Room. As part of this effort, ERCOT engineers worked with Powertech Labs to design and implement a system of reporting system performance parameters from dynamic simulations to the ERCOT EMS (Fig. 6.3).

This tool is intended to facilitate reporting and archiving of relevant results from contingency analysis with dynamic simulations and would allow monitoring of impact of large generation loss and transmission fault events. The reported metrics include:

(a) Loss of generation events.

　(i) Minimum frequency (in Hz).
　(ii) Amount of generation tripped (in MW).

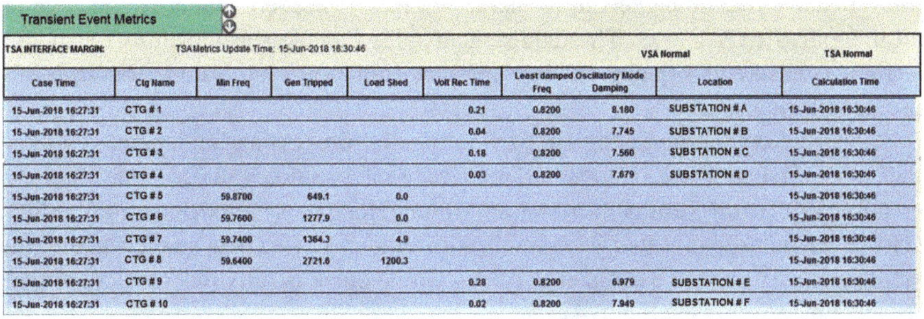

Transient Event Metrics									
TSA INTERFACE MARGIN:	TSA Metrics Update Time: 15-Jun-2018 18:30:46					VSA Normal		TSA Normal	
Case Time	Ctg Name	Min Freq	Gen Tripped	Load Shed	Volt Rec Time	Least damped Oscillatory Mode Freq	Damping	Location	Calculation Time
15-Jun-2018 16:27:31	CTG #1				0.21	0.8200	8.180	SUBSTATION #A	15-Jun-2018 16:30:46
15-Jun-2018 16:27:31	CTG #2				0.04	0.8200	7.745	SUBSTATION #B	15-Jun-2018 16:30:46
15-Jun-2018 16:27:31	CTG #3				0.18	0.8200	7.560	SUBSTATION #C	15-Jun-2018 16:30:46
15-Jun-2018 16:27:31	CTG #4				0.03	0.8200	7.679	SUBSTATION #D	15-Jun-2018 16:30:46
15-Jun-2018 16:27:31	CTG #5	59.8700	649.1	0.0					15-Jun-2018 16:30:46
15-Jun-2018 16:27:31	CTG #6	59.7600	1277.9	0.0					15-Jun-2018 16:30:46
15-Jun-2018 16:27:31	CTG #7	59.7400	1364.3	4.9					15-Jun-2018 16:30:46
15-Jun-2018 16:27:31	CTG #8	59.6400	2721.8	1200.3					15-Jun-2018 16:30:46
15-Jun-2018 16:27:31	CTG #9				0.28	0.8200	6.979	SUBSTATION #E	15-Jun-2018 16:30:46
15-Jun-2018 16:27:31	CTG #10				0.02	0.8200	7.949	SUBSTATION #F	15-Jun-2018 16:30:46

Fig. 6.3 Transient Event Metrics display in EMS

(b) Transmission fault events.

   (i) Slowest voltage recovery (largest time; in seconds).
   (ii) Location of the slowest voltage recovery.

(c) Generally reported items.

   (i) Amount of load shed, if any (in MW).
   (ii) Frequency and damping of the least damped oscillatory mode.

6.3.3 Looking Ahead

There are difficulties with implementing a dynamic simulation tool in the Control Room to work in concert with the EMS. Such tools require the use of mathematical models to represent the dynamic behavior of the included devices in the study. While a large number of devices (steam/gas turbines, generators, static and rotating exciters, power system stabilizers, etc.) can be represented through standardized models (published by organizations like IEEE), new technologies like the Type-3 and Type-4 wind turbines and solar generation inverters do not have standardized modeling implementations. "Generic" model specifications exist for these technologies, but these do not always provide a faithful representation of device behavior—which is critical for analysis in areas like the Panhandle with low system strength. For these devices, ERCOT has elected to use user-created models provided by the manufacturer. However, such models are applicable only to the particular software platform for which they are created. Since ERCOT uses different tools for dynamic simulation studies in Planning and Operations horizons, acquiring user-created dynamic model representations for all such tools is necessary.

ERCOT has now put into place, requirements [10] on generators to provide any user-created dynamic models in formats (multiple, if needed) that are applicable to all the tools ERCOT uses. The requirements also include specifications for model testing to ensure the quality of performance and parity of features between all provided formats.

Advanced computing techniques using parallel processing implemented within the tools used at ERCOT together with internal processes to facilitate data set creation will render results in minimum times. The data sets for dynamic stability studies can be used repeatedly to simulate different events. These studies would run simultaneously since the evaluation of any single event is fully independent of that for any other event (see Fig. 6.4).

With these requirements and tools in place, ERCOT is now in a position to implement full dynamic simulation studies in the Control Room to calculate GTLs for the defined GTCs and assess system performance at a high level of fidelity.

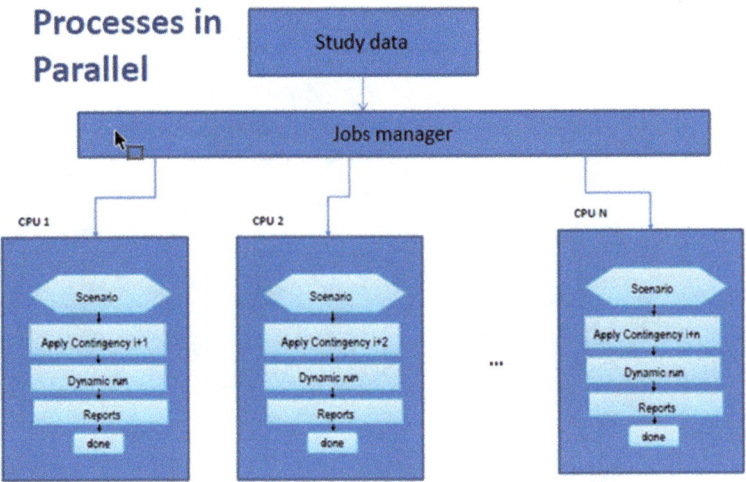

Fig. 6.4 Simultaneous event simulation in a dynamic study

References

1. ERCOT, *Quick Facts—June 2020* (ERCOT, 2020)
2. S.-H. Huang, J. Schmall, J. Conto, J. Adams, Z. Yang, C. Carter, *Voltage Control Challenges on Weak Grids with High Penetration of Wind Generation: ERCOT Experience* (IEEE PES General Meeting, San Diego, 2012)
3. Y. Zhang, J. Schmall, S.-H. Huang, J. Conto, J. Billo, and E. Rehman, "Evaluating System Strength for Large-Scale Wind Plant Integration" Proceedings of 2014 IEEE PES General Meeting, July 27–31, 2014, National Harbor, MD
4. ERCOT, "System Strength Assessment of the Panhandle System PSCAD Study", 2016. [Online]. http://www.ercot.com/content/news/presentations/2016/Panhandle%20System%20Strength%20Study%20Feb%2023%202016%20(Public).pdf
5. ERCOT, "Panhandle and South Texas Stability and System Strength Assessment", 2018. [Online]. http://www.ercot.com/content/wcm/lists/144927/Panhandle_and_South_Texas_Stability_and_System_Strength_Assessment_March....pdf
6. Y. Zhang, S. Rajagopalan, and J. Conto, "Practical Voltage Stability Analysis" Proceedings of 2010 IEEE PES General Meeting
7. ERCOT, "Panhandle Renewable Energy Zone (PREZ) Study report", 2014. [Online]. http://www.ercot.com/content/news/presentations/2014/Panhandle%20Renewable%20Energy%20Zone%20Study%20Report.pdf
8. VSAT, Version 17 User Manual, Powertech Labs Inc, April 2017
9. DSA Manager, Powertech Labs Inc.
10. ERCOT, Dynamics Working Group Procedure Manual Revision 14, 2020. [Online]. http://www.ercot.com/content/wcm/key_documents_lists/27301/DWG_Procedure_Manual_Revision_14_ROS_Approved_20200206.docx

Chapter 7
Use of Voltage Stability Assessment and Transient Stability Assessment Tools at PJM Interconnection

Dean G. Manno and Jason M. Sexauer

7.1 Introduction

PJM is the regional transmission organization (RTO) for 13 states and the District of Columbia, covering an area of the United States that includes the Mid-Atlantic region and parts of the Midwest. As part of its mission, PJM monitors the bulk electric system and system operating limits, including voltage stability, transient stability, and dynamic stability limitations.

Historically, PJM Interconnection, a regional transmission organization, used offline approaches to calculate stability limits. Those methods, developed before the modern digital age, presented some impediments to achieving the greatest efficiency. More recently however PJM has adopted some powerful tools and methodologies to help monitor and maintain stability. They involve online, real-time applications for frequently analyzing the system, identifying limits, and keeping operators informed about changing conditions. PJM's Real-Time Transfer Limit Calculator focuses on the relationship between generators and load and the impacts on voltage stability while the Transient Stability Analysis tool, in general, looks at the impact generators have on transient and dynamic stability.

7.2 PJM Tools for Maintaining Voltage Stability

Maintaining a stable voltage is essential to the reliability of the bulk electric system. Key factors that can affect stability include the length of high-voltage transmission

---

D. G. Manno · J. M. Sexauer (✉)
PJM Interconnection, Audubon, PA, USA
e-mail: Dean.Manno@pjm.com; Jason.Sexauer@pjm.com

lines that carry power to load centers and the impacts that generators can have on the flow of power. It is important for operators to know the limits under which the system will perform reliably, especially under changing conditions.

7.2.1 Background: Voltage Stability and Instability

Voltage stability is defined as "the ability of a power system to maintain steady voltages at all busses in the system after being subjected to a disturbance from a given initial operating condition." [1] Such disturbances could include the sudden shutdown of a power plant, for example, or the loss of a transmission line. Voltage instability is most common when system load is far (in terms of electrical distance) from a strongly networked system of generation.

After a system disturbance, loads respond in a variety of ways, including motor slip corrections, transformer tap changes, distribution voltage regulation, and thermostat responses. These load responses may attempt to draw more real power and reactive power through the transmission system. The transmission lines are heavily inductive, meaning the transmission line's inductive reactance is often 7–10 times more than the line's resistance. As more real and reactive power is transmitted through the lines, the lines themselves demand more reactive power to serve the load. This worsens voltage drops along the transmission lines, causes lower voltage at the receiving end, and triggers more load adjustments that pull more power. Since the receiving-end voltage decreases, the power increase to support the load comes in the form of increased current through the transmission lines (P = IV). When this effect snowballs out of control, the system experiences a voltage collapse. A voltage collapse is defined as "the process by which the sequence of events accompanying voltage instability leads to a blackout or abnormally low voltages in a significant part of the power system." [1].

Figure 7.1 shows the voltage at the receiving end of a system (V_R) as a function of the real power transferred across the system in megawatts (MW). This is called the

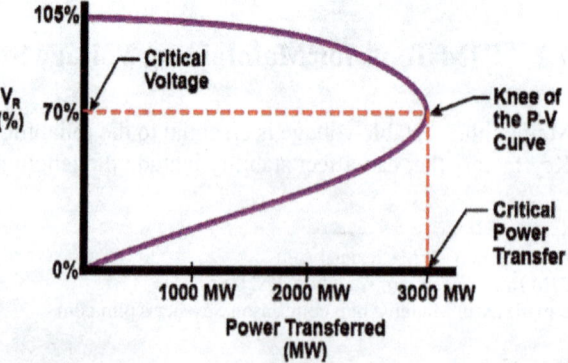

Fig. 7.1 System P-V Curve—Voltage at the receiving end of a system (V_R) as a function of the real power transferred across the system (MW)

system's "P-V Curve." As the system transfers more real power to the receiving end, the receiving-end voltage decreases. This is represented by the top half of the curve (stable equilibrium points). At the "Critical Power Transfer," the receiving voltage hits a "Critical Voltage" level and voltage collapse ensues (saddle-node bifurcation points). This is represented as the "Knee of the P-V Curve."

The system cannot deliver any amount of power past the Critical Power Transfer level. The underside of the curve consists of theoretical operating points (unstable equilibrium points), where the power delivered to the load is dominated by high currents and low voltages. Operating points on the underside of the curve are infeasible. Disturbances affect the curve by shifting the Critical Power Transfer point left to a lower value.

Transformer tap adjustments, generator excitation, and reactive devices may be able to provide reactive support to boost voltages and extend the horizontal length of the P-V Curve. However, these devices may reach their limits before restoring the voltage to acceptable levels. Location of generators, reactive devices, and load tap changers (LTCs) relative to the load also play a factor. Due to the dominantly inductive characteristic of transmission lines, reactive power cannot be transmitted long distances. If reactive support is not located close to areas where voltage issues are occurring, attempting to transmit the reactive power to the voltage deficient location may cause more severe voltage drops across the delivering transmission lines.

7.2.2 PJM's History with Voltage Stability and Inter-Regional Operating Limits

Around the 1960s, America's electric power systems began to be transformed by "mine-mouth" generation plants. Instead of building coal plants close to city loads and transporting coal from mines to the plants, electric generation companies found it more profitable to build coal-fired generation plants next to coal mines and build transmission lines to deliver the electric power to the distant loads. This economic model was the dominant factor that impacted PJM's generation mix from the 1960s to the 2000s. During this time, PJM's most economic generation plants were coal-fired units in the Appalachian Mountain areas of western Pennsylvania and West Virginia. As PJM expanded in the 2000s, it inherited more mine-mouth, coal-fired generation plants along the New River. This laid the foundation for voltage stability concerns in the PJM footprint.

PJM's most economic generation was in its western territory: Ohio, New River Valley, West Virginia, and western Pennsylvania. PJM's major load centers were in its eastern territory: Philadelphia, New Jersey, Baltimore, and Washington, D.C. The network connecting these two regions consisted of few transmission lines that navigated through the Appalachian Mountains. As PJM attempted to deliver power long distances from west to east, voltage stability issues occurred because of the voltage drops across these long lines. If too much power was being delivered, particularly during periods of high demand, the backbone of the system was not able to maintain adequate voltage levels after a disturbance. These voltage stability issues were observed before the system reached any thermal constraints. PJM needed to be able to track and control line flows based on pre-contingency and post-contingency voltage issues.

7.2.3 Implementation

Unlike controlling for thermal issues, which have easy-to-identify limits, controlling for reactive issues demands more-complicated calculations to determine the limits. These reactive limits need to consider nearby voltage levels, load consumption, power factors, reactive device support, out of service transmission facilities, generation reactive outputs, and reactive consumption of transmission lines.

Thus, it is difficult to define a voltage limit in terms of the power flow across one specific line. This led to the need to develop a voltage stability analysis tool for PJM to define and monitor dynamic voltage limits. PJM's focus is not on trying to define local voltage issues, but instead on finding wide-area voltage instability threats on the backbone transmission network.

PJM's voltage stability analysis tool is called the Real-Time Transfer Limit Calculator (RTTLC). It is designed to associate voltage violations with power flow on specified transmission lines. PJM's RTTLC was implemented in the early 2000s. The limits calculated by this tool are based on the system voltage collapse point, bus voltage drop point limits, and/or bus voltage limits under N-1 conditions; the inputs are from PJM's Energy Management System, which also includes the State Estimator and Security Analysis tools in the real-time mode (Fig. 7.2).

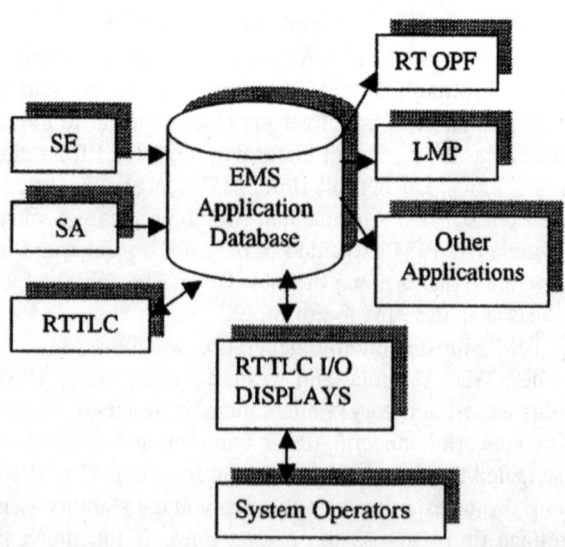

Fig. 7.2 Framework for Real-Time Transfer Limit Calculation (RTTLC). The limits calculated by RTTLC are inputs for PJM's Energy Management System, alongside the State Estimator and Security Analysis tools [2]

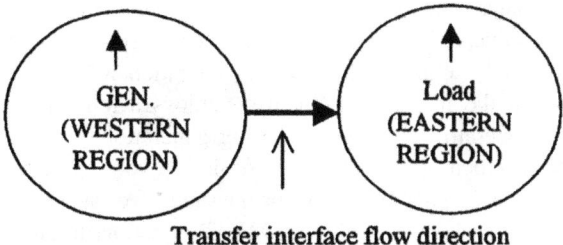

Transfer interface flow direction

Fig. 7.3 When first implemented in the early 2000s, the PJMs RTTLC voltage stability analysis tool would define western region generation as the source, eastern region load as the sink, and the EHV network between these regions as the interface [2]

Steps in PJM's Real-Time Transfer Limit Calculation (RTTLC):

1. Define an area that contains generation (the "source").
2. Define a different area that contains load (the "sink").
3. Define the group of lines that deliver the power from the source to the sink (the "interface") (Fig. 7.3).

    (a) The interface does not have to include all the lines that connect the source to the sink, but it should include the most important lines at the highest voltage. In PJM's case, only extra high voltage (EHV) lines of 345 kV and above are defined as the interface between the source and the sink.

4. Define which busses to monitor for voltage violations.

    (a) Usually, these are the highest voltage busses along the interface or in the sink. In PJM's case, only busses of 345 kV and above are monitored.

5. Define N-1 contingencies to analyze.
6. Receive latest solved power flow case from EMS State Estimator.
7. Raise base case sink load and source generation in steps (approximately 500 MW).
8. Solve using Continuation Power Flow engine.
9. Monitor defined busses for voltage violation or system non-converges.
10. Repeat steps 7 through 10 for the base case and each defined N-1 contingency.
11. Compute the maximum pre-contingency transfer for each contingency.
12. Determine the limiting contingency by identifying the one with the smallest MW pre-contingency transfer. This flow level becomes a defined limit for voltage stability violations for that interface [2].

RTTLC can process a case and provide updated limits every 5 min. When a specific power flow level on an interface is identified as a point when a voltage violation would occur, PJM can control the power flow on the interface to stay below the defined limit. The flow limit of the interface is inserted as a constraint into PJM's Security Constrained Economic Dispatch (SCED) engine. The SCED engine will help operators dispatch generation assets to make sure flows on the interface stay below the defined value, as well as control for other constraints on the system.

PJM's RTTLC only tells operators what generation MW output adjustments can reduce flow on the interface. PJM also deploys a tool that provides "non-cost" solutions to operators before dispatching generation. The Voltage Stability Analysis & Enhancement tool (VSA&E) from Bigwood Systems Inc. runs in parallel with the RTTLC. This tool performs the same steps as the RTTLC but takes an additional step by finding sensitivities of MVAR injections throughout the system on the interface transfer levels. This allows VSA&E to recommend adjustments in capacitor banks switching, LTCs, or generator reactive outputs in order to increase the maximum power transfer before voltage issues are observed and generation dispatch is necessary. This helps ensure that PJM is dispatching the most economic generation available.

Both PJM's RTTLC and Bigwood Systems' VSA&E tools also have offline study modes. PJM runs day-ahead peak case studies for each day. These study cases are run through the offline RTTLC to determine if transfer limits are to be expected the next day. Then the case is run through offline VSA&E to maximize transfer limit capabilities before generation dispatch is necessary. These day-ahead study case interface limits are used as constraints in the day-ahead market for dispatch control.

7.2.4 PJM Interfaces Today

When first implemented, PJM only had three defined interfaces: Western, Central, and Eastern. Through these interfaces, PJM calculated the maximum transfer capability from western regions to eastern regions. Today, PJM monitors 10 transfer interfaces. These are not necessarily limited to west-to-east transfers. PJM now also monitors transfers from north to south, south to north, and east to west.

PJM no longer experiences as many west-to-east interface issues as when it first implemented the RTTLC. There has been a boom in natural gas generation plants in PJM's eastern regions in the past 10 years because of increased drilling activity in the Marcellus Shale region and because of decreased natural gas prices. The increase in new generators in eastern PJM and reinforcements in the transmission system have reduced reactive-interface congestion in the RTO. Operators, nevertheless, continue to use the RTTLC and VSA&E tools every day in the Operations Center to help monitor voltage issues that may arise.

7.3 PJM Tools for Transient Stability Analysis

PJM has implemented the Transient Stability Analysis using Powertech Labs Inc.'s Transient Stability Analysis & Control (TSA&C) tool. This is an online, real-time tool that performs a stability analysis every 5 min. Previously such studies were performed manually about once a year. Transient Stability Analysis (TSA)[1] came about because offline stability analysis was often too conservative, resulting in higher costs to ratepayers for the dispatch of higher-cost generation than if more-precise limits were known. Offline analysis also was too limiting because no offline study could foresee every possible operating condition.

Since 2006, PJM has employed Transient Stability Analysis in a real-time setting to determine stability limits and to dispatch generation to help maintain stability [3]. The tool leverages data from both PJM's Energy Management System State Estimator and the dynamic modeling data from the System Dynamics Working Group to identify insecure contingencies and determine limits to mitigate these risks. The tool also can be run offline to determine stability limits for upcoming transmission facility outages in the operations planning time horizon.

7.3.1 Process for Mitigating Transient Insecurity in the Control Room

The PJM Transient Stability Analysis system (TSA) runs approximately every 5 min. If, during the execution, an insecure contingency is observed—a potential loss of transmission that could cause a generator to be unstable—an alert is sent to the Intelligent Event Processor (IEP), which is an alarm system monitored by PJM operators. Upon seeing the alarm, PJM operators will use the Transient Stability Analysis & Control tool (TSA&C) to identify the insecurity, what units are causing it, and what mitigating actions are possible. Typical results from the tool would include increasing voltage support from the generator, which is reactive power, expressed in megavars (MVARs), or reducing generator output, expressed in megawatts (MWs).

However, before taking any action, the operators will verify the reasonableness of the results. Dispatchers will verify that the contingency is valid for the current operating state. They also will verify that the contingency is near the generators which have been identified as causing the issue. Next, they will validate the State Estimator (SE) solution that generated the result, which means they will ensure that the SE MWs and MVARs for the unit are reasonable and accurate. They also will identify the change that caused the insecurity: this could be the result of a transmission line tripping off or a change in the operational state of a generator. Where possible, the operator will compare the TSA limit against the offline planning limit posted in PJM's operations manuals.

---

[1] Throughout this chapter, the term "TSA&C" will be used to discuss the specific software product developed by Powertech Labs Inc., whereas "TSA" will be used to discuss the broader ecosystem, usage, and business processes that employ the TSA&C tool. These two concepts are closely connected, so at times the terms could conceivably be used interchangeably.

When the operators have taken all these steps to verify and validate the insecurity, they will manually dispatch the affected generator, using the recommended reductions from TSA&C. It also may be necessary for dampening issues—when there is instability, or oscillation, among various generating units—to run an offline TSA study, using the study mode to determine a limitation. If the limitation is expected to be in effect for multiple days, PJM will act to ensure the restriction is reflected in the PJM Day-Ahead Market. There are several ways to do this. One approach is to have the generator bid into the market only up to their restricted output. A second approach is to build a thermal surrogate, which makes the stability limit appear as a thermal limit, that the market software can use to restrict the units.

7.3.2 Training and Job Responsibilities

The first line of defense for real-time operation of the PJM TSA is the staff of real-time dispatchers. They monitor TSA for post-contingency insecurities and impose restrictions as recommended by the tool. They are trained on the tool once every 3 years. Training includes both a classroom presentation on the theory and practical processes of TSA as well as simulated insecurities in the training environment.

If issues arise that are beyond the capabilities of the real-time operators, PJM maintains on-call support for TSA. The on-call support team is composed of back-office engineers who maintain the TSA tool, perform the offline studies for upcoming outages, and maintain the PJM manuals that list offline limits. On-call staff receive on-the-job-training and go through a formal testing process. To build experience with TSA, the support staff starts off performing offline studies for upcoming scheduled outages. They also perform PJM manual maintenance studies before being assigned to real-time, on-call duty.

For technical issues that have an impact on the operation of the TSA&C software, PJM employs an on-site, round-the-clock IT support group, which is capable of restarting servers and troubleshooting the application. In the event they are unable to resolve a software issue, they call out to the TSA on-call group mentioned previously. In the event of extended application failure, the last-known stability limits are used. For emerging issues during an application outage, the offline limits from the PJM manuals are used.

7.3.3 Evolution of Use

Prior to real-time TSA, stability limits were determined using offline studies. Most limits came from planning studies, which used worst-case assumptions to determine limits for predetermined transmission facility outages. If an operational scenario arose that was not covered by the offline limits, PJM electric system planners were enlisted to perform a study and determine limits, a process which was arduous because they would have to build a case mirroring the configuration.

When PJM first introduced TSA to the real-time environment, there was some understandable trepidation among PJM operators and transmission owners. In general, TSA allows PJM to operate the bulk electric system to less-conservative limits that more closely reflect the system's actual capabilities than the offline planning studies. This means that if the new tool incorrectly models or analyzes the situation, a unit could run the risk of being operated beyond its safe capability. As such, PJM undertook an extensive benchmarking period in which the limits from the offline studies were validated against the real-time tool. This helped raise confidence in TSA. During this period, TSA was only used in areas where the benchmarking was complete and the transmission owner (TO) was comfortable with its use. For other areas, the more-conservative offline limits were used.

The next major evolutionary step was to start monitoring for dampening issues. Initially, TSA only monitored transient (first-swing) issues. Dampening was not at first considered because it is a phenomenon involving interactions between large swatches of generators in the system. At first, the real-time model did not contain enough dynamic modeling to accurately perform dampening analysis. As additional stability areas were added to the tool—clusters of units in the same general geographic area—PJM's modeling of areas outside of its footprint improved, non-priority generator dynamic models were added and verified, and accurate analysis of dampening became possible.

The most recent innovation in TSA has been the introduction of nodal modeling. Historically, TSA has used an SE export based on the industry-standard PSS/E bus-branch model format. This format occasionally incorrectly reflects what the system would look like after a contingency. This includes instances of post-contingency bus splits, modeling breaker failure schemes, and other related issues. Newer versions of the TSA tool have, instead, applied node-breaker modeling, which more accurately simulates post-contingency topology of the system. Node-breaker modeling is the preferred representation in operational power flow modeling. As Powertech continued to evolve its TSA&C product to serve real-time systems, it also began to support node-breaker modeling. PJM worked with the company to enhance TSA&C to support PJM's specific nodal modeling. This has improved the experience in the control room, as there are fewer situations where contingencies are being incorrectly modeled and where special workarounds need to be developed.

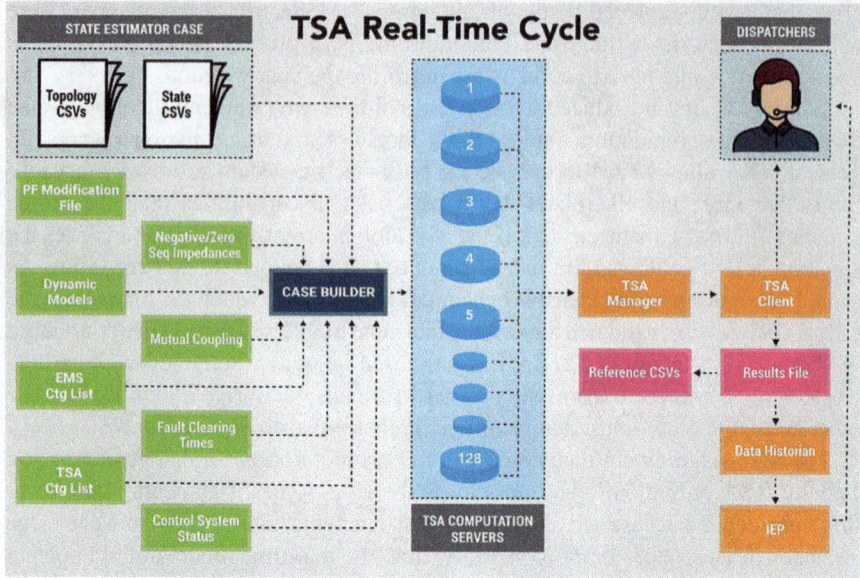

7.3.4 Key Data Inputs

PJM executes a TSA run approximately every 5 min in a manner that is similar to the way contingency analysis is typically run in a transmission control room. This repeating cycle takes the latest available information about generator and load withdrawals, system topology, control system statuses, and other factors, and ensures that generators will remain secure following the contingency loss of a single element, per the NERC TPL-001-4 standard [4].

7.3.4.1 State Estimator Case

The primary source of data for the TSA run is based on the steady-state positive-sequence load-flow generated by the Energy Management System's (EMS) State Estimator (SE). This provides TSA with many of the key assumptions it needs related to the "fast-moving" parameters of the analysis, including current generator and load withdrawals, statuses, and system topology. The SE converges approximately every minute. However, PJM's TSA system solves a case approximately every 5 mins. As a result, TSA takes only every fifth case.

The SE case is exported as a series of about 30 CSV files. The CSV files are broken into "state" files, which contain the "fast-moving" parameters enumerated previously, and "topology" files, which contain "slow-moving" parameters like impedances, connectivity, and unit VAR curves. The design intent was to send only the state files every SE run and the topology files only when their data changed. However, in practice, PJM found it easier to send all the data at the same time so that historic cases can be reconstructed without trying to divine the topology file effective at a given time. This package of about 30 CSV files has become a sort of de facto standard at PJM, with several internal and external applications using the format. It has become preferred over the Common Information Model (CIM) format for some uses because it results in smaller, easier to manage files and an extendable nature while being backward compatible.

7.3.4.2 Power Flow Modification File

Before the SE case is solved and transient stability analyzed, there is a power flow modification file that is used to modify the real-time SE case to be more acceptable for transient analysis. PJM uses this file to make changes to unit modeling to better reflect their dynamic model, as well as to add in dynamic models for DC facilities not represented in the EMS. TSA support personnel can also use the file to change modeling, especially of the external footprint where equivalence models are used, when a specific outage condition is causing TSA&C to produce inaccurate results.

7.3.4.3 Contingency List

As the application of transient analysis is largely a post-contingency phenomenon, it is important to maintain a list of contingencies to study. PJM analyzes about 1500 contingencies in TSA, in contrast to about 7000 contingencies that are run in the EMS for steady-state contingency analysis. In general, TSA only monitors contingencies within about 3 busses of generators with known stability issues as determined by offline planning studies. The contingencies monitored are both bolted three-phase faults for loss of a single transmission element and single-line-to-ground faults for a breaker failure.

The contingency definitions for the bolted three-phase faults are maintained in PJM's EMS. The file is exported to TSA in an .xml format proprietary to the EMS. PJM has internal scripts that convert this file into a more standard format for use in TSA&C. While the bolted three-phase fault contingency type closely mirrors the sort of contingency analyzed by our EMS, there are some key additional details that are needed which are added by various support files:

- Fault clearing time: The steady-state version of a contingency only contains the elements operated for a fault, but not how long the fault persists before being cleared. The fault clearing time is a key element in stability analysis. As a result, this data must be added. PJM maintains a list of company/voltage level defaults as well as specific clearances per equipment and per contingency as needed.
- Fault location: The steady-state version of a contingency does not care where the fault is applied, as it is concerned only with the post-clearing topology

of the system. However, stability analysis is concerned with the pre-clearing, fault-on, and post-clearing topology of the system. TSA takes each steady-state contingency and creates multiple TSA contingencies with fault locations assumed at each terminal of the transmission elements. For example, a simple line contingency will be broken into two contingencies: One with the fault assumed at the near end of the line and a second with the fault assumed at the far end of the line. A more complicated example would be a transmission line with a load-tap in the middle. This would create six contingencies: two for each side of the two transmission line segments, and two for each the high and low sides of the transformer.

The contingency definition for the single-line-to-ground faults are maintained outside the EMS, only in TSA, as this is a contingency type that is not applicable to steady-state analysis. They are stored in the contingency format native to TSA&C.

7.3.4.4 Dynamic Model Data

Another key data input required for transient analysis is a mathematical model of the physical parameters and control systems associated with transmission system components, primarily generators. This involves things such as generator physical parameters like inertia, governor models, automatic voltage regulator models, and power system stabilizer models.

The industry standard for this type of data is the PSS/E dyr file, which contains a generator's bus identification, unit identification, model name as described in accompanying reference materials, and a list of numeric parameters for the model of generator. This information is submitted to PJM's planning department pursuant to MOD-26 and MOD-27 standards [5, 6] for the System Dynamics Working Group (SDWG) planning dynamic model. PJM system planners forward this information to PJM operators, who map the planning model identifiers to EMS model identifiers via a special header row added to the dynamic model file. This special header row also allows for PJM operations to account for modeling differences between the planning and operations models, such as a different number of units (for example, cross-compound steam unit expressed with separate high- and low-pressure generators in the planning model, expressed as a single unit in the operations model) and the presence or absence of a generator step-up transformer from either model.

7.3.4.5 Additional Impedance Data

The SE case described previously only contains the positive-sequence impedance. However, because single-line-to-ground faults are also considered in TSA, negative- and zero-sequence data also is required. Further, the mutual coupling between transmission lines in shared right-of-way should also be considered. This information is maintained in PJM's offline short-circuit model and mapped into TSA.

7.3.4.6 Control System Status Data

In offline planning studies, the status of automatic voltage regulators and power system stabilizers are assumed to be in their default state based on how the plant typically operates. However, operationally, these devices may be unavailable because of plant maintenance or operational conditions. Thus, a file is provided to TSA with the real-time status of these devices. Generation operators report this information to PJM's outage tracking tool. The information is bridged out of this tool, manually reviewed, and then used by TSA.

7.3.5 Cycle Execution Process

The key input for the TSA process, the SE case, is delivered to a file share via PJM's Service Oriented Architecture. When TSA recognizes a new file has been delivered, it merges with the other data inputs cited previously and transforms the file into a format proprietary to Powertech's TSA&C technology.

The case is then delivered to 128 processes spread across four servers, which analyze the transient response for a part of the contingency list, resulting in each process analyzing about 13 contingencies. Powertech has implemented early-termination and post-filtering logic, which helps improve the computation time.

If an insecure contingency is observed, TSA&C will run a tool called Preventative Control Measure (PCM) which determines what generator action (increasing MVAR and/or reducing MW) is needed to make the contingency secure.

The results are then fed back to a manager process, which reassembles the data so that it can be viewed by PJM operators via a viewer. PJM operates two data centers, and identical TSA environments are installed at both however only one is running at a given time. PJM periodically rotates operations between the data centers to ensure the functionality of both.

7.3.6 Key Outputs and Integrations

The output of most importance from TSA is the identification of any insecure contingency. PJM defines an insecure contingency as a contingency that results in transient instability (it is first-swing unstable) and/or a contingency that exhibits less than 3% damping. When an insecure contingency is found, the units that are causing the insecurity also are typically identified via PCM. This information is stored in the data historian. The data historian also feeds the Intelligent Event Processor (IEP), a dispatcher alarming tool that notifies the dispatcher that a TSA insecurity has been observed and action is required to remedy the issue. Finally, detailed information regarding the insecurity is gleaned from the TSA&C logs and emailed to TSA support personnel, and archived to a set of .csv files for further analysis.

7.3.7 Study Mode

In addition to the real-time version of TSA, PJM also utilizes TSA in study mode to facilitate offline analysis of system stability for upcoming equipment maintenance outages. Typically, PJM will perform stability analysis for an outage with known stability issues about 3 days before the outage is scheduled to start. Further, as part of the standard outage approval process, a system-wide stability analysis is performed on the outage approval and day-ahead cases to ensure transient security. This process allows PJM to determine preliminary stability limits before the operating day so that generator restrictions can be communicated to the generation owner, and the day-ahead energy market can accurately account for the restriction on the unit.

The study model takes most of the same inputs cited for real-time. The key difference is the SE case is partially replaced with a power flow case from the offline study. As mentioned previously, the SE case is provided as a set of about 30 CSV files divided into "state" and "topology" files. The study mode uses the "topology" files from the last SE run but takes "state" files generated from the offline power flow, which comes from the EMS study package. These state files were explicitly designed so they could replace the state files from the SE export in this manner.

This hybrid approach allows PJM the ability to perform the study setup in the EMS using the tooling and infrastructure used for steady-state operations planning studies. This means the training required for study engineers and operators is minimal, as most of the case building and setup can be done using a toolchain they are familiar with instead of learning a new toolchain specific to TSA&C.

One of the other features of the PJM study mode is the ability to filter the contingencies analyzed to specific "stability areas." In the outage management software, PJM tracks transmission outages that are likely to cause stability issues based on offline planning studies and manuals. When an outage is studied for approval, it is flagged as needing a stability study and the stability area. When engineers perform the stability study, they can restrict the contingency list to just this area, which significantly speeds the execution time. This is important because the study system has about a quarter the computational capability of the real-time system.

References

1. P. Kundur et al., Definition and classification of power system stability IEEE/CIGRE joint task force on stability terms and definitions. IEEE Trans. Power Syst. **19**(3), 1387–1401 (2004). https://doi.org/10.1109/TPWRS.2004.825981
2. J. Tong, Real time transfer limit calculations, in 2000 Power Engineering Society Summer Meeting (Cat. No.00CH37134), Seattle, WA, 2000, pp. 1297–1302, vol. 2. https://doi.org/10.1109/PESS.2000.867576
3. J. Tong, L. Wang, Design of a DSA tool for real time system operations. Powercon (2006)
4. NERC TPL-001-4, https://www.nerc.com/_layouts/15/PrintStandard.aspx?standardnumber=TPL-001-4&title=Transmission%20System%20Planning%20Performance%20Requirements&jurisdiction=null

5. NERC MOD-026-1, https://www.nerc.com/_layouts/15/PrintStandard.aspx?standardnumber=MOD-026-1&title=Verification%20of%20Models%20and%20Data%20for%20Generator%20Excitation%20Control%20System%20or%20Plant%20Volt/Var%20Control%20Functions&jurisdiction=United%20States
6. NERC MOD-027-1, https://www.nerc.com/_layouts/15/PrintStandard.aspx?standardnumber=MOD-027-1&title=Verification%20of%20Models%20and%20Data%20for%20Turbine/Governor%20and%20Load%20Control%20or%20Active%20Power/Frequency%20Control%20Functions&jurisdiction=United%20States

Chapter 8
Online and Offline Stability Assessment Development at National Grid, UK

Fan Li, Martin Bradley, and Frederic Howell

8.1 Introduction

Traditionally stability assessment is carried out at the day-ahead stage by the operational planning team at National Grid. Critical faults on a constraint boundary are analysed and a transfer limit established with appropriate post-fault actions identified. Stability studies are carried out only for the peak demand period once a day in a daily security assessment, assuming that the high demand scenario is the 'worst case' for stability, just like for thermal assessment.

With the fast increase of wind generation and other renewable energy sources in the GBSO system, assessing system stability in operational time scales is becoming more and more difficult and resource-demanding if relying on the traditional offline planning analysis. The highly uncertain nature of renewable energy has further complicated scenarios that should be covered by an operational planning stability assessment. It has become more acute than before that a single peak demand snapshot stability assessment is insufficient to ensure the system security, all the time, every day.

For power system analysis and operational planning, although many fast and simplified approaches have been developed intending to assess system stability 'directly', e.g. from its 'transient energy function', the detailed time simulation remains as the most accurate yet most flexible approach used by the industry. Time simulation can accommodate all types of complex dynamic control system models.

F. Li (✉) · M. Bradley
National Grid, St. Catherine's Lodge, Wokingham, UK
e-mail: Fan.Li@nationalgrid.com; Martin.Bradley@nationalgrid.com

F. Howell
Powertech Labs, Inc., Surrey, BC, Canada
e-mail: Frederic.Howell@powertechlabs.com

© Springer Nature Switzerland AG 2021
S. (NDR) Nuthalapati, *Use of Voltage Stability Assessment and Transient Stability Assessment Tools in Grid Operations*, Power Electronics and Power Systems, https://doi.org/10.1007/978-3-030-67482-3_8

However, time simulation for a real power system is computationally demanding and time-consuming. At National Grid, an offline assessment tool was used, which however is restricted to a small group of contingency cases and focused on the limit finding of a few known constraint boundaries.

In order to meet challenges set by the Government renewable energy target and deliver secure and economic system operation to meet the Future Energy Scenarios, the operational planning and system control teams in National Grid have identified the following requirements on the capability for stability assessment:

- Develop an online automatic system for fast stability assessment in real-time, which will still allow time to readjust generator-system intertripping and other special protection schemes as the last line of defence.
- System stability should be assessed not only for the known stability constraint boundaries but also for the 'new' areas and 'unknown' problems, which are due to fast-growing renewable generation and associated system changes introduced by new control technology and power electronics.
- Stability analyses in both offline study and online assessment need to be consistent in meeting the requirement of the GB Security and Quality of Supply Standards (SQSS).
- An offline system is required to support 'what-if' studies for any selected lead-time in the range spanning from the hour-ahead dispatch horizon to the Year-ahead outage planning stage.
- Both online and offline assessments should meet their respective performance criteria. In the online application, the assessment should be done in line with the state estimation cycle in real-time; and in an offline application both base case assessment and transfer limit derivations should be completed in a practical yet acceptable time scale. There should be no need for any manual intervention in the execution of multiple contingency analyses. Efficiency improvements should be delivered by these new capabilities to the operational planning and control processes.

This chapter presents a review of the development of on- and offline stability assessment systems at National Grid. Considerations for the design from National Grid operational requirements are outlined; implementation aspects are summarised. Some examples of experience from end-users since the commissioning of these new assessment systems are collected here, where their benefits to the business process improvement have been demonstrated.

8.2 Solutions to Meet National Grid Requirements

8.2.1 Solution to 'Identify Unknown Stability Problems'

This requirement has been identified specifically for National Grid processes. The common practice for system stability in most utilities is focused on a small group of known critical contingencies and their associated constraint limits. This practice was also the approach adopted traditionally in the National Grid operational planning process. However, with a fast-developing system incorporating renewable generation and new control systems, this approach adopted in the offline planning study is no longer sufficient for system operation. The control room is exposed to the increasing risk that critical stability contingencies are deviating from those of offline studies as the operating point fluctuates in real-time. 'Engineering judgement' from experienced control engineers is no longer as solid as before. They would like to have a computer system that can analyse the stability risks over the entire system in real-time without the reliance on any presumptions about the system. The priority to them is not the limit of a 'known' constraint, but rather is the risk in the areas that are traditionally 'unknown' in the context of stability. This 'safety-net' system was what National Grid did not have in its process before.

For system stability, the GB SQSS requires that the GB HV transmission system, either intact or with planned outages, under any 'prevailing' operating conditions, should not be unstable, i.e. exhibit either pole-slip (synchronous generators lose synchronism with the system) or unacceptable sustained poor damping, when the system is subject to a credible system fault. This is termed as a 'secured event of a fault outage'.

The SQSS further clarifies that system faults are those on single or 'credible double' circuits, or bus bars or mesh corners where that part of the system is designed to be secured against these types of faults. Switch faults are also considered at certain locations where faults will severely impact on the integrity and security of the main interconnected transmission system, particularly in the area where a significant amount of generation would be disconnected because of such a fault.

From the definitions of security in the SQSS, the requirement to cover for the stability risks arising from 'unknown' system faults therefore can be interpreted as the need for screening all system credible faults, which includes both circuit and bus bar faults. Therefore, any arbitrary combination of circuit trips and cascade events can be ignored.

For circuit faults, National Grid offline studies have the contingency cases for static load flow studies to assess both thermal and voltage violations. However, at National Grid both thermal overloading and voltage violation are evaluated at the time when the Delayed Auto Re-closure (DAR) and other automatic post-fault actions are completed, which is typically after 3 mins of the occurrence of the fault (namely 'Time Phase 2'). Therefore, these contingency models cannot be used for transient stability assessment. The typical time scale for transient (electromechanical) stability is commonly considered between a few milliseconds to a few seconds. This transient time phase is before any DAR and automatic tap changing and switching actions. Also, all credible faults for the 'static' load flow study will have to be refined with details of the primary protection switch sequence included.

For stability studies, it is accepted as a common practice that a balanced three-phase fault in general is the worst case for short circuit faults [Kundur, 1994, Chapter 13] [1].

Fault locations are also critical for the transient stability assessment. Since, in general, it is difficult to pinpoint which location is the 'worst case', faults at all circuit ends are studied. This approach inevitably would at least double the number of contingency cases compared to load flow studies.

To avoid intensive effort to implement detailed protection schedules, generic fault clearing times are used. The SQSS requires a safety margin to be included such that the fastest protection is assumed to have failed and the fault is cleared by the second fastest protection. For the National Grid HV transmission system, it is considered sufficient to assume that a generic fault is cleared at the near-side to the fault in 80 and 120 ms at the far-side.

In manual offline planning studies, bus bar faults used to be created ad hoc based on the individual planner's experience. This was an area where oversight and human errors might occur, and cause consistency in the implementation of the SQSS to vary. Bus bar faults however cannot be predefined as for circuit faults, because running arrangements can change all the time for the purposes of outages, fault levels and flow controls. Bus bar fault switching sequences must be created 'on-the-fly' from the current system snapshot. This task is undertaken by the Data Preparation Tool (DPT) that creates the base case system snapshot, where the switching sequences for all bus faults are created via its internal tracing algorithm. The nodal-branch model in the load flow calculation has been extended to include circuit breakers as zero-impedance dummy lines. 'Nodes' have been mapped to bus sections that are connected by sectioning or coupling circuit breakers. A new user interface has been developed which allows users to create bar faults either individually or collectively for an area or a particular substation, with ease.

8.2.2 Time Simulation Approach

The GB SQSS requires that transient stability is measured by

- No pole-slip.
- No unacceptable damping.

Both criteria are presented intuitively to power system engineers in terms of time domain system responses on rotor angle swing curves.

For a human operator, pole-slip can be observed directly from the rotor angle swing curve, where if the post-fault swing is not dying away and the system does not settle to a new steady state a group of synchronous machines is understood to have 'pole slipped' against others in the system.

Damping is loosely defined on the overall rotor angle curves where the approximate exponential decay time constant is less than 12 s.

In order to obtain these time domain responses accurately, and comply with the Grid Code on the use of generator data and models, detailed dynamic controller models need to be included in the system for transient stability assessment. A time simulation approach therefore is the only option for National Grid, where any 'direct methods' with inevitable simplifications have been considered only suitable for initial critical case sifting.

The Powertech DSATools™ suite has been selected for the National Grid as the basis to develop its operational Stability Assessment tools. The Transient Security Assessment Tool (TSAT) provides fast time simulation and can be deployed on multiple computational servers. It can assess both transient and damping based on system time responses. User-defined dynamic models can be developed and supported in TSAT. Both TSAT and DSA managers are customised to meet National Grid's requirements, which will be described in detail in the sections to follow.

8.2.3 Solutions to Meeting the Requirements of the SQSS

In order to comply with the GB SQSS, the commercial off-the-shelf version of TSAT would need some 'bespoke' development and customisation.

For the detection of pole-slip, in the manual process it is observed by engineers visually, based on a 20 s-simulation window. With the automatic execution of multiple contingencies, it is impractical to manually examine all cases one-by-one. In TSAT, there are several ways to identify transient instability. A Stability Margin (SM) index has been developed by Powertech to quantify a pole-slip reliably. Based on the Extended Equal Area criterion it is more accurate than using thresholds that are arbitrarily defined on rotor angle and angular speed (slips).

For damping assessment, in addition to its existing measure on damping ratio, a Damping Decay Time Constant is introduced into TSAT to allow the National Grid engineers to assess directly whether the system decay time is within 12 s, as required by the SQSS.

A system snapshot is captured by a DPT and imported into TSAT as a resolved load flow solution in the commonly used PSS/E data format. However, in order to support bus bar fault analysis, circuit breakers are included in the load flow model as zero-impedance lines, where a sufficient level of nodal visibility is reserved via these circuit breakers. Out-of-service pieces of equipment are also kept to preserve the topology of the intact system for referencing in case of re-energising. Load flow data are further enhanced to allow for equipment names in addition to 'bus numbers' to make the tool more user friendly. In addition to the power flow file, additional 'Bus-node mapping' and 'bus fault clearing data' files are exported from the DPT

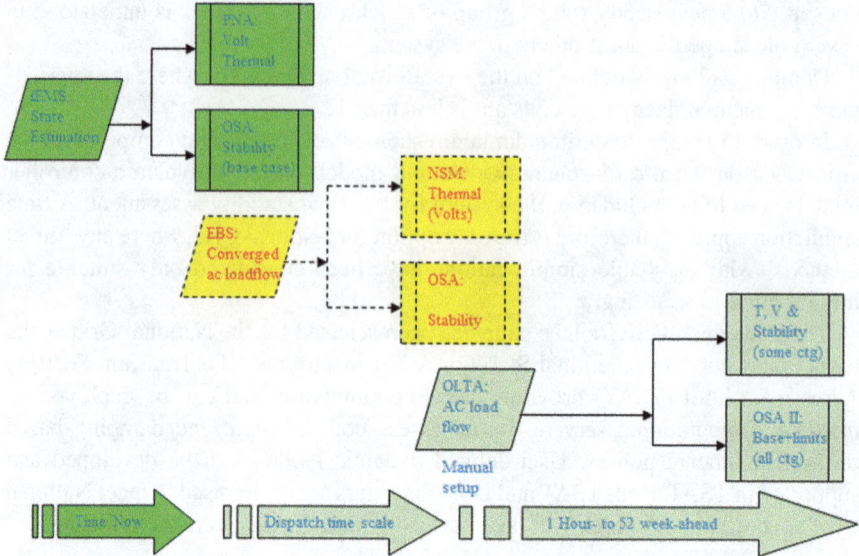

Fig. 8.1 NG system operation tools and process time scales

to preserve the bus-node topology mapping and support for the 'on-the-fly' creation of bus fault analysis.

There are more than 1000 credible single and double circuit faults on the GB HV transmission system. For stability assessment to cover for all 'unknown' possibilities, the number of contingencies is more than doubled by considering fault locations at all ends of these circuits (some circuits have three or even four ends). Together with the selected bar faults, for a base case assessment, there are more than 2000 contingency cases that would require to be scanned in every assessment execution cycle. In the National Grid planning process, for each one of these cases, it requires a full 20 s simulation to assess both transient and damping stability accurately. DSATools supports the multiple computational server architecture where the contingency cases can be distributed among these servers and analysed in parallel.

Developments that are specific for each Online Stability Assessment (OSA) or its offline (OFSA) applications are covered in detail separately in sections below.

In Fig. 8.1, the following terms are used

- iEMS: Integrated Electricity Management System (National Grid's SCADA system);
- PNA: Power Network Analysis (security assessment for voltage and thermal criteria);
- EBS: Electricity Balancing System (the market management system);
- NSM: Network Security Mode (in EBS);
- OSA: Online Stability Assessment;
- OSA II: it has been developed into Offline Stability Assessment, OFSA;

- OLTA: Offline Transmission Analysis;
- ctg: short hand for contingency cases;

8.3 Development for Online Stability Assessment—OSA

The Power Network Analysis (PNA) module in the National Grid Integrated EMS (iEMS) system does not have a stability analysis function. In order to meet the requirement from the Control Room, an online stability assessment (OSA) capacity is developed to support real-time operation, based on the state estimation (SE) from the iEMS system.

SE from iEMS provides OSA with the network configurations together with the snapshot of the current operating point as the base case to initiate simulations for stability contingency assessment. This implementation removes the requirement of supporting a separate network model in OSA. Both network model and its associated scenario (state) variables are updated in real-time every SE cycle. An intact baseline network model is stored in OSA only for reference purposes in case of the need to 'inage' (restore to service) a piece of equipment in some contingency or to energise generation in transfer analysis. This full system model is updated regularly by the OSA support team.

The generator-to-system tripping schemes (i.e. 'intertripping' schemes) are implemented identically to their real scheme logical models, and their status of selections are imported from the iEMS Data Historian to OSA. These intertripping schemes will 'fire' in the contingency cases when the monitored circuits are tripped during the simulation. This implementation removes the need for constant manual checking of intertripping availabilities and arming status in a static fault contingency model, and hence is suitable for the online implementation.

The DPT and iEMS-OSA data conversion interface has been developed and integrated with the XA/21 SCADA system by GE.

Dynamic controller models are converted from the National Grid Offline Transmission Analysis (OLTA) program, which are not required to be updated in real-time. These are converted as part of the OSA project and updated when the models in OLTA are updated. Dynamic models required by National Grid transient stability studies are detailed generator controller models, including both AVRs and governors. In addition to these traditional generator and controller models, various types of SVCs, HVDCs, TCSC, and wind turbines are also modelled. The detailed dynamic models contain a large number of state variables, which is the dominant factor for the performance of online implementation. In order to meet the Real-Time performance requirement (i.e. more than two thousand 20 s-simulations to be completed in 15 min), a sufficient number of computational servers are deployed and configured for performance. Contingency cases have also been refined to remove those post-fault islanded devices that could cause numerical convergence problems.

A 'Swing Margin' calculation has been developed in TSAT to quantify the loss of synchronism between any system separations and can handle the reference

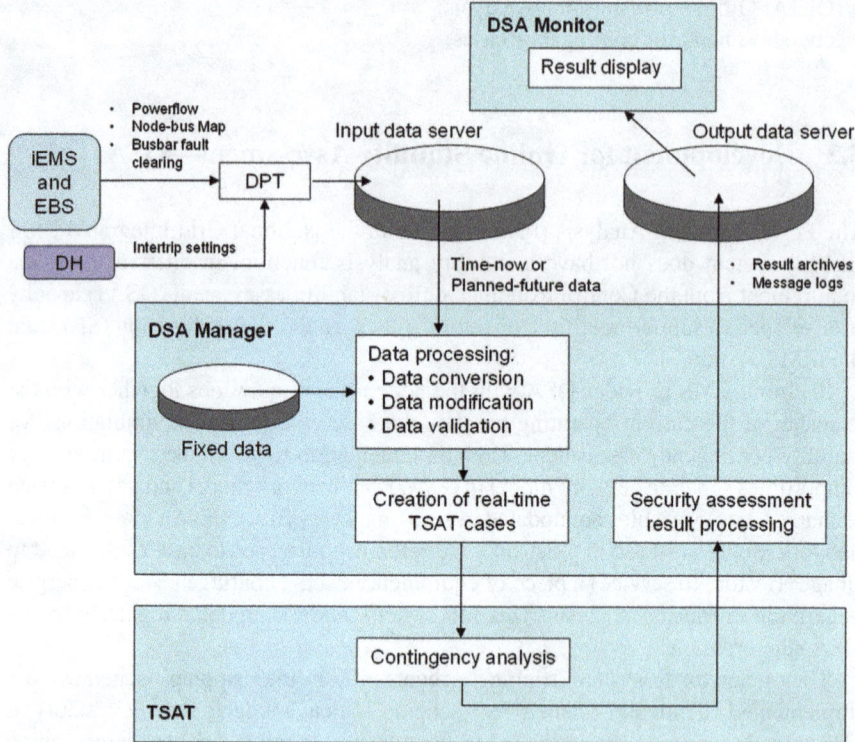

Fig. 8.2 NG OSA architecture

framework of each AC island, which is the functionality essential for automatic online assessment. The display reference framework has been further enhanced in the OSA project for viewing assessment results consistently across the GB system. Weighted Mean Rotor (w.m.r.) angle output has been added to TSAT for convenience in plotting all contingencies against this single invariant reference framework in OSA, regardless of synchronous machine outages and variations of fault locations.

The conceptual technique model (CTM) of the OSA is as shown in Fig. 8.2.

OSA, as the enhancement to the iEMS PNA suite, is classified as a 'business critical system'. It is implemented on the critical national infrastructure (CNI) network with duplicated data synchronisation resilience and a full disaster recovery (DR) environment, which is shown as in Figs. 8.3 and 8.4.

OSA is designed with two operational modes. The online Real-Time mode runs automatically in line with the iEMS state estimation cycle time, nominally 15 min. Within this cycle time it scans more than 2000 contingency cases, covering both credible circuit faults and user-selected bus bar and switch faults. Insecure cases are displayed in tabular form in the online summary, and cases over predefined warning

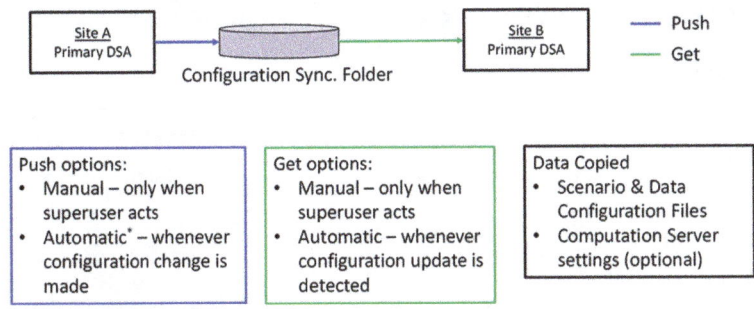

Fig. 8.3 OSA synchronisation schematic

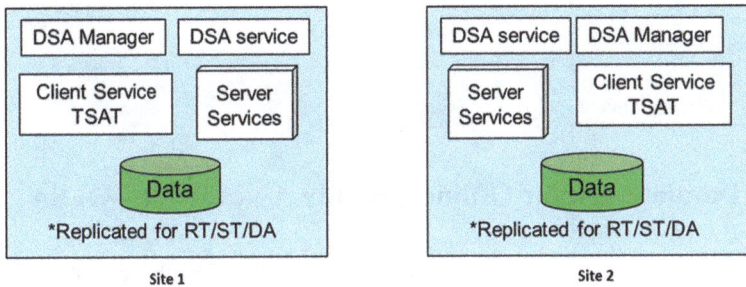

Fig. 8.4 Site-Site overview

limits can be further interrogated via OSA's What-if (WI) study mode. The WI mode can capture cases from either iEMS directly or from the reload of the user-selected historical OSA analysis. WI mode shares computational resource with the system Disaster Recovery (DR) environment and is capable of being dynamically reconfigured during the mode change to share the support to both RT and WI in the same environment. Because the DR environment is constructed with the same level of performance capacity as RT environment, it can process the full set of contingency cases for the WI study rapidly. However, normal users are restricted to load flow and contingency modifications in a WI study. If users want to take a small number of contingency cases for further studies, they can export these OSA cases to the offline TSAT to run simulation with the full control of configuration parameters and all other modelling data. The change made by users to the offline TSAT will not affect the original case back in the OSA (Fig. 8.5).

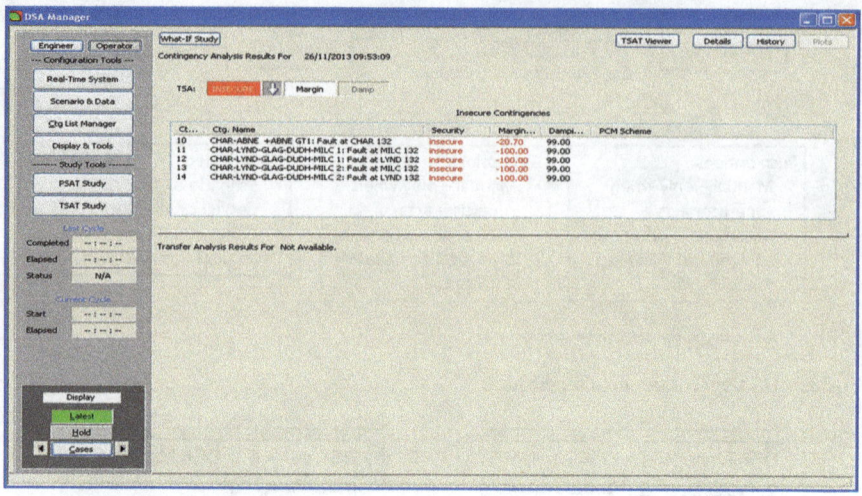

Fig. 8.5 OSA real-time display

8.4 Development for Offline Stability Assessment—OFSA

The National Grid offline transmission analysis (OLTA) model is established using DIgSILENT PowerFactory. While capable of performing stability analysis, OLTA is not practical to support routine assessment of stability for all credible (2000+) faults on the GB system in the same way as it does for voltage and thermal conditions.

From a system operation point of view, OSA provides the last safety-net in real-time however it does not have lead-time to allow adjustments to any generation re-dispatch. Planned generation dispatch, together with post-fault actions, provides the most economical solution to manage system constraints. It is clear that an automated boundary transfer limit analysis will be more suitable and valuable when it is carried out at the operational planning stage than in real-time. Therefore, there is a requirement to develop an offline stability assessment capacity to provide accurate and up-to-date stability constraints to the dispatch program at National Grid. As a common practice, while the NG dispatch program can carry out security constrained optimal load flow and incorporating directly both thermal and voltage constraint in its optimisation formulation, it relies on the offline stability study to provide the 'group constraints' in terms of MW boundary flow limits for the stability constraints. A single peak demand assessment cannot sufficiently capture the variations in stability constraints through a day; therefore, there is a business requirement to develop offline stability transfer analysis capacity that can cover the entire dispatch horizon.

Further to the Operational Planning requirement above, in the National Grid business process, stability assessment is required for a 'planned future' that spans from day-ahead dispatch, current year operation planning to year-ahead outage planning stages to allow them to assess all credible faults over the entire GB HV transmission system.

OFSA, following the successful development of OSA, is developed to allow Operational Planning and other National Grid planning teams to study stability

- for all credible system faults; and,
- for transfer analysis and stability limit derivation.

which can support a 'planned future' covering all time scales from the dispatch to 10-year-ahead system development planning process.

OFSA takes a system snapshot from OLTA, where the base case network and operational scenario of the time concerned are captured. OLTA is a 'time driven' model in that the system evolution with time is managed by 'schemes'. A 'study case' can be set to any time in the future, and it will activate all network reinforcement schemes that are applicable to that time. In this way, users at National Grid from different planning teams can create their own models in OLTA and then export to OFSA as the base cases for further stability studies.

In OFSA its dynamic models are the same as those in OSA and are converted from OLTA. With different stages of network expansion, there will be generators commissioned or decommissioned. All the dynamic models associated with these generators are stored in the same master dynamic model data set, but their usage is determined by the network expansion schemes where the mapping to the connection bus bars is represented properly. In this way, there is only a single copy of the master dynamic model parameter set that will need to be supported for OFSA.

For contingencies, Transfer Analysis and Boundary definitions must be managed by each planning team for their appropriate network models of their planning time scales. The TSAT Transfer Analysis function allows combination of different generation and demand scaling methods, e.g. scaling in proportion to unit MW output ('sharing'), or by order ('Merit Order'). These methods can be used in combination and applied to different dispatch groups, e.g. for generators in England and Wales and those in Scotland; and for synchronous units and for wind farms, etc. In the OFSA project, TSAT has been enhanced to allow re-energisation or switching off generators in the Merit Order dispatch.

For the base case and each transfer scenario, TSAT studies can be configured to use their own dynamic models, contingency cases, simulation parameters, and boundary flow displays. All transfer analysis will be derived from the same base case. Any change to the base case load flow will be applied to all transfer analysis.

Unlike in OSA, the intertripping schemes are not modelled with their logic models. Instead, they are simply modelled as part of the switching sequence in contingency events, because there is no information for generation, nor their intertrip arming status available at the planning time scales. Operational Tripping Schemes (OTS) can be optimised by setting permutations as different contingencies and scenarios in the same assessment case.

An OLTA-OFSA interface has been developed by DIgSILENT, where in addition to the data conversion that has been described for OSA, two other major enhancements have been included to support Transfer Analysis (TA).

- Tap dependent impedance for transfer scaling load flow re-calculation.

  Unlike OSA where only the base case is analysed, in OFSA the base case is scaled and successive load flows are calculated to derive the limit. Taps are calculated and the associated transformer impedance values are adjusted per solved tap positions. Other state variables in these TA load flows are also re-calculated according to the user's requirement.
- Preservation of generator connectivity topology for re-energising.

  In TA, generators that are available but switched off in the base case can be required to be dispatched for the limit derivation. These generators will need to be connected to the generator transformer, then the busbars that are energised to the HV network. The topology between generators and the HV system must be preserved for the generators that are currently 'off' in the enhanced nodal-branch model.

Powertech have enhanced

- PSAT for options of the load flow MW mismatch distribution, and rebalance options of the generation scaling in TA;
- TSAT Merit Order dispatch, to allow for busbar energising options;
- TSAT by converting several new types of controller models from OLTA, including HVDC, TCSC, and generic STATCOM models.

Powertech have also customised DSA Manager for OFSA, where

- Contingency selections can be customised for each base case and TA;
- Boundary MW flow definition and monitoring options have been enhanced;
- The Transfer Results Display is enhanced with multiple pages of display which can be customised;
- Transfer Constraints can be selected for analysis;
- Users can carry out either individual or both of the base case and transfer analysis;
- Boundary flow summary of all selected constraints are loaded automatically and can be viewed in the corresponding offline PSAT project.

Example output for the OFSA display is shown in Fig. 8.6.

Because OFSA is designed to cover for all National Grid planning time scales, it is hosted on the company's business local area network. There are two 'production environments' and a 'fix-on-fail' test environment. The two production environments are independent of each other and there is no data synchronisation between them. The performance requirement for OFSA is that execution of all base case contingencies together with up to 10 TAs should be completed in 30 mins, with its current level of dynamic modelling.

8 Online and Offline Stability Assessment Development at National Grid, UK

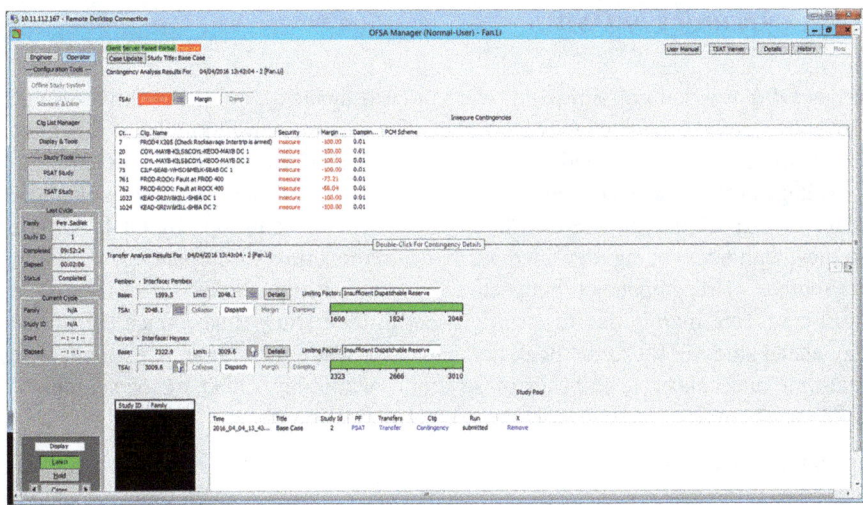

Fig. 8.6 OFSA display

8.5 Experiences of OSA

8.5.1 Robustness of OSA System

The robustness of OSA is underpinned by its DPT and the performance of SE.

SE from iEMS provides OSA with the network model together with the snapshot of the current operating point as the base case to initialise simulations for stability contingency assessment. Although SE resolves inconsistencies and gaps in the SCADA metered data, its result can be different from a valid load flow solution, due to its functional and algorithmic differences from that of a load flow. Because of the insufficient metering from some CCGTs and embedded generators, and lack of information on switching status in the LV network, sometimes large MW and MVAr flows can be circulating in parts of the interconnected LV network. Generation output from some parallel balancing units can have one unit with its output doubled, while the other being 0 MW. These circulating powers can exceed the equipment ratings and cause non-convergence of a load flow. Overloading a generator at the initial steady state can prevent its dynamic controller models being initialised properly, leading to instability in the simulation. The iEMS support team at National Grid resolves these problems by

- Improving the metering and observability in the PNA model;
- Carefully prioritising metered and estimated data at the sites in question;
- Identifying and separating the problematic equivalent LV interconnectors in the PNA model.

8.5.2 Accuracy of OSA

When OSA was commissioned, it was validated against the OLTA offline results. For any insecure cases that OSA reported, they were reviewed in OLTA. Since the OLTA scenario is usually modified from a peak demand study, it is difficult to line up its voltage profiles across the entire system with that from the iEMS state estimation. After careful alignment, most of these cases could be confirmed by OLTA offline studies, with a few being identified attributed to modelling discrepancies.

Sources of discrepancies between online and offline stability assessments, in most cases, are mainly due to misalignment of their study times, since the offline day-ahead study is set to the peak demand time, while the online assessment is in real-time and can be a snapshot of any time of the day. Other major sources of differences are found as follows:

- Online and offline voltage profiles.

  Since the system dispatch in the GB balancing mechanism only dispatches MW generation, while the reactive dispatch is carried out separately as part of the system control process, there is inevitably a difference in reactive power flows between the offline model and the online metered values. The assumptions on reactive power compensation in the offline model may not exactly match the deployment of compensation in real-time.
- Generation and demand pattern changes.

  The day-ahead offline study models a peak demand scenario using the forecasted demand and generation physical notifications. Both demand and generation mixes can change in real-time, especially when a significant share of generation is provided from renewable sources.
- Demand apportionment and Voltage Dependent Load (VDL) assumptions.

  - The reactive demand forecasted at the day-ahead stage is based on certain approximations. In the 'apportionment' process of setting up an offline OLTA model, these forecasted MVAr values from the Grid Supply Points (GSPs, typically 132 kV) are usually allocated to Bulk Supply Points (BSPs, typically 33 kV) in prorate to BSP active power demands. Because the reactive demand can be either positive or negative, this process can introduce errors.
  - Usually, in an offline model, certain assumptions are made for the voltage dependency of loads. The VDL model characteristics are static and will not be adjusted for different demand scenarios over time of a day.

In addition to these 'true' differences between predicted and actual network conditions, there are also modelling differences between offline OLTA and online OSA.

- All dynamic models in OSA are derived from those in OLTA, which are updated manually by the support team. Inevitably OSA dynamic models can lag behind those in OLTA. Regular review and a rigorous business process of triggering an

update whenever the OLTA model is updated are in place to minimise this update lag.
- The base case can be distorted when it is exported from PNA/SE to the PSAT in OSA. As mentioned before there are algorithmic differences between SE and power flow computation. A converged SE may not guarantee a converged power flow. Solutions to this issue have been discussed in Section 8.5.1.
- Differences in the power flow can also be introduced by computation parameters. Ideally, a converged SE solution should be able to be resolved in a power flow with all P-V buses and tap changers 'fixed' to the imported MW and MVAr values and tap ratios. However, not all modelling information is captured in the iEMS-OSA exporting process, e.g. voltage control modelling is not exported. Therefore, information on how the voltage controllers are controlling the 'remote buses' (e.g. V set point, droops, etc.) are lost. In the PSS/E data format, these remote bus data are partial and inconsistent if they are used with the exported model without these voltage controllers. The solution to this problem is to ignore these partial data (e.g. target volts) of the remote-controlled buses while maintaining the voltage at the terminal buses to their solved values.

8.5.3 Statistics of OSA Performance

During the 'Trial Use' period of the OSA Project in 2013–2014 there were 25 insecure cases reported. These insecure cases included both transient stability and poor damping cases. Among these, 23 cases have been confirmed as genuine insecure cases with OLTA. Two cases had differences due to some modelling and parameter issues in some small embedded synchronous units. These data issues were resolved with the support team.

Some of these insecure cases were not able to be identified during the day-ahead planning. These were due to the time when they occurred, which were not at any studied 'Cardinal Points' (i.e. the turning points on the daily demand curve). Furthermore, most of these cases were not in areas where National Grid would normally assess for stability. OSA has identified 'unknown' stability problems and allowed the system operator to re-secure the system in real-time operation, preventing system instability had these faults happened. Some of these examples are given in the next section (Fig. 8.7).

8.5.4 Examples of Problems Identified by OSA

Among the OSA reported cases (which were subsequently confirmed by OLTA), many have been identified as the 'unknown' new problems to the system operation and planning teams. These new 'unknown' cases are a consequence of the increasing

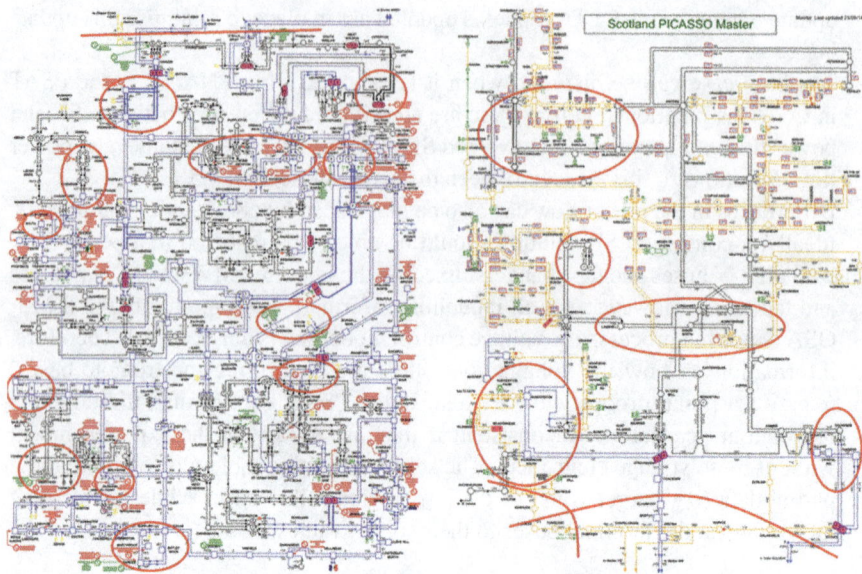

Fig. 8.7 Cases that OSA validated against OLTA (More information of constraint boundaries can be found in National Grid 10 Year Statement [2])

renewable energy and rapid reduction of synchronous generation. Some examples are shown below.

- In addition to the well-known B6 constraint at its peak transfer level, some new damping issues have been identified in both North East (NE) England and South Wales by OSA. These cases happened with some planned outages in the period between midnight and early morning, when demand was low and no planning study was done. After these cases were identified, control engineers readjusted the voltage control plan by selecting different voltage control circuits and raised the voltage profile slightly to re-secure the system and avoid additional constraint cost. In the NE case restrictions on nuclear generation were avoided during those planned outages. This case has also highlighted a new challenge the GBSO is facing in South West England where high voltage is becoming a prominent problem. During the low demand period the target volts are desirably set lower, but as a consequence synchronous machines are operated towards their MVAr stability limits and system damping is reduced. The system operator would have to manage a delicate balance between high voltage control and the need for system damping.
- In most cases, OSA was found to revise constraint limits to be more stringent than in operational planning timescales and, as a consequence, the system was secured by the 'safety-net' in real-time, but with higher costs. Occasionally OSA could identify higher transfer limits and brings cost savings directly. This is particularly the case when the Anglo-Scottish B6 transfer is operated at its

stability limit. With the fast-growing wind generation capacity in Scotland, since 2013 there have been several occasions when excess wind generation was restricted in Scotland due to the wind being stronger in real-time than was forecasted at the operational planning stage. Some additional safety margin has since been introduced into the day-ahead planning process to accommodate these uncertainties. With the commissioning of OSA, the Electricity National Control Centre (ENCC) is able to readjust the constraint limits in real-time, to remove unnecessary restrictions and transfer more "green" energy across the B6 boundary from Scotland to England/Wales.

8.6 Experiences of OFSA

8.6.1 Accuracy of OFSA

With the base case model and scenario data being identical, OFSA assessment results line up closely with those from OLTA. However, some sources of differences have been identified, which are not controllable by users:

– There is the "update lag" previously mentioned, whereby dynamic controller models are updated in OLTA first and then in OFSA.
– The effect of mutual coupling is not considered for parallel lines in the DSATools/PSAT but is considered in the PowerFactory/OLTA model for the National Grid TO network.
– Differences exist in treating static power injections in the simulation. In PowerFactory static generators are modelled as 'constant current' VDL, while in TSAT these are modelled as constant impedance VDL. With a 'generation netting' mechanism and the introduction of new global load type of 'VDL Net', this difference is reduced.
– Synchronous machine models [1] in PowerFactory and TSAT are slightly different.

The focus of user acceptance testing for numerical validation of OFSA has been on transfer boundary limits. Results were obtained for major stability constraint boundaries for a snapshot captured in March 2016 (see Table 8.1). In this case, it showed good agreement between OFSA and OLTA in the constraint limits for all boundaries. There were low levels of wind generation that were not controlled by any wind turbine dynamic models. The impact of the static generator model difference therefore was limited in this case. These static generators would have more noticeable impact on the B6 boundary results. The major difference is from the synchronous machine models in this case.

Some operational guidance has been issued to account for these algorithmic differences between DSATools and PowerFactory.

Table 8.1 OFSA constraint limits validated against OLTA in the user acceptance tests (More information of constraint boundaries can be found in National Grid 10-Year Statement [2])

Constraints	OLTA limits (MW)	OFSA limits (MW)	Limited by
B5	2975	2979	Transient
B6	2657	2628	Damping
B4	1666	1637	Transient
SW	1922	1907	Transient

8.6.2 Additional Considerations in Transfer Analysis (TA)

During the Trial Use of OFSA, it is found that if the base case is far from the limit of a constraint, the TA generation scaling can lead to pessimistic results. During the automatic generation dispatch, reactive compensation usually is not adjusted. Voltage can be depressed significantly with a heavy increase of constraint boundary power flows. A poor voltage profile will not only impact on the stability results but also could cause the dispatch scaling process to fail in Transfer Analysis.

To avoid this problem, an operational guidance is issued to advise users to start the TA close to the estimated constraint limit; also, additional transient voltage criteria have been adopted to treat the non-converged case to the insecure voltage collapse; therefore, the simulation can continue from these voltage collapsed cases rather than stopping at an error.

8.7 Conclusion

NGESO is required by the SQSS to operate the system securely for defined contingencies. Before the development of the OSA system, in the operational planning process power system engineers could only inspect network topology from experience, and use OLTA to study certain well-known problems for potential stability issues at the day-ahead stage. Only a small number of stability trips could be studied for a daily peak demand scenario due to the manual process and the limitations in offline tools. As NGESO did not have any tool to assess stability automatically across the whole GB system during the operational planning, it was not practical to test all credible faults on the system and assess stability throughout a day in real-time. OSA has met the challenge of increasing uncertainties faced by Control Engineers and enabled them to update the post-fault actions if insecure conditions arise in real-time; OFSA, on the other hand, makes it possible for operational plans to identify stability issues across all credible contingencies before they are handed over to the control phase. The transfer analysis in OFSA establishes power transfer limits for all active constraints across the GB system and allows time for re-dispatch of generation to economically mitigate these constraint risks.

The derivation of stability limits and optimisation of constraint management used to be an intensive manual process. Introducing an (automated) OFSA system also helps improve efficiencies in the operational planning process.

Some further developments have now been placed on the National Grid Dynamic Security Assessor (DSA) road map.

OFSA will further be developed to provide consistent updates on stability constraint limits of the system, from every phase of the planning process to the final real-time control. This development will provide a profiled group constraint to the dispatch optimiser for all leading times across the dispatch horizon. Currently, the dispatch program works on a constant day-ahead stability limit, which is derived from a single daily peak demand scenario.

If constraint limits can be refined in OSA, it will increase the Control Room situational awareness and allow the system operator to see the 'head room' from the current operating point, and use that as an approximation for the next state estimation cycle interpolating between 'time-now' and the next dispatch point. This assessment will be a rolling process updated in every SE cycle, as a refinement to the assessments from OFSA, which are calculated only once (albeit for every dispatch point) at the day-ahead stage.

References

1. P. Kundur, Power System Stability and Control (1994)
2. National Grid Electricity Ten Year Statement 2018, https://www.nationalgrideso.com/insights/electricity-ten-year-statement-etys

Chapter 9
Use of Voltage Stability Monitoring and Transient Stability Monitoring Tools at the Nordic Power System Operators: Introduction of Synchrophasor Applications in the Control Room

Kjetil Uhlen, Dinh Thuc Duong, and David Karlsen

9.1 Introduction

With the green shift the electrical power system will be increasingly dynamic. Operation and control are challenged by the increasing amounts of variable renewable energy sources, distributed generation, and electrification of transport. Development of microgrids, on one end of the scale, and mega scale system integration through HVDC grids on the other end, represent new operational challenges. Innovations with fundamental improvements to control center monitoring and control systems are needed for dealing with these challenges.

Phasor Measurement Units (PMUs) and Wide Area Monitoring Systems (WAMS) are expected to be important elements of the future tools for operation and control. The development of IT platforms and WAMS applications are promising but testing and deployment in the control room environment have so far been limited. Introducing WAMS to the operators in the control rooms is therefore an important and necessary step to gain a new understanding of these challenges and of how to further develop control center monitoring and control systems.

On this background, a research and demonstration project was initiated by Statnett—the Norwegian Transmission System operator (TSO)—in order to gain

---

K. Uhlen (✉)
Norwegian University of Science and Technology, NTNU, IEL, Trondheim, Norway
e-mail: Kjetil.uhlen@ntnu.no

D. T. Duong
Lede AS, Porsgrunn, Norway
e-mail: ThucDinh.Duong@lede.no

D. Karlsen
Statnett SF, Oslo, Norway
e-mail: David.karlsen@statnett.no

operational experiences. Through implementation and deployment of selected WAMS applications, operators at Statnett control centers have been given access to prototype tools for online monitoring and assessment of power system dynamics and stability.

This chapter describes the WAMS platform and applications that were implemented and presents some of the results and experiences gained. The focus is on two applications that were developed for online voltage stability and transient stability monitoring.

9.2 Objectives

Still today, most control centers rely on SCADA information where measurements from RTUs are updated less than once per second. This makes it almost impossible for the operators to observe and understand the majority of dynamic responses in the power grid. A main motivation for WAMS is the need for higher resolution information and increased awareness of dynamic phenomena. The objective of the SPANDEx[1] project was to implement and test WAMS applications in the control centers, but also to further develop and test new applications by utilizing the National Smart Grid Laboratory[2] at NTNU/SINTEF. Further, the project aimed at assessing critical ICT issues, and finally based on the results and feedback from the operators, the project will develop a roadmap towards full integration of PMUs and use of synchrophasors in the future SCADA/EMS solutions.

9.3 WAMS Platform

In order to implement and test various monitoring functions in the control room, a suitable and dedicated application platform is needed. The following basic requirements were specified for the platform to cover the needs of the R&D project:

- Easy installation and implementation of WAMS applications.
- Accessible for external research partners, project partners, and operators at several control centers.
- Ability to receive data from current PMU infrastructure.
- A front-end solution (viewing tool) in order to easily present results and findings.

---

[1] The project: Synchrophasor/PMU Application Integration and Data Exchange (SPANDEx) is a collaboration between Statnett SF, NTNU, SINTEF, and GE Grid Solutions and received funding from the Research Council of Norway between 2016 and 2019.

[2] See ntnu.edu/smartgrid.

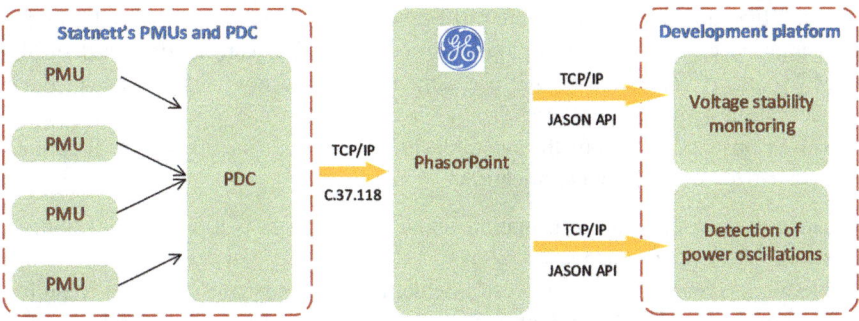

Fig. 9.1 Control room integration platform for development and demonstration of WAMS applications

- A user interface with graphics and functionalities that operators would expect from a control center application.

While the first requirements could be met by developing an open tailor-made tool, the last suggests choosing a commercial software tool, first of all in order to ensure a professional user interface and alarm handling functionality. The choice was to use GE's *e-terraPhasorpoint* as the basic platform on a secure server. To comply with cybersecurity requirements, access to the server is through two remote desktop connections. The possibility to implement and test new WAMS applications was solved by a custom-made interface.

By the end of 2019, Statnett has nearly 150 PMUs in about 50 substations, reporting more than 1000 synchrophasors. The infrastructure uses the IEEE C37.118 communication protocol where the PDC during this project was configured to send PMU streams to the "WAMS Platform" on a secure server as illustrated in Fig. 9.1.

This server was originally designed to receive a limited amount of PMU streams, but as the control center operators got more involved, there was soon a demand for more PMUs. The server currently receives data from more than 504 phasors, coming from about 30 substations.

9.4 Application Development

The WAMS applications were chosen based on maturity of development and relevance for the operators. The main criterion was that they must provide additional and useful information on disturbances and dynamic properties, which today is not available from the present SCADA/EMS.

In order to make the selections, it was necessary to examine and understand the most pressing problems viewed from the operators in the control rooms, and which applications they believe are most relevant. In addition to decide on the choice of applications, the candidate locations were identified, i.e., in which parts of the network the various problems were most pressing or likely to be observed. Finally, decisions on placement and deployment of new PMUs were made based on the need for better observability. All assessments and decisions were taken in close collaboration with the operators.

Based on feedback from the operators the following functions were identified and prioritized as preferred application:

- General detection and information about disturbances that are easily obtained from PMU measurements.
- Network islanding detection—early detection of incidents that result in islanding of parts of the grid—and providing critical information about stability and voltage quality in the isolated parts.
- Power oscillation monitoring—estimating frequency and damping of critical (low damped) electro-mechanical modes.
- Power oscillation (transient stability) monitoring—fast detection of oscillations and their nature and severity following disturbances.
- Voltage stability monitoring—estimating maximum loadability on selected corridors and load areas.

The first three functions were readily available in *e-terraPhasorpoint*, but the functions for online voltage stability and transient stability monitoring, as illustrated in Fig. 9.1, had to be implemented as separate applications.

The following sections describe the development and use of voltage stability and transient stability monitoring tools in more detail.

9.4.1 Online Transient Stability Monitoring

The Nordic power grid experiences from time to time low damped electro-mechanical oscillations that require mitigating actions. These can be low-frequency inter-area modes, but also stability problems of more local character can be critical—especially in areas with a surplus of variable generation. It is crucial that the operators get early warnings about low damped modes before the magnitude of power oscillations become critically high.

By continually estimating frequency and damping of critical (low damped) electro-mechanical modes, the operators can get precise information about the problem and where and how to take action [1].

A lot of research has been undertaken and several applications have been implemented for power oscillation monitoring based on normal operation measurements—often referred to as ambient data. One of these applications is readily available in *e-terraPhasorpoint* and therefore also used for continuous power oscillation monitoring in this project.

Low damped system modes do not necessarily lead to critically large power oscillations. However, large power oscillations may occur even if the inherent damping of system modes are ok! This is often the case following larger disturbances and the system is on the verge of transient instability.

Therefore, in addition to the "traditional" power oscillation monitoring function, a new algorithm for detection and characterization of power oscillations is being tested in a prototype tool. The algorithm and implementation of the "*PowerApp*" are described below:

The idea behind this tool is to detect system disturbances from the stream of synchrophasor data, and then immediately capturing in a shorter time window the time responses following a disturbance. The information is presented as separate time plots on the operators' monitor, and in parallel a simple analysis is performed on the captured data to provide basic information of the response, such as magnitude and dominant frequency of oscillations and an estimate of the damping ratio of the dominant mode. The monitored signals can be frequency, voltage angle, or active/reactive power computed from voltage and current phasors.

In this way, the information obtained from the *PowerApp* tool serves the following main purposes:

– It gives the operator a very early alert or alarm about disturbances.
– the generated time plot following a disturbance facilitates an easy visual inspection of the severity of the disturbance.
– If the disturbance leads to sustained or damped oscillations, the tool provides valuable information about the characteristic of the response.

We choose to call this a tool for *online transient stability monitoring*. This is mainly to distinguish it from other power oscillation monitoring algorithms that first of all attempt to identify the small-signal properties of the power system dynamics. This tool focuses on the detection of larger disturbances and, as quickly as possible, to alert operators and present information about the disturbance. It does not (at least not yet) provide information associated with classical transient stability assessment, such as critical clearing times or low voltage ride-through capabilities.

The algorithm and prototype implementations are presented below. More details on the algorithm are described in [2].

The algorithm behind the tool we refer to as the "*PowerApp*" consists of three stages as illustrated in Fig. 9.2.

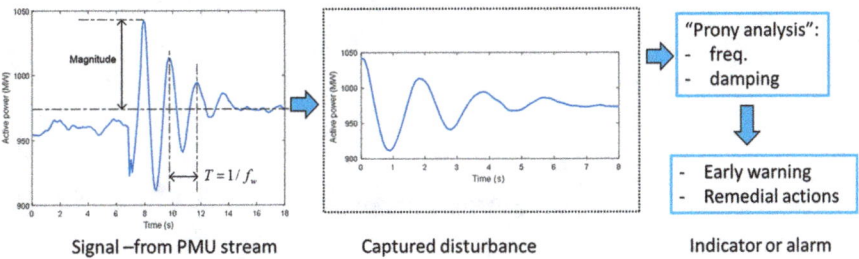

Fig. 9.2 Illustration of the algorithm implemented in *PowerApp*

1. *Signal from PMU stream*: The first stage is a function that automatically detects disturbances in the monitored signals. There are several possible ways to do this, but the one that is illustrated here is a simple one that continuously monitors the maximum and minimum points of the signal. When the difference between a maximum and the next minimum within a defined (short) time period exceeds a threshold, this is considered a disturbance and the difference is considered as the peak-to-peak magnitude of the disturbance. Figure 9.3 shows an example of how the peak-to-peak magnitude of fluctuations is observed in a monitored signal.
2. *Captured disturbance*: The time instant when the detected magnitude is larger than a predefined threshold is defined as the starting point of a disturbance, which then triggers the capturing and presentation of the time response following the disturbance.
3. *"Prony analysis"*: The captured signal is analyzed. If the disturbance results in an oscillatory response—which is often the case—the captured ringdown in the time window is analyzed by the Prony method [3] or an alternative algorithm, in order to estimate the oscillation frequency and damping ratio of the dominant mode. This is used as an indicator of the severity of the disturbance. Figure 9.4 illustrates the process and how useful information is extracted from a longer time series.

At present, this is implemented as a pure monitoring tool, and so any remedial actions resulting from the information must be initiated by the operator. However, in the future one may be able to develop automatic actions based on the indicators.

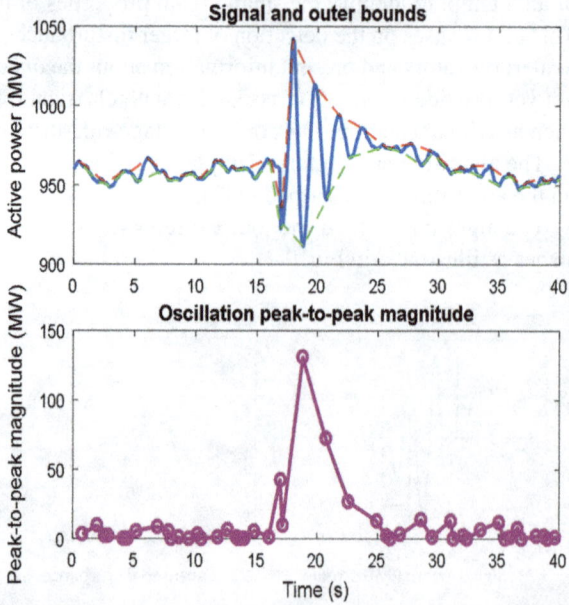

Fig. 9.3 Observation of peak-to-peak magnitude of oscillations in a monitored signal

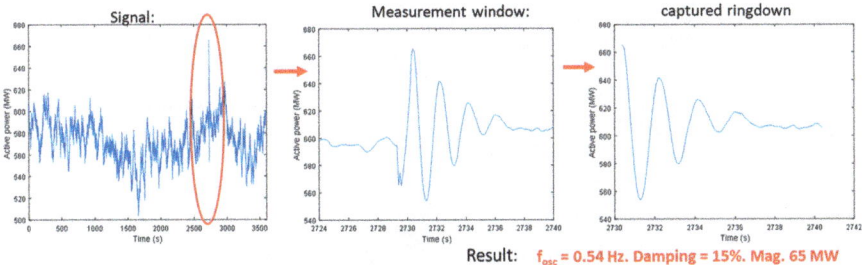

Fig. 9.4 Example of capturing and extraction of information from a signal

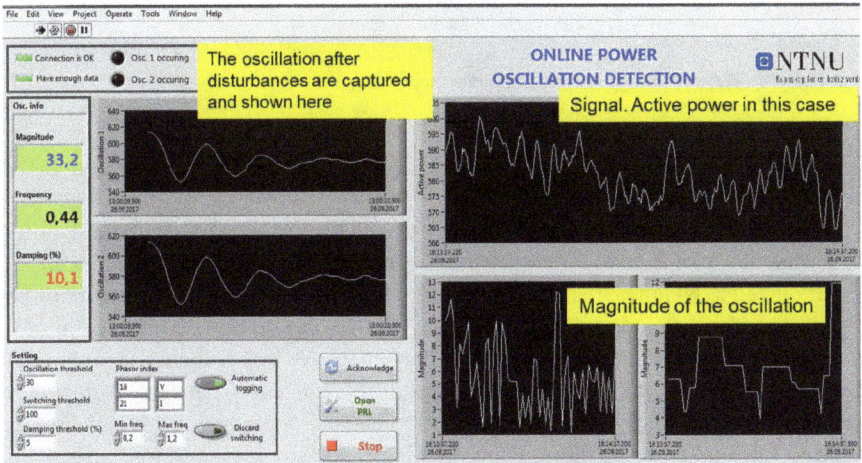

Fig. 9.5 *"PowerApp"* development version implemented in *NI LabView*

The implementation and testing of the method were done in two steps. First, a research version was developed in order to test various displays and detection algorithms. This was implemented in the *NI LabView* environment and using a software [4] developed in earlier research projects for real-time synchrophasor applications. The user interface is shown in Fig. 9.5.

Second, the prototype version used by Statnett consists of two separate applications as shown in Fig. 9.6; where one is the online monitoring tool and the other is for post fault analysis with the possibility to retrieve and study previous events.

9.4.2 Online Voltage Stability Monitoring

Voltage control problems, including the risk of voltage instability, are a concern in several areas of the Nordic transmission and subtransmission grid where there is heavy power transfer to areas dominated by loads. In particular, if there is a limited

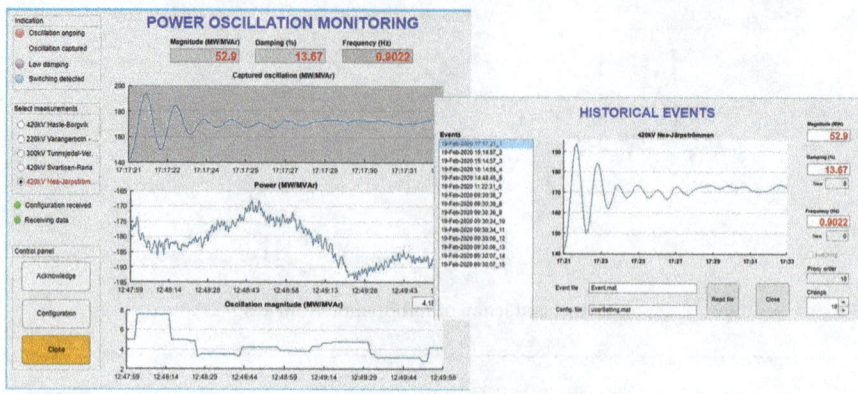

Fig. 9.6 *PowerApp* prototype version implemented at Statnett

number of lines (typically two or three) connecting to an area, this is vulnerable, for example, during periods of maintenance.

Normally, the system operator will perform analysis—including voltage stability assessments—to ensure that the system is operated according to the applied security criterion (e.g., N-1). In real-time however you are not always guaranteed to be within the planned operational limits. That may be in normal operation or in post contingency situations. Incidents, such as line outages or loss of local generation can lead to critically low voltages or risk of instability. Changes happen fast—sometimes several incidents occur almost at the same time—and operators themselves are not used to run (online or offline) voltage stability analysis. It is therefore a need for simple measures giving online information about the margin to instability in order to take fast and correct actions when operation becomes critical.

This was a main motivation when deciding to develop the tool for online voltage stability monitoring at Statnett.

What is expected from a tool for online voltage stability monitoring? The answer depends on the nature of the voltage problems in the grid, but certainly also on the skills of the operators. In our case, the following information was regarded most important:

– First, to know the exact operating point at given locations (voltage and power transfer) at any time.
– Next, to get an estimate of the maximum loadability and thereby the operational margin—measured as an active, reactive, or apparent power margin.
– Finally, and a much more difficult problem, what will be the power margin following a certain contingency?

Figure 9.7 shows one way of visualizing this information. Various algorithms for online voltage stability monitoring have been reported in the literature. Many of these are based on "impedance matching," that is, algorithms that estimate the grid

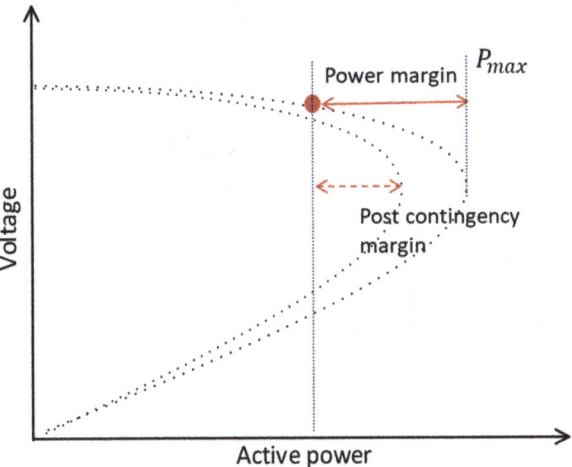

Fig. 9.7 Example visualization of operating point and power margins

impedance (Thevenin impedance or short circuit impedance) and compare this with the load impedance at the point of interest.

In practice however it has been found very difficult to obtain a robust and accurate estimate of the Thevenin (grid) impedance unless you are very close to maximum loadability (the "nose point"). It should also be noted that the point of instability and the maximum loadability is not necessarily the same and that even in the ideal case impedance matching may not provide a correct estimate of the power margin [5, 6].

In this work, we have used a different method which has proved more robust. The chosen algorithm is based on sensitivities that are calculated from voltage and current phasors and subsequent estimation of maximum loadability in the monitored load areas [6, 7].

Below is a brief explanation of the indicator that we call the *S-Z sensitivity* and its application:

The PMU (or PMUs) located at a transmission corridor connecting to load area—or directly on a feeder to a load area—sends a stream of voltage and current synchrophasors to the algorithm, which computes the apparent power, S_L, and the magnitude of the load impedance, Z_L:

$$S_L = \left| \vec{V} \cdot \vec{I}^* \right| \quad Z_L = \frac{|\vec{V}|}{|\vec{I}|} \tag{9.1}$$

The algorithm filters these signals and computes the rate of change, denoted as dS_L and dZ_L. The *S-Z sensitivity* indicator is now simply the ratio of these two:

$$S - Z \text{ sensitivity} : \zeta = \frac{dS_L}{dZ_L} \tag{9.2}$$

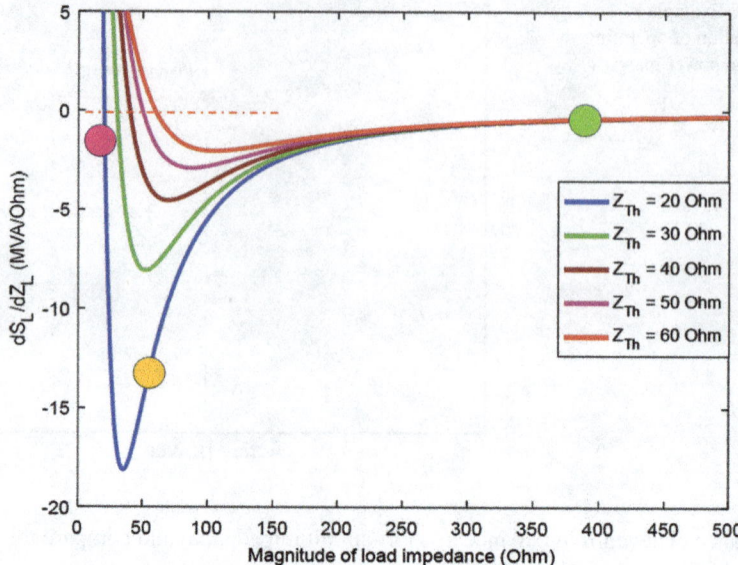

Fig. 9.8 Visualizing the *S-Z sensitivity*

The importance of this indicator is easily understood as follows: If the load demand is increasing, Z_L is decreasing and dZ_L will be negative. Simultaneously, if the system is stable we would expect that more power is fed to the load so that S_L is increasing and dS_L is positive. In the opposite case, if the system is at or beyond the point of maximum loadability, dS_L will be zero or negative. Thus, we can state that the system has reached the point of maximum loadability if the *S-Z sensitivity* is equal to or larger than zero.

A good way of presenting the indicator is shown in Fig. 9.8. Here, the online computed value of the *S-Z* indicator is plotted against the magnitude of the load impedance. In that plot, we have also drawn a number of fixed curves that indicate how the indicator would move if the strength of the (ideal) grid remains unchanged. That is, the curves represent trajectories at constant Thenevin impedance.

For example, if the load is low the indicator will stay in the area around the green dot, and since the sensitivity is low when everything is ok, the dot will not jump around very much. In the opposite case, when you are close to instability (the red dot), the sensitivity approaches zero and you are quite sure to have reached a critical operating point. The most interesting area is probably where the yellow dot is; Here, the operators should be alerted that you are approaching a critical condition.

There is one more interesting feature of the S-Z indicator. It can be used to give a quite robust estimate of the Thevenin impedance. If the grid is considered by its Thevenin equivalent, voltage E_{Th} behind the impedance, Z_{Th}, the following equation can be derived [6, 7]:

$$\frac{dS_L}{dZ_L} = \frac{E_{Th}^2 \left(Z_{Th}^2 - Z_L^2\right)}{\left(Z_L^2 + Z_{Th}^2 + 2Z_L Z_{Th} \cos\theta\right)^2} \tag{9.3}$$

By rearranging the terms, one end up with a second-order equation to solve for the Thevenin impedance, Z_{Th}:

Define $\zeta = \frac{dS_L}{dZ_L}$

$$\left(I^2 - \zeta\right) Z_{Th}^2 - 2\zeta \cos\theta Z_L Z_{Th} - Z_L^2 \left(I^2 + \zeta^2\right) = 0 \tag{9.4}$$

In this equation, θ is the angle between the load impedance and the Thevenin impedance. This angle is still unknown, but it can be estimated with quite good accuracy with prior knowledge of the grid. For example, in a high voltage transmission grid, you would know that the X/R—ratio is high, and this can be used to set an initial (low) value of $\cos\theta$.

When Z_{Th} is computed, the solid blue line can be drawn, as shown by the snapshots in Fig. 9.9. This can be used for two purposes: In the rightmost snapshot the blue line crosses the red dot, and this verifies that you have a good estimate of the grid impedance. It indicates the trajectory of the indicator as the load changes, and it can further be used to make other visualizations, for example, voltage versus power (nose curve types of the plot) that the operator might be more familiar with.

The rightmost snapshot shows a case where the blue line does not exactly cross the red dot. This might be an indication that the indicator is less reliable, or that the assumed value of $\cos\theta$ is slightly wrong.

As for the *PowerApp*, the implementation and testing of the method were done in two steps. First, a research version "*Vapp*" was developed in order to test various displays and alarms as shown in Fig. 9.10. This was also implemented in the *NI LabView* environment. The user interface includes four plots:

– The upper left shows the operating point, voltage versus active power, and plots the associated "nose curve" based on the estimated Thevenin impedance. The dashed lines indicate trajectories drawn with other values of the Thevenin

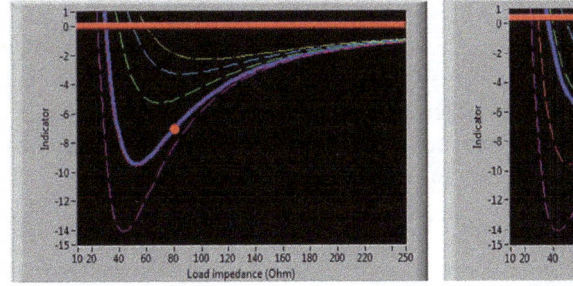
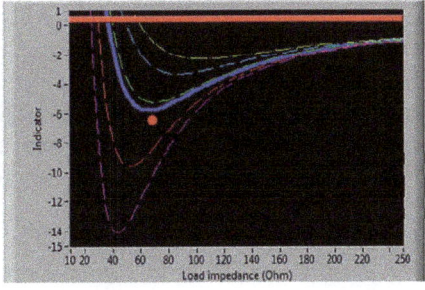

Fig. 9.9 Visualization of the S-Z indicator and the corresponding "constant-Thevenin impedance" line

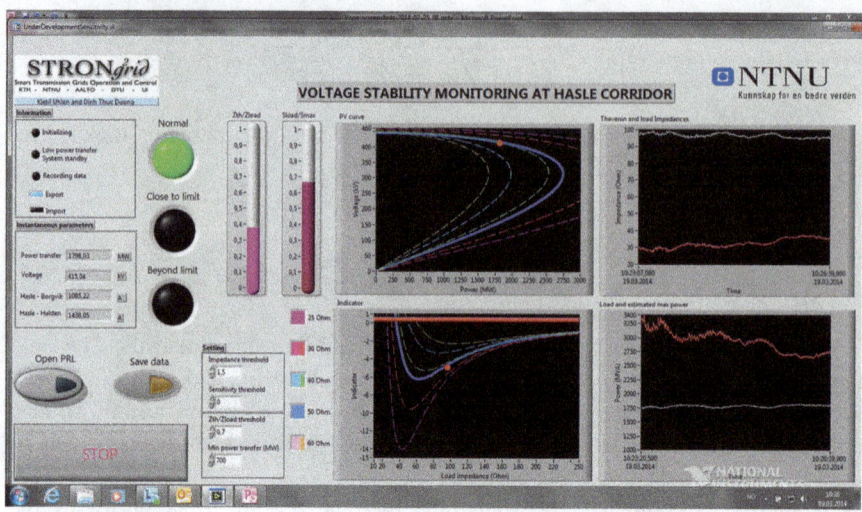

Fig. 9.10 The research version user interface of the *Vapp*

impedance. This provides information of what would be the margin if the strength of the grid changes.
- Upper right figure shows trend plots of the magnitudes of the load and Thevenin impedances, respectively.
- Lower left shows the S-Z indicator as described above.
- Finally, the lower right figure shows trend plots of the actual power transfer and the estimated maximum power.

The prototype version used by Statnett consists of two separate applications, as shown in Fig. 9.11; where one is the online monitoring tool and the other is for post fault analysis with the possibility to retrieve and study previous events. In the prototype version, you have the possibility to select upto five locations to monitor, but in an industrial version this is of course only limited by the availability of measurements and computational issues. So far there are only two live plots available. That is, the online voltage versus power curve (nose curve) with indication of the operating point, and a trend curve showing the magnitude of the load impedance and estimated Thevenin impedance.

Up to present (luckily, one might say), we have seen no cases where the monitored areas have experienced a voltage collapse. The screen-dump in Fig. 9.12 serves to illustrate one incident where the grid was somewhat weakened, and a sudden load increase brought an area quite close to maximum loadability.

9 Use of Voltage Stability Monitoring and Transient Stability Monitoring...

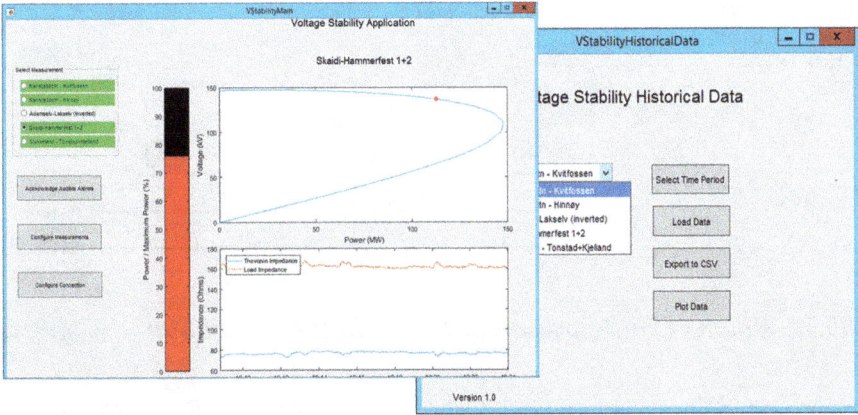

Fig. 9.11 Prototype voltage stability application. Picture in front shows the online application user interface. In back the user interface for post-event analysis

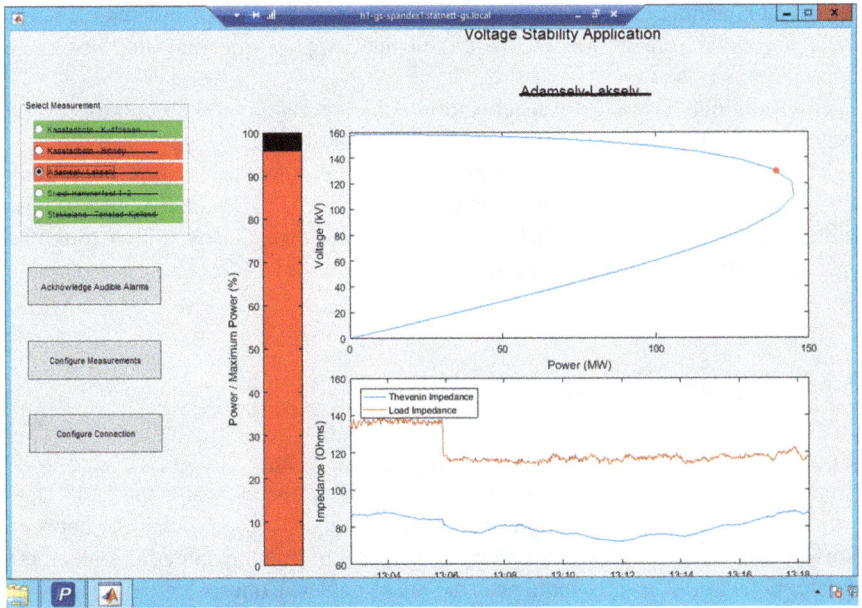

Fig. 9.12 Example showing a case where there was a sudden load increase and consequently a very low power margin

9.5 Other Examples of WAMS

In this section we discuss a few more general observations and experiences gained from having the prototype wide area monitoring system at the control center [8].

Fig. 9.13 PMU measurements from substations

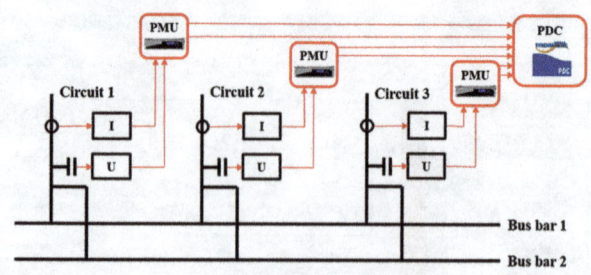

9.5.1 Robust Configuration of PMUs in the Substation

WAMS applications are most useful when system stability is at risk, e.g., during extreme power flow conditions or following contingencies (outages or short circuit faults). Therefore, it is crucial that the required stream of information is always available. One illustrative example was during a storm, where two lines tripped and caused an islanding event. One of the substations had PMU measurements, but the frequency measurement was on the circuit that tripped—meaning that important information was lost.

By installing a PMU for each circuit, the operators get access to several frequencies and voltage measurements from one substation.

The solution shown in Fig. 9.13 ensures that the power calculation of a circuit always uses the correct voltage measurement. The PMUs sample three-phase currents and voltages from each circuit. No matter which power line trips, the operators will always have access to correct frequency measurement.

9.5.2 Access to Voltage Phase Angle

When reconnecting a power line, the voltage phase angle difference across the circuit breaker must normally be less than 30°. The operators use the EMS state estimator to check the phase angles before attempting to close. The border between Norway and Sweden/Finland has a length of 1600 km, and the operating state of the synchronous grid naturally influences the phase angles. Concerning power lines on corridors between the Nordic TSOs, synchrophasors information has shown that the phase angle calculations by the EMS are often incorrect, mainly because the grid configuration at the other end is unknown. One comparison showed that the state estimator reported a phase angle difference of 14° while PMUs measured 45°. In this case, the line would have failed to reconnect, and without the information from the synchrophasors the reason for this would not have been obvious.

9.5.3 Monitoring HVDC

During the winter of 2014/2015, several commutation failures were detected on HVDC converters located in southern Norway. In June 2017, the WAMS implementation at Statnett identified a commutation failure based on detection of rapid changes in voltage phase angle from PMUs located close to the converter station. This event had system-wide consequences, as the fault on the HVDC connection in the south led to a sudden loss of load and increase of system frequency. This again caused a frequency sensitive load to trip in the north of Norway. The load that tripped is located some 1600 km away from the HVDC link.

The event demonstrates clearly how the measured responses in frequency throughout the system can be used to improve situational awareness and to identify where faults have occurred. Figure 9.14 shows three disturbances. At fault 1 (at 31 s) the frequency at substation 2 (south) starts tp increase first. This is the measurement from the PMU closest to the HVDC converter. The frequency rise indicates a loss of load, or in this case, a loss of export on the HVDC link. The second disturbance (at 39 s) is when the frequency sensitive load in the north of Norway tripped. Substation 1 is located close to this load and the grid is relatively weaker in that area. This can be observed by the local frequency oscillations in the north, and that the impact on the other frequency measurements are too small to be noticed. At the third event (at 41 s), the HVDC link goes back to full export, again causing the frequency to oscillate between the north and south.

Closer investigation of the PMU measurements at substation 2, revealed that the HVDC link had ramped from full export to zero in less than a few hundred milliseconds. After exactly 10 s, the HVDC ramped back to full export (Fig. 9.15). By use of WAMS, the control center was able to provide the personnel at the

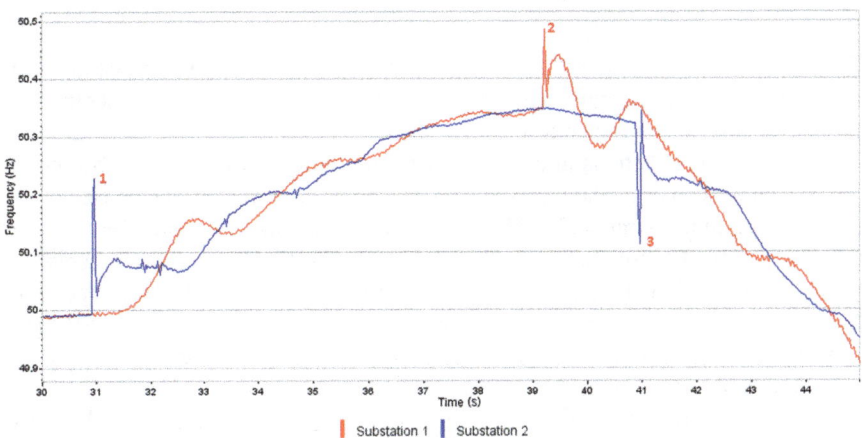

Fig. 9.14 Response in system frequency at two distant locations after HVDC fault

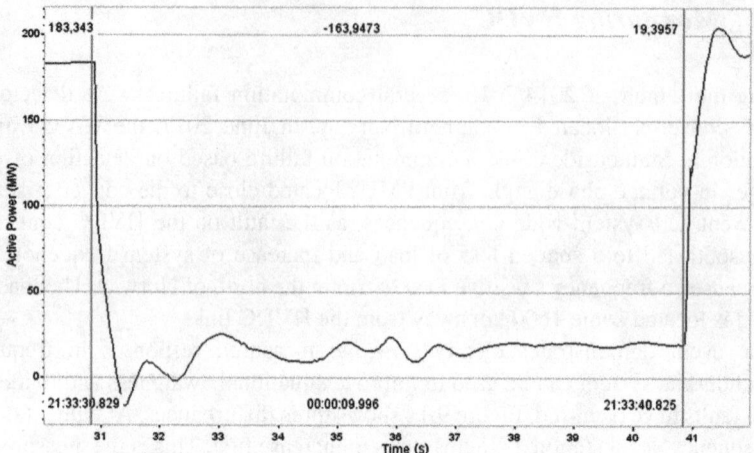

Fig. 9.15 Power flow on one of the AC lines feeding the HVDC converter station

substation with this additional information, which would not be possible to obtain from SCADA/EMS.

Note on frequency measurements: The "spikes" observed in frequency measurements from PMUs close to the disturbances are likely due to almost instantaneous changes in voltage angle. One could say that this is an error or a weakness of the frequency measurement algorithm, but on the other hand, these spikes give a clear signal that there has been a disturbance electrically close to the monitored node.

9.6 Concluding Remarks

PMU data has obvious advantages over SCADA measurements when dealing with dynamic events, i.e., events that put system stability at risk. As the operation of power systems becomes increasingly complex, such events are likely to occur more often. Time to take actions in order to maintain secure operation is also becoming shorter with less inertia and exposed to ever larger and more frequent variations in power flows. In this context, WAMS is expected to be more and more useful for the operators, as well as for the design of automatic control and protection systems.

Through implementation and prototype testing of WAMS at Statnett, operators have had access to new tools for online stability monitoring. The results from this project regarding applications and operator experiences contribute to accelerate the full-scale deployment of this technology.

The results have provided valuable insight into how PMUs should be deployed in the substations, how the WAMS applications should be visualized in the control room, and some of the advantages of monitoring the power system with synchrophasors information have been acknowledged. The importance of involving the control room operators and draw on their experiences should not be underestimated. Their knowledge and experience trigger ideas for new applications.

Obviously important for further developments is to thoroughly address also the ICT related challenges, such as data management, communication requirements, and cybersecurity threats.

References

1. J. Seppänen, Methods for monitoring electromechanical oscillations in power systems, PhD thesis, Aalto University, 2017
2. D.T. Duong, K. Uhlen, An empirical method for online detection of power oscillations in power systems. Presented at the 2018 IEEE Innovative Smart Grid Technologies—Asia (ISGT Asia) conference, Singapore, 2018
3. J.F. Hauer, Application of prony analysis to the determination of modal content and equivalent models for measured power system response. IEEE Trans. Power Syst. **6**(3) (1991)
4. L. Vanfretti, V. H. Aarstrand, M. S. Almas, V. S. Perić, J. O. Gjerde, A software development toolkit for real-time synchrophasor applications. in *IEEE Powertech 2013*, Grenoble, France, 2013
5. C.D. Vournas, C. Lambrou, P. Mandoulidis, Voltage stability monitoring from a transmission bus PMU. IEEE Trans. Power Syst. **32**(4), 3266–3274 (2017). https://doi.org/10.1109/TPWRS.2016.2629495
6. D.T. Duong, Online voltage stability monitoring and coordinated secondary voltage control, PhD thesis, NTNU, 2016
7. D.T. Duong, K. Uhlen, S-Z sensitivity: an indicator and a tool for online voltage stability monitoring. Presented at the panel session Real-time Voltage Control and Stability Monitoring. Submission Number 18PESGM0443 at the IEEE PES General Meeting, Portland, OR, USA, August 8, 2018
8. D. Karlsen, K. Uhlen, L.K. Vormedal, Introducing PMU-based applications in the control room setting, Paper C2-124, CIGRE Session 2018, Paris

Chapter 10
Voltage Stability and Transient Stability Assessment Tools to Manage the National Electricity Market in Australia

Stephen Boroczky and Lochana Perera

10.1 Introduction to the National Electricity Market

Australia's power systems have developed from a number of independent regional systems, which evolved as population and industry developed in dispersed coastal areas.

Subsequent interconnection in these coastal systems in the east and south of Australia has led to the formation of the national grid, which is often referred to as the National Electricity Market (NEM). The NEM, like all other systems in Australia, operates at 50 Hz at various high voltages that reflect the independent regional origins of the interconnected system.

The NEM AC transmission system has evolved into a system that predominantly runs along the Australian coast, stretching from Cairns in far north Queensland down to New South Wales and Victoria, and west to Port Lincoln in South Australia for more than 4000 km (2500 miles), along with an additional DC interconnection to the island of Tasmania. As such, it is a loosely meshed, long, thin network that represents one of the longest interconnected power systems in the world. Coupled with the ever-increasing penetration of solar PV, Distributed Energy Resources and low inertia, this is presenting its own unique operational challenges, particularly in the various stability domains.

The Australian Energy Market Operator (AEMO) is the independent system operator and is responsible for the management of the NEM transmission system. Faced with the challenges of operating such a network within its technical limit, AEMO increasingly relies on real-time security assessment tools to confidently operate the NEM closer to this technical envelope.

S. Boroczky (✉) · L. Perera
Australian Energy Market Operator, Melbourne, Victoria, Australia
e-mail: Stephen.Boroczky@aemo.com.au; Lochana.Perera@aemo.com.au

Because of the way the NEM power grid has evolved, the weak points of the system are generally near the interregional interconnectors, but some intra-regional issues remain.

Existing processes of mapping out the NEM technical envelope by offline methodologies, backed up with real-time analytics are serving AEMO well. However, with the ever-changing landscape of the NEM, some recent incidents have highlighted that these traditional methods of assessing transient and voltage stability may not be as appropriate as they have been in the past. New improved and more sophisticated methods may be needed in the future.

10.2 Power System Stability in the NEM

AEMO has operated the NEM power system without significant challenges in power system stability related issues for many years. The recent accelerated uptake of renewable energy sources such as wind power plants, grid-scale and distributed PV power plants and energy storage devices has changed the power system landscape very rapidly, resulting in several challenges in maintaining power system stability [1, 2], security and reliability. In addition to the change of generation spread, the change of demand response, characterised by a reduction in traditional induction machine load and an increasing share of inverter-controlled drive systems and air-conditioning loads have altered the power system demand behaviour significantly.

The current power system managed in the NEM is a low inertia, low system strength, less synchronous generator dominated and low visibility entity, adversely affected by a large proportion of volatile demand profile with reduced load relief supplied by a large quantity of distributed PV and embedded generators. Emerging challenges of power system operation as a result of rapid transformations taking place has underscored the importance of the stability analysis tools for the monitoring and real-time decision-making in the NEM more than ever before.

10.2.1 Managing Stability Issues in the NEM

The National Electricity Rules (NER) [3] requires AEMO to be responsible for the management and secure operation of the NEM power system. AEMO takes every effort to operate the power system inside a stable and secure technical operating envelope. In this regard, AEMO defines the power system operating envelope using power system limits or limit equations. Limit equations are provided by Transmission Network Service Providers (TNSP) or AEMO. All the TNSP developed limit equations are also extensively reviewed by the AEMO.

Due-diligence studies are undertaken to independently assess and verify the stability limit equations by performing stability simulations. The limit equations are formulated as linear combinations of power system variables and applied as constraints in a least cost-based generator dispatch algorithm. This is solved by a linear programming optimisation algorithm known as the National Electricity Market Dispatch Engine (NEMDE). NEMDE synthesises a least cost-based generator and load dispatch outcome, that ensures power system limits that are encoded as linear constraints into the least cost-based optimisation algorithm are enforced.

NEMDE outputs are used in the NEM to control and direct generator and load dispatch. Traditionally, AEMO enforced via NEMDE thermal, voltage stability, transient stability and oscillatory stability limits as well as generator ramping limits and Frequency Control Ancillary Service (FCAS)-based limits [4]. In line with the deteriorating system strength of the NEM, AEMO now has to maintain system strength of the power system by using system strength constraints. System strength constraints ensure that sufficient synchronous generation is in operation to maintain the short circuit levels in local power system regions [5].

A power system can be operated in extremely large number of operating conditions in terms of generation dispatch patterns, reactive power device utilisation, load profiles, network topology and switching patterns. Thousands of power system studies with varying power system operating conditions are, in general, undertaken to understand any particular power system stability phenomenon. For example, the transient stability limit between Victoria (VIC) and New South Wales (NSW) is defined via four AC interconnectors. It is developed by undertaking transient stability simulations of the worst-case credible contingency,[1] transient stability limit searching and multivariable linear regression analysis of potential power system variables of thousands of power system snapshots covering an extremely large number of power system operating conditions.

As shown in Fig. 10.1, the limit equation is developed by fitting a multivariable linear equation or equations to cover at least 95% of the stability limits of all the power system operating conditions studied. A safety margin known as an operating

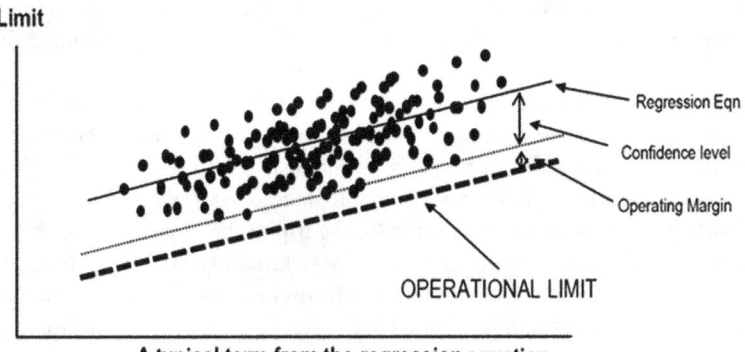

Fig. 10.1 Formulation of NEMDE constraint equation by linear regression

---

[1] Usually a two-phase to ground fault and trip of Hazelwood—South Morang 500 kV line.

margin is also applied on top of the 95% confidence stability equation developed to cover measurement and modelling imperfections associated with the process and short-term variations in load and generator outputs.

Automated software tools are extensively used for all aspects of limit equation development. Voltage stability limit equations and oscillatory stability limits of the power system are developed using similar methodologies. Currently, transient stability limit equations are developed using power system simulations undertaken using Power System Simulator for Engineering (PSS/E) of Siemens PTI. The transient stability limits are developed to maintain rotor angle stability of synchronous generators following a large disturbance such as a trip of a large generator, load or fault and trip of a major transmission line. Voltage stability limit equations are also developed using PSS/E and stability criteria of maintaining acceptable voltages at all buses with a reactive power reserve margin of at least 1% of the maximum three-phase fault levels at all buses in the power system.

10.2.2 Description of NEM Stability Issues

The NEM power system operation is constrained by several stability limits in its interconnected regions. The following stability limits are illustrated in Fig. 10.2 which shows the nature and relative location of these limits.

- Power flow to New South Wales from Queensland is limited by the transient stability limits for the fault and trip of Armidale—Dumaresq 330 kV line or for the trip of Boyne Island potline (approximately 400 MW) and the oscillatory stability limits.
- Power flow north from New South Wales to Queensland is limited by the voltage stability limit for the trip of Liddell-Muswellbrook 330 kV line.
- Power flow north from Victoria to New South Wales is limited by the transient stability limits for a fault and trip of the Hazel Wood—South Morang 500 kV line [6].
- Power flow south from New South Wales to Victoria is limited by voltage stability limits for the contingent loss of the largest Victorian generator or the Basslink DC interconnector between Tasmania and Victoria.
- Victoria to South Australia on Heywood interconnector is limited by voltage stability and transient stability limits for the trip of the largest generator in South Australia or the fault and trip of South East—Tailem Bend 275 kV line.
- Furthermore, the DC interconnector Murraylink power flow from Victoria towards South Australia is limited by a voltage stability limit for the loss of Bendigo—Kerang 220 kV line which is a major line supplying the DC interconnector.
- Power export to Victoria from South Australia is limited by an oscillatory stability limit.

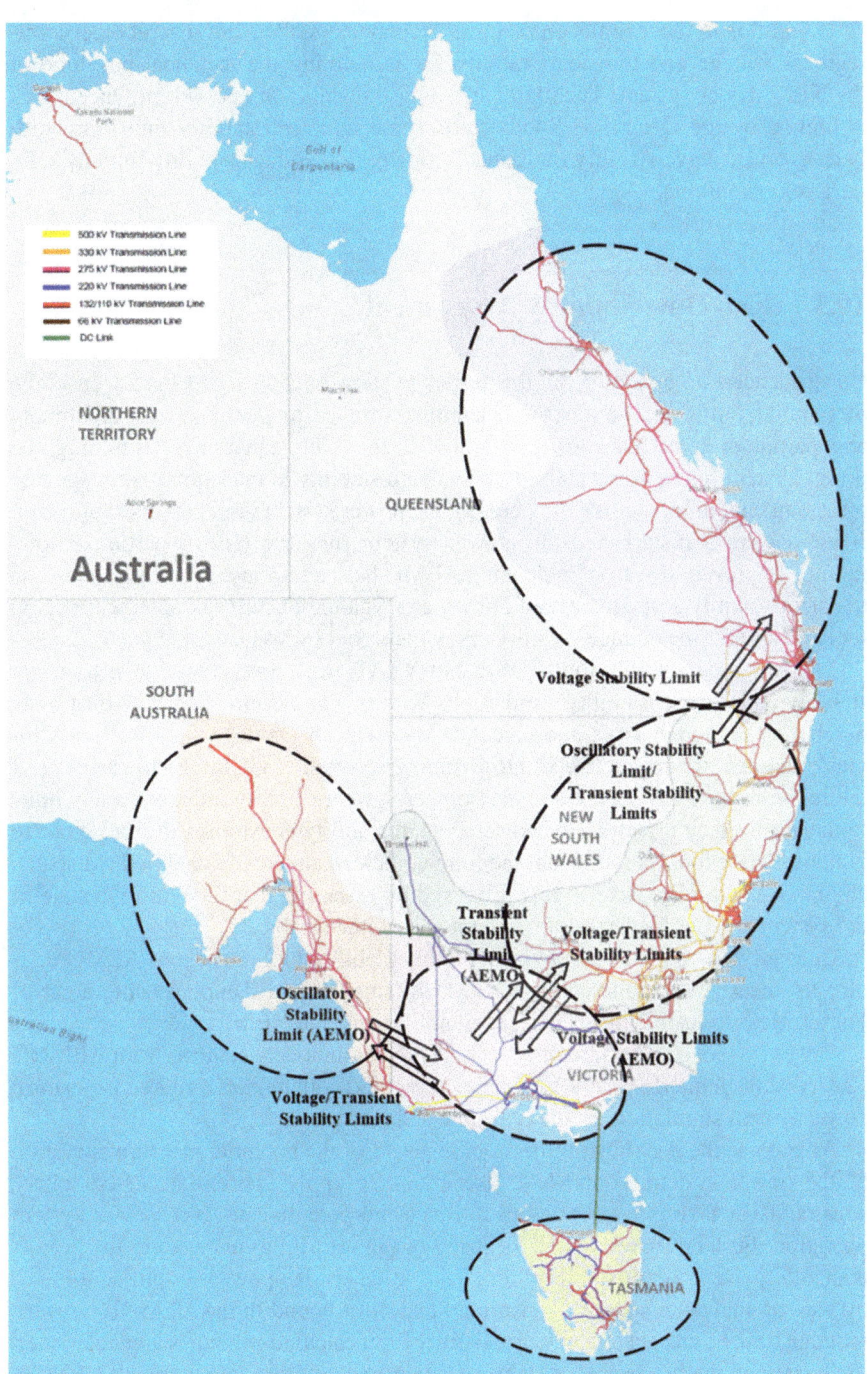

Fig. 10.2 Map illustrating the NEM with its regions depicting typical stability issues and their locations

In addition to those limits on major interconnectors, there are several other intra-regional voltage and transient stability limits defining the technical envelope of the NEM power system. The extremely long transmission network in Queensland is limited by the Central to South Queensland transient stability limit, far north Queensland voltage stability limit and Tarong region voltage stability limit as main intra-regional limits.

10.3 Real-Time Stability Assessment

Consequences of operation of the power system outside its technical envelope of stability limits can be adverse, including damage to power system equipment and partial or total blackout of regions. If the limit equation constraining the power system for transient stability or voltage stability is inadequate for a specific operating condition, the risk of operating the power system insecurely is significant. If there were a disturbance, the power system runs the risk of partial or total failure. To prevent that from happening AEMO has an in-house developed real-time transient stability assessment tool known as Dynamic Security Assessment (DSA) tool and a real-time voltage stability assessment tool (VSAT).

The real-time stability tools, DSA and VSAT, respectively undertake transient stability and voltage stability simulations with power system snapshots over a set number of interconnectors and calculate transient and voltage stability limits for major defined contingencies. Control room is provided, via the EMS displays of the real-time stability tool, calculated power system stability interconnector limits to make sure they are aware of power system stability constraints that are close to binding. These tools complement the limit equations already defined in the dispatch engine and basically perform due diligence by assessing that the real-time state of the system is secure in terms of transient and voltage stability.

Currently, there is no real-time oscillatory stability tool in use, but AEMO does directly measure the inherent power system small signal damping using a set of Phasor Measurement Units (PMU) scattered around the NEM.

Power system state estimator snapshots are continuously sourced from AEMO's Energy Management System (EMS) and combined with models required to perform power system simulations in the real-time stability tools.

As soon as the AEMO control room is aware of the potential insecure operation of the power system, controllers take action to apply appropriate discretionary constraints or take other actions to mitigate the potential insecure power system operation. Further analysis is also carried out in AEMO to understand the performance of power system stability limits and to determine power system congestion by way of analysing stability constraints that have bound in the NEMDE in every month. Offline transient and voltage stability limit calculation studies are undertaken to understand the headroom of stability constraints at time slots in which NEMDE has indicated that the power system has reached its limits. These offline studies enable AEMO to understand if the offline developed stability limits are optimistic or pessimistic, enabling it to continuously refine the operating envelope of the power system.

Currently, DSA implements six scenarios, consisting of over 60 simulated contingencies that represent the transient stability issues highlighted in Fig. 10.2. Figure 10.2 also highlights some of the 25 scenarios VSAT runs to monitor the interface limits of concern and identify the limiting contingency for each scenario.

Any DSA or VSAT alarm that cannot be explained is extensively investigated by AEMO Operations to understand the real causes of the alarm. Based on these investigations, AEMO ascertains the issues associated with current voltage and transient stability limits, more appropriate settings for the limits such as critical contingencies, modelling issues and control system settings. The limits or issues that compromise power system security are promptly flagged and fed back to TNSPs or to AEMO's own departments who undertake stability limit determination, to modify or improve the constraint equations that monitor the operating envelope of the power system.

Statistical analysis is also undertaken to analyse the headroom between the real-time stability tool-based limits and offline study-based limits. If a certain limit has a large headroom most of the time, it shows that the power system can be operated beyond the offline calculated limit. Such limits are further evaluated by AEMO using offline stability studies and flagged to the TNSPs for further refinements and improvements.

10.3.1 The Scenario Definition

The scenario forms the basis for both DSA and VSAT to describe the interface limit to be determined, and the source and sink definitions that describe what needs to be adjusted to vary the transfer along that interface.

To determine a scenario limit, the transfer is increased along the interface by increasing source generation and increasing sink load (or reducing sink generation).

If a scenario is found to be unstable in the base case (current state of the system) both DSA and VSAT will reduce the transfer to determine at which point the case becomes stable again. This advises the control room operators the necessary transfer reduction to guarantee stability.

Scenarios that are chosen are based on the areas identified by the planning models as the weak points in the system. These normally end up being around the regional interconnectors. The scenario definitions mimic the constraint equations developed with the offline tools as described above. That means the scenarios in DSA and VSAT are confirming the accuracy of the constraint equations and, where there is a deviation, will determine what the true limit is under each power system condition.

10.3.2 DSA: Transient Stability Assessment

The Dynamic Security Assessment (DSA) is a bespoke application developed by AEMO with a commercial power system simulation engine at its core and orchestrated by python scripts [7, 8]. Figure 10.3 shows how the different DSA subsystems interface with each other, both on the EMS and on the DSA server itself.

The EMS application prepares the case by exporting the current real-time powerflow model from the state estimator, together with scenario and contingency definitions which are passed to the DSA server. Here, PSS/E powerflows are created that represent the state of the network as determined by the state estimator. Dynamic models are added and mapped to the powerflow data, as well as the scenario data.

Python scripts coordinate the analysis of each case. Efficient use of the multicore environment is made by allocating a single contingency simulation in each scenario to a single core. This queuing system maximises the use of each core and avoids the need to swap multiple processes if too many simulations were to be scheduled at once.

Once the base case stability is determined, the scenarios define what transaction analysis is performed by increasing (or decreasing) the transfer across the interface of interest. A binary search algorithm is deployed to determine the stability limit. The stability limit is the level of transfer at which the scenario becomes unstable for at least one contingency. This information conveys to the operator the transfer limit and the current operating margin to this limit.

If the scenario is unstable in its base case, (i.e. in the "current" state of the network), a reverse limit search is performed, advising the operators how far the

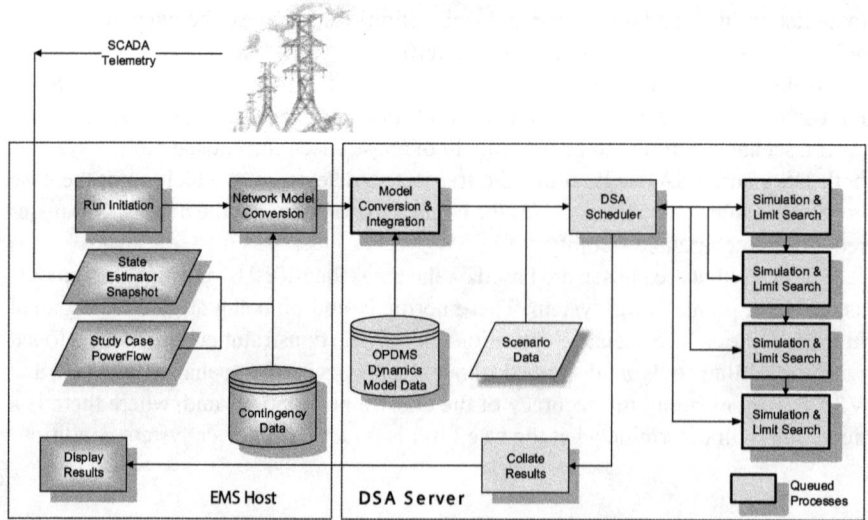

Fig. 10.3 DSA architecture showing the various stages of analysis in the process

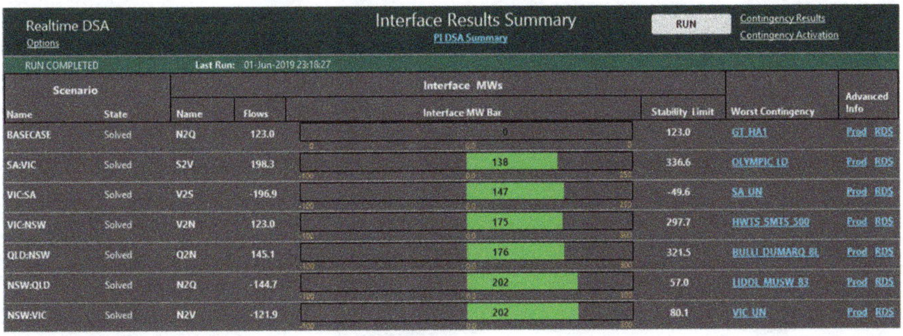
Realtime DSA
Options
Interface Results Summary
PI DSA Summary
RUN
Contingency Results
Contingency Activation
RUN COMPLETED
Last Run: 01-Jun-2019 23:18:27
Scenario
Interface MWs
Name
State
Name
Flows
Interface MW Bar
Stability Limit
Worst Contingency
Advanced Info
BASECASE Solved N2Q 123.0 0 123.0 GT HA1 Prod RDS
SA:VIC Solved S2V 198.3 138 336.6 OLYMPIC LO Prod RDS
VIC:SA Solved V2S -196.9 147 -49.6 SA UN Prod RDS
VIC:NSW Solved V2N 123.0 175 297.7 HWTS SMTS 500 Prod RDS
QLD:NSW Solved Q2N 145.1 176 321.5 BULLI DUMARQ 8L Prod RDS
NSW:QLD Solved N2Q -144.7 202 57.0 LIDDL MUSW 83 Prod RDS
NSW:VIC Solved N2V -121.9 202 80.1 VIC UN Prod RDS

Fig. 10.4 Screenshot of main DSA page showing the different scenarios along with the computed interface transfer limits

transfer should be reduced to become secure again. In this instance, a discretionary constraint is placed in the market to bring the transfer back to within this secure limit.

Figure 10.4 shows an example of DSA results that can be viewed in the EMS. The dominant information on this screenshot is the calculated transfer limit, the operating margin to that limit and the limiting contingency.

10.3.3 Stability Criterion

The stability of an individual simulation is determined by a special angle spread model that is deployed as a model within PSS/E. It attempts to measure the amount of *swing* between any two generators on the system. By ignoring the initial angle, the model ignores any bias that could be introduced by machines that are on the extremities of the system. Figure 10.5 shows how the maximum swing between any two machines is determined.

A few of the key features of this model include:

- A simulation is declared as unstable when the difference between any two rotor angle swings (i.e. ignoring the offset) exceeds the threshold.
- Conversely, an early termination criterion terminates the simulation early if the rotor angle swings are not growing significantly.
- It is aware of islands being created dynamically as a result of the contingency or RAS action. It will only compare generators that are within the same island.
- It can exclude specific generators from being monitored. This avoids the problems of using an angle from power electronic devices, such as voltage source converters and wind farms, where the machine angle has no real meaning.

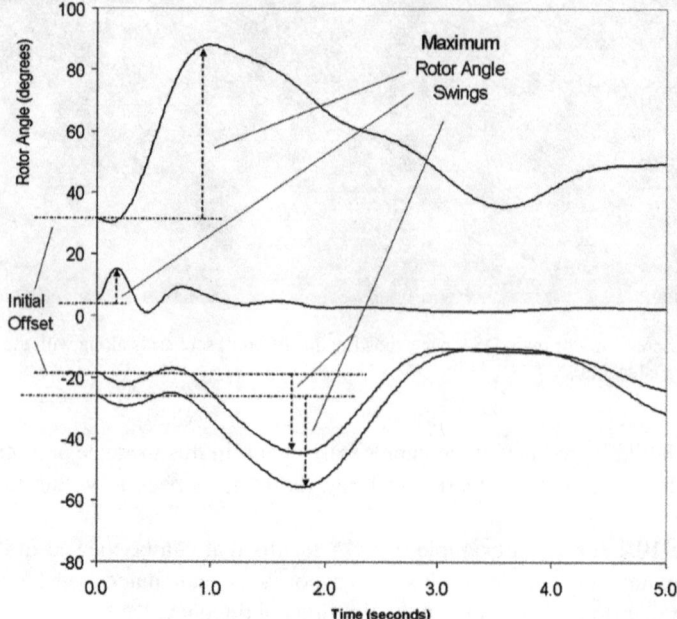

Fig. 10.5 How DSA determines the stability of a simulation. If the machine rotor angles diverge past a threshold then the case is declared as unstable

10.3.4 Typical DSA Issues

By far the most frequent issues AEMO experiences with DSA can be attributed to dynamic model robustness. In order to accurately represent equipment connected to the power system, most of the models used by AEMO are user-defined models. Whilst a model can work reasonably well for tailored planning studies, when thrust into an operational environment, a model may start misbehaving. A model needs to be able to cope with operational conditions outside the equipment's normal operating range. For example, a governor needs to be able to initialise if a generator happens to be outputting more than its rated capacity, or if it is operating outside its normal capability curve.

This issue has been exacerbated with the increasing penetration of wind and solar farms. The models are more complex and take additional inputs such as wind speed or irradiance. If the given wind speed or irradiance is not quite appropriate for the output of the farm, the models can have problems initialising.

Other issues come about during the limit search. As generation is scaled in the source, the wind speed or irradiance may no longer be appropriate for a farm's scaled output, even though it is still within its capability. Similarly, the scaling needs to respect the ratings of individual turbines or inverter groups. Unless windspeed, irradiance, number of turbines and inverter groups are adjusted as generator output is scaled, there is a risk of initialisation failure.

10.4 VSAT: Real-Time Voltage Stability Assessment

VSAT (depicted in Fig. 10.6) is essentially a powerflow-based steady-state analysis that primarily performs PV-analysis along an interface definition. The 25 scenarios target the known weak spots in the NEM, some of which are identified in Fig. 10.2, and mimic the constraints already applied in NEMDE. Figure 10.7 shows an example EMS screenshot showing some of the VSAT scenarios and their limiting contingency at the time.

10.4.1 Impact of HVDC Frequency Modulation

Basslink is the HVDC interconnector between the southern mainland and the island of Tasmania. With a capability of 500 MW, this link represents about a third of the entire load in Tasmania. The link also has a frequency modulation block so that frequency response can be shared between the mainland and Tasmania. Essentially,

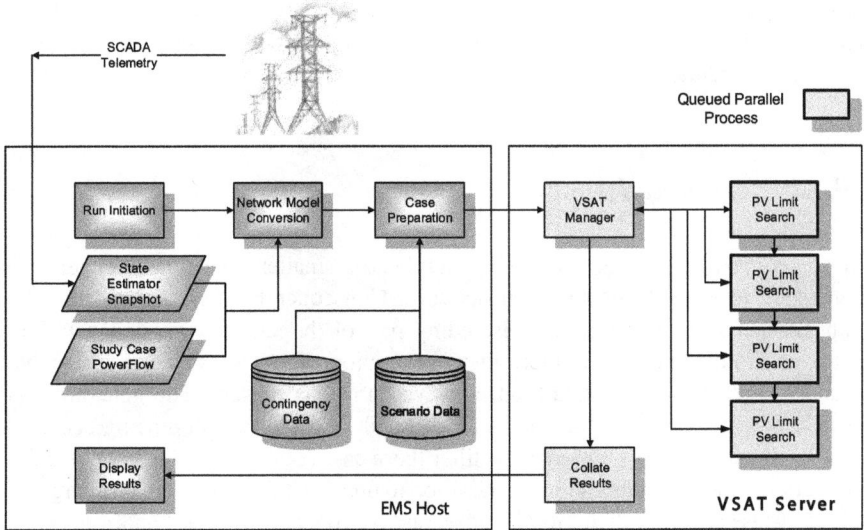

Fig. 10.6 VSAT architecture is very similar to DSA architecture. It is slightly simpler as it avoids the need to combine dynamic data models. VSAT schedules the PV limit search process locally as well as on remote servers

Fig. 10.7 VSAT limit display showing some of the 25 scenario limits determine by VSAT. From an operational perspective, the interface flow limits are of the most interest on this display

Basslink will reduce transfer north for frequency lowering events in Tasmania. It is important to model this even for voltage stability as it can significantly affect the flows through the corridors leading into Basslink. Currently, Basslink is modelled as equivalent load in the mainland and Tasmania in VSAT. To model the frequency response from Basslink some dummy generators have been added in the model at the Basslink terminals. These generators are assigned with governor response to factor in the frequency behaviour of Basslink.

10.4.2 Islanding Issues

In the NEM there are an increasing number of smaller intermittent generators embedded in the sub-transmission network. Consequently, a contingency in the transmission system may end up islanding part of the sub-transmission network. If this island contains some intermittent generation, VSAT tries to create a viable island as both generation and load, however small, is present. The generation is usually insufficient to support the local load and the island ends up being declared as voltage unstable. It is important to filter these cases out because there are usually protection schemes on this sort of generation to prevent it from forming an energised island. These can be difficult to diagnose, and it is only with careful analysis of the instability causing contingency can one determine the real issue. In these situations, the normal practice is to disable the contingency until the conditions causing this issue has passed.

10.4.3 Scaling Issues

The scaling of wind and solar farms can pose its own issues, particularly if they are situated on the end of a radial corridor. In determining the scenario stability limit, one needs to ensure that these wind and solar farms are not scaled so far as to inadvertently introduce a local voltage stability issue along the corridor that is unrelated to the scenario.

10.5 Power System Events

Whilst operationally every effort is made to avoid the onset of system events, nevertheless, they do provide the opportunity to formulate a better understanding of the evolving nature of the power system in terms of its dynamic behaviour. The following two events not only illustrate how the dynamic performance of the NEM system is changing but also the increasing level of complexity required to accurately represent its behaviour. The learnings from these events have helped to apply strategies that mitigate future occurrences of such system events.

10.5.1 South Australian Black System Incident

In the afternoon of 28 September 2016 an extreme weather condition including high winds, thunderstorms, lightning strikes and heavy rainfall in South Australia have resulted in multiple transmission line faults, loss of major 275 kV transmission lines, loss of a large amount of generation and the trip of the major interconnector between South Australia (SA) and Victoria (VIC) and a total black out of the South Australia power system [9]. The events leading to the black system occurred in a short time from 16:16 to 16:18 h. The system restoration was initiated at 17:23 h by restoring the interconnector between South Australia and Victoria and 80–90% of the electricity supply was restored by midnight on 28 September 2016. Consequently, the last remaining segment of the transmission system (excluding transmission lines with major tower damage) was energised at 21:00 h on 30 September 2016.

An extensive and thorough investigation was undertaken by AEMO on the black system event in South Australia and several conclusions were reached. AEMO has examined the sequence of faults that occurred in the system, the generator performance to the sequence of events, the performance of the associated protection and control systems and compared all the available high-speed electrical measurements with PSSE- and PSCAD-based simulation of the event sequence leading to the system black out event.

It has come to light that six voltage dips over an 88 s period have resulted in a significant power reduction (approximately 456 MW) of South Australian wind farms because of a protective feature of windfarms previously unknown to AEMO. This protective feature resulted in a significant amount of power reduction of wind farms if they were subjected to more than a preset number of voltage dips within a 2-min period. The ensuing power deficit was then transferred to the Heywood AC interconnector between South Australia and Victoria and rapidly growing angular difference between South Australian generators and Victorian generators has activated the loss of synchronisation protection and tripping of the double circuit interconnector and isolating the South Australian power system from the rest of the NEM.

As a result of approximately 900 MW loss of generation from the interconnector, the remaining generators in South Australia were unable to maintain South Australian demand, causing the system frequency to collapse faster than South Australia's under frequency load shedding (UFLS) scheme (very high rate of change of frequency) could activate, subsequently resulting in a total blackout condition.

During the restoration phase, neither of the two black start generators in South Australia were successful in restoring power to a major generator and therefore the system restoration was carried out using the VIC-SA AC interconnector.

From this investigation, AEMO has learned that:

1. Non-credible contingencies with previously unknown consequences can occur in the power system, moving the power system into unknown territory,
2. Conventional approaches used in power system stability analysis may be broadened to account for the impact of severe and multiple weather-related contingencies including islanding conditions,
3. Power system modelling should be included with all critical protection settings,
4. Power system restoration should be studied more extensively and tested more comprehensively and
5. The requirement should be established for a minimum number of synchronous generators in SA.

AEMO has also taken steps in the medium term and in the long term to address the high Rate of Change of Frequency (RoCoF), challenges related to robust and reliable detection and measurement of high RoCoF events, emergency control schemes to enable successful islanded operation of the power system, the effects of distributed and energy storage in delivering fast frequency response in the transforming power system dominated by asynchronous generators and diminishing synchronous generators. The investigation of the above incident has also highlighted the importance of using detailed Electromagnetic Transient (EMT) type models to better understand the voltage and transient stability and performance of power electronic control-based equipment in the renewable generator dominated power systems. In this respect, AEMO has undertaken to broaden the scope of Generating System Model Guidelines to include more detailed EMT type models and to include not just the dispatchable generators, but also the embedded generation, the voltage supporting and controlling equipment and the protection systems. AEMO has also undertaken to study both detailed EMT-based and RMS-based models to investigate power system stability issues in especially low inertia and low system strength regions of the NEM power system [5].

10.5.2 Multiple NEM Region Separation Event

At 13:12 h on 25 August 2018, [10] the two 330 kV AC interconnectors Dumaresq-Bulli Creek lines 8L and 8M connecting Queensland (QLD) and New South Wales (NSW) tripped, separating QLD region and NSW region following a direct lightning strike on the tower structure supporting the two circuits and double back flash over on one phase of each circuit during an extreme weather event including thunderstorms, rain and damaging winds. The double back flash over has simultaneously caused a phase to ground fault on one phase (phase C) of each circuit and loss of synchronism between QLD and NSW. The single-phase reclosing action of the interconnector did not close the circuit breakers following reclosing action because the above simultaneous single fault trip of both circuits exceeded the voltage, angle and frequency thresholds checked (synchro check) prior to the closing of the circuit breakers and therefore opened all six circuit breakers of the two lines.

After the separation, the frequency of QLD rose to 50.9 Hz as a result of supply surplus (power flow just before the event was from QLD to NSW). The remaining power system experienced a supply deficit resulting in a low frequency and triggering of the APD Portland Tripping (EAPT) scheme (which also trips the SA-VIC AC interconnector) separating South Australia region from Victoria (VIC) and New South Wales (NSW) regions.

The separated South Australian (SA) region also had a supply surplus (power flow just before the event was from South Australia to Victoria) post-separation, causing South Australian frequency to rise. The event has caused the VIC and NSW island to experience a power deficit and a frequency of 49 Hz, triggering under frequency load shedding (UFLS) schemes in Victoria, interrupting approximately 1000 MW of customer and industrial loads in Victoria and NSW. Frequency controller on Basslink DC-link has also responded to the under-frequency condition in Victoria increasing its power flow from Tasmania to Victoria from 500 to 630 MW and causing the trip of 81 MW of contracted load under Tasmania's under frequency load shedding scheme. The incident created three independently operating islands in the power system. The restoration started at 13:35 h by connecting South Australia into VIC + NSW island and at 14:20 h by connecting QLD into SA + VIC + NSW island.

The incident investigation suggested that the power system would have responded much better had it have more primary frequency response during the event than was available. It is also recommended that AEMO reclassify the simultaneous trip of 8L and 8M lines as credible when necessary and apply new constraints. Following extensive transient stability studies undertaken by AEMO,

TransGrid (NSW TNSP) and PowerLink (QLD TNSP) using both PSSE and PSCAD, AEMO has constrained the QLD to NSW interconnector at 850 MW and modified synchro-check conditions of auto-reclosing logic of the circuit breakers of 8L and 8M for the simultaneous loss of two phases across both 330 kV lines 8L and 8M when such a contingency is declared credible during extreme weather events.

10.6 Transformation of the NEM and Future Impacts

Recent years have seen the emergence of a large number of asynchronous plants such as grid-scale solar PV plants, wind power plants and grid-scale storage devices in the NEM, especially in the weak and remote parts of the power system. At the same time, the NEM has experienced a huge decline of dispatchable, high inertia, base load synchronous generators such as Northern Power Station (520 MW) and Hazel Wood Power Station (1600 MW) retirements. Subsequent decline in system inertia and lower fault current contributions have introduced several new transient and voltage stability issues into the NEM.

Voltage collapse occurs at power transfers well below the nominal levels in weaker networks. Large voltage and current oscillations could result in dynamically changing load impedance conditions. Furthermore, the lack of system inertia affects the rate of change of frequency (RoCoF) in the power system increasing the RoCoF in many contingencies thus causing the power system to be highly vulnerable to frequency variations. Increased distributed generation such as household PV and storage as well as changing customer behaviour has reduced the visibility of the power system for the power system operators. Decreasing day time customer demand due to the explosion of distributed PV, increasing light load operation of long transmission lines and increasing peakiness of the NEM load profile have worsened the task of managing high voltage in the power system during low demand conditions, especially in South Australia and Victoria. Real-time transient stability tools and voltage stability tools are widely considered as some of the important instruments of supporting the control rooms of the modern twenty-first century NEM. Continuous and real-time assessment of transient and voltage stability issues in the NEM support the power system operators by increasingly identifying and predicting possible stability issues in the power system. New contingencies, emergency control schemes and new limits are also required to be studied by real-time stability tools with the emergence of a widely dispersed asynchronous generator fleet.

Victoria's remote north-west transmission network has evolved into a renewable energy hub in recent years, exposing the power system to new stability issues such as low inertia and system strength. AEMO has recently undertaken extensive electromagnetic transient simulations of this part of the network dominated by asynchronous generators using PSCAD to understand and to define new limits to prevent large signal oscillations of voltage, active power and reactive power following a large disturbance such as the fault and trip of a major transmission line, a generator or a load.

AEMO has now provided several limits to define the operating envelope of the asynchronous generators in north-west Victoria to manage large disturbance voltage/power oscillations and system strength. AEMO's current real-time transient stability assessment tool, DSA uses PSSE which models the NEM power system using positive sequence, RMS models. Inability to represent the key components responsible for weak grid instability such as Phase Locked Loop (PLL) of power electronic converters, inability to represent full three-phase power system and inability to model electrical transients are some of the shortcomings of the current real-time transient stability tools used in AEMO. However, the complexity of modelling, the time and computing power required to undertake EMT type simulations in real-time stability tools is immense at present, still justifies the use of simplified RMS-based models in real-time stability assessment and offline assessment of EMT-based complex models.

The use of both EMT type tools (such as PSCAD) and RMS type tools (such as PSS/E) have been proven essential in several power system incident investigations undertaken by AEMO in the recent past, mainly attributable to the continuously evolving power system flooded with modern power electronic converter fed generators and other equipment.

AEMO will continue to rely on a combination of offline analytical assessment that is validated by the real-time stability tools. As computing performance improves, it may be possible to introduce EMT phasor style simulations into AEMO's real-time stability suite. With the extreme complexity of these models, one would need to be careful not to introduce further initialisation issues that could render the application impractical.

This evolving landscape is demanding increased modelling complexity for AEMO to remain confident in its assessment of the power system's technical capability and to maintain the security of the system. It will be driving the future use of voltage stability and transient stability analytical tools for the NEM.

References

1. P. Kundur, Definition and classification of power system stability. IEEE Trans. Power Syst. **19**(2), 1387–1401 (2004)
2. Australian Energy Market Operator, *Power System Stability Guidelines* [Online] (2012), https://www.aemo.com.au/-/media/files/electricity/nem/security_and_reliability/congestion-information/2016/power-system-stability-guidelines.pdf
3. Australian Energy Market Commission, *National Electricity Rules* [Online] (2020), https://www.aemc.gov.au/regulation/energy-rules/national-electricity-rules/current
4. Australian Energy Market Operator, *FCAS Model in NEMDE-Scaling, Enablement and Co-optimisation of FCAS Offers in Central Dispatch* [Online] (2017), https://aemo.com.au/-/media/Files/Electricity/NEM/Security_and_Reliability/Dispatch/Policy_and_Process/2017/FCAS-Model-in-NEMDE.pdf

5. Australian Energy Market Operator, *System Security Market Frameworks Review - System Strength Impact Assessment Guidelines* [Online] (2019), https://aemo.com.au/energy-systems/electricity/national-electricity-market-nem/system-operations/system-security-market-frameworks-review
6. Australian Energy Market Operator, *Limit Advice* [Online] (2020), https://aemo.com.au/en/energy-systems/electricity/national-electricity-market-nem/system-operations/congestion-information-resource/limits-advice
7. S.C. Savulescu, *Real-time stability assessment in modern power system control centers* (John Wiley & Sons, Hoboken, 2009)
8. Working Group 601 of Study Committee C4, *Review of on-line dynamic security assessment tool and techniques*, CIGRE Technical Brochure (2007)
9. Australian Energy Market Operator, *Black System South Australia 28 September 2016* [Online] (2017), http://www.aemo.com.au/-/media/Files/Electricity/NEM/Market_Notices_and_Events/Power_System_Incident_Reports/2017/Integrated-Final-Report-SA-Black-System-28-September-2016.pdf
10. Australian Energy Market Operator, *Final Report—Queensland and South Australia system separation on 25 August 2018* [Online] (2019), http://www.aemo.com.au/-/media/Files/Electricity/NEM/Market_Notices_and_Events/Power_System_Incident_Reports/2018/Qld%2D%2D-SA-Separation-25-August-2018-Incident-Report.pdf

Chapter 11
Use of Online Transient Stability and Voltage Stability Assessment Tools at State Grid China

Shi Bonian, Yan Jianfeng, and Jin Yiding

11.1 Introduction

The major characteristics of China's power supply and consumption are the reverse distribution of power generation (in the north and west area) and load center (in the south and east area). In order to meet the urgent needs of clean energy delivery, heavy load center power supply, energy conservation, and emission reduction, State Grid has been vigorously developing UHVAC/UHVDC technologies for long-distance and large-capacity power transmission [1, 2]. Till the end of 2016, six 1000 kV UHVACs and five ±800 kV UHVDCs have been deployed within State Grid, being the world's only one power utility with parallel operation of UHVAC and UHVDC, as shown in Fig. 11.1.

At the same time, the coupling influence between AC and DC transmission and the reciprocal influence between power exporting region and power importing region are getting even stronger, and thus leading the power grid dynamics to become more complex [3–6].

Take one hydropower dominated exporting power grid as an example, as shown in Fig. 11.2. This is a provincial power grid located in central China and deployed

---

S. Bonian (✉)
R&D Center, Beijing Sifang Automation Company, Beijing, China
e-mail: shibonian@sf-auto.com

Y. Jianfeng
State Grid Simulation Center, China EPRI, State Grid Corporation of China, Beijing, China
e-mail: yanjf@epri.sgcc.com.cn

J. Yiding
System Operation Department, National Dispatching and Control Center, State Grid Corporation of China, Beijing, China
e-mail: jin-yiding@sgcc.com.cn

© Springer Nature Switzerland AG 2021
S. (NDR) Nuthalapati, *Use of Voltage Stability Assessment and Transient Stability Assessment Tools in Grid Operations*, Power Electronics and Power Systems, https://doi.org/10.1007/978-3-030-67482-3_11

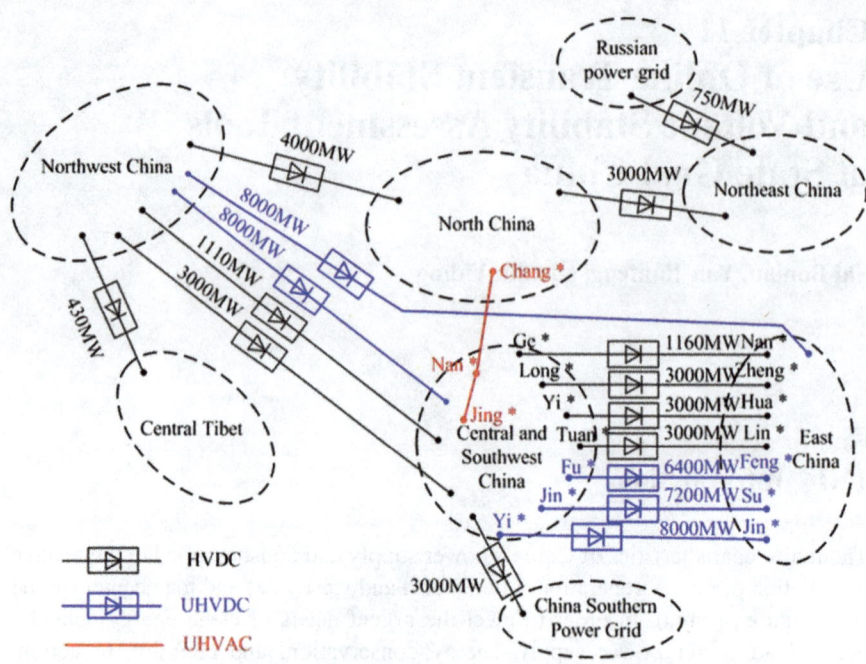

Fig. 11.1 Regional HVDC/UHVDC/UHVAC Interconnection of SGCC (until 2016)

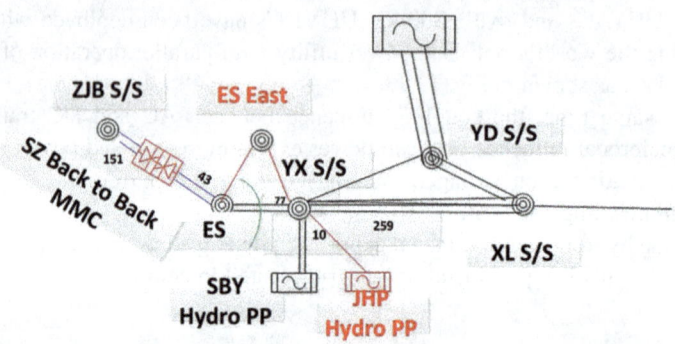

Fig. 11.2 Diagram of a hydro power exporting network with MMC interconnection

with a number of hydropower plants. At the end of 2019, a new 1250 MW, ±420 kV MMC (Modular Multilevel Converter) HVDC project, SZ Back to Back, was put into commission, which blocks the original power flow transfer path, and results in the transient instability problem of the hydropower exporting grid. When the transmission power on SZ B2B MMC decreases from 800 MW to 600 MW, more local hydro generation will be supplied to maintain the same power export on the XL corridor. For the same N-1 fault on the line YD-YX, the JHP generator will lose synchronism with increased local output, as shown in Fig. 11.3.

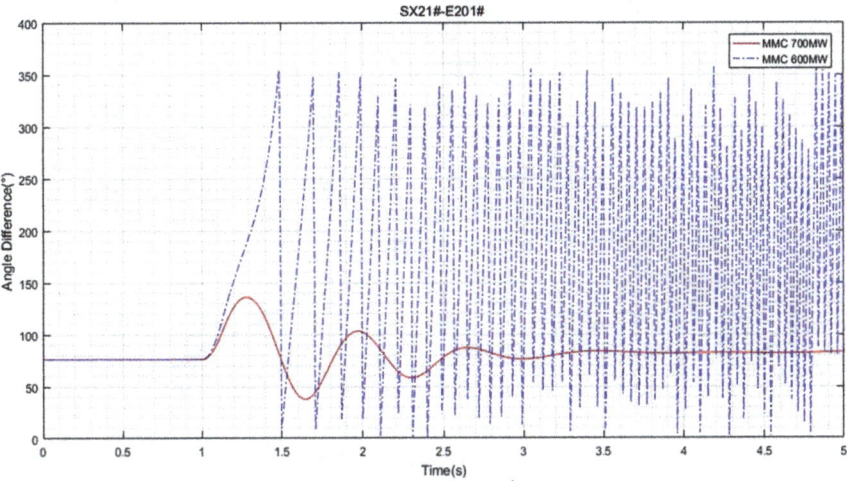

Fig. 11.3 Comparison of generator angle curves for different MMC transmission power

Besides the transient angle instability problem, with the increased deployment of UHVDC projects, the number of conventional thermal power plants at the receiving end grid is decreased, and voltage support capacity for the receiving end grid also weakened thus leading to the severe voltage instability problems such as:

1. As some conventional thermal power plants at the receiving end grid are replaced by the HVDC power import, power grid voltage regulation capability deteriorates. According to the HVDC design principle, under normal operating conditions, the reactive power exchange between the HVDC converter station and the AC grid is zero. During normal operation mode switching, the HVDC converter station will display reverse voltage regulation capability compared with conventional generators, as shown in Fig. 11.4, where the converter station will consume more reactive power with voltage decrease, while the conventional generators will generate far more reactive power. With AC bus voltage reduction of 1%, the HVDC converter station will absorb 50 MVar from the AC grid, while the same size of conventional generators can provide at least 300 MVar or more.
2. A large amount of reactive power will be absorbed during DC fault, which leads to the problem of dynamic voltage stability. Operation practice shows that when an UHVDC commutation failure occurs, the inverter side will absorb the reactive power as high as 4000–5000 MVar. Figure 11.5 illustrates the measured reactive power consumed by the converter station during an UHVDC commutation failure.

Take the SD AC/DC hybrid receiving end power grid in North China as an example, as shown in Fig. 11.6. The infeed DC section consists of ±800 kV ZLT-QZ UHVDC, ±800 kV EKZ-LY UHVDC, and ±660 kV YCD-JZD HVDC, with

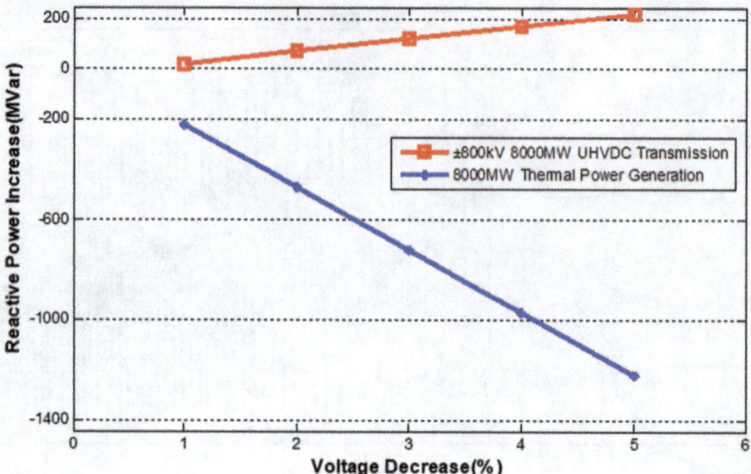

Fig. 11.4 Comparison of reactive voltage regulation capability between HVDC converter station and conventional generator

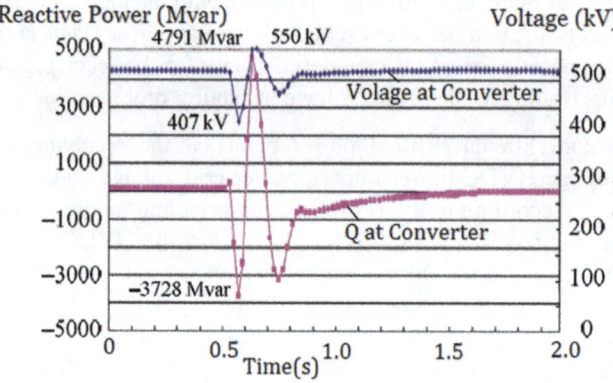

Fig. 11.5 Measured voltage and reactive power of converter station during an UHVDC commutation failure

total rated power of 24,000 MW. The AC receiving section consists of four 1000 kV UHVAC lines and four 500 kV HVAC lines.

Under the calculating condition as SD local load level of 60,000 MW, with AC receiving section importing 9000 MW, UHVDC ZLT importing 7000 MW, UHVDC EKZ importing 4500 MW, HVDC YCD importing 4000 MW, and generation output from renewable energy resource (such as PV plants and Wind farms) being 13,000 MW, considering the inverter LVRT (low voltage ride through) capability of PV plants and wind farms, when N-2 faults occur on lines such as Q-L double lines, voltage collapse will be inevitable in SD power grid, as shown in Fig. 11.7.

11 Use of Online Transient Stability and Voltage Stability Assessment Tools... 221

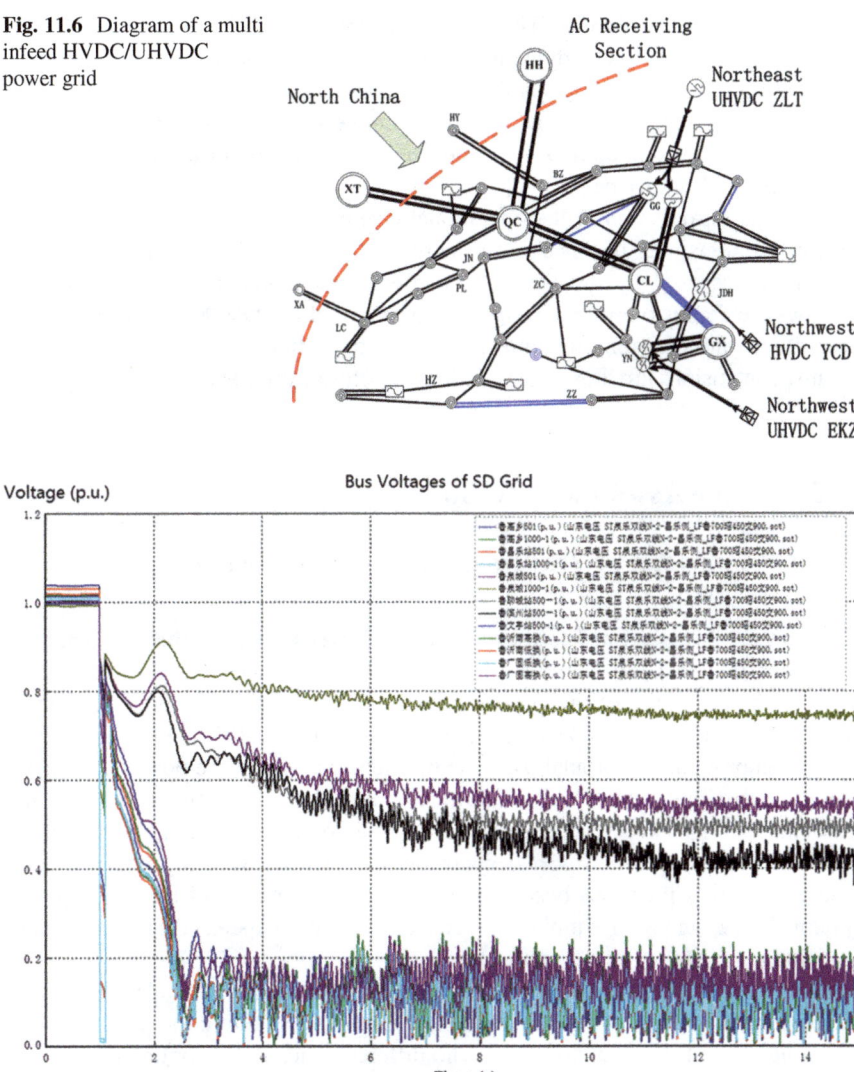

Fig. 11.6 Diagram of a multi infeed HVDC/UHVDC power grid

Fig. 11.7 Voltage instability of SD power grid under N-2 fault

The even more complex dynamics observed within today's State Grid, including fluctuating renewable energy, bidirectional power flows caused by demand responses and storage devices, hybrid HVAC/HVDC parallel systems with heavy power transfer, and increased applications of advanced protection and control systems for power electronic devices, will easily and frequently lead to short period power supply imbalance. In some extreme cases, cascading failures and large-scale blackouts may occur if the disturbances are not evaluated thoroughly and not mitigated in a timely manner. Therefore, to address the above challenges, it is of critical importance to assess dynamic security and operational risks of the system in near real-time, such as online stability assessment.

Since 2009, State Grid has deployed the online stability assessment system on the newly developed unified D5000 platform, for all the provincial level and above dispatching and control centers [7].

After the introduction of the operational challenges faced by the SGCC in this section, the remainder of this chapter is organized as follows. Section 2 introduces the algorithms used by the D5000 online assessment system, and then Sect. 3 provides the details of system design and structure of the D5000 online system. In Sect. 4, the online system implementation and some study cases will be presented. Finally, conclusions are drawn in Sect. 5, with future work identified.

11.2 Online Assessment Technique

11.2.1 Algorithm of Transient Stability Assessment

Transient stability is defined as the ability of the power system to return to its normal conditions after a large disturbance. In this chapter, transient stability specially refers to the capability of the synchronous generator to maintain the synchronism after large disturbance such as permanent short-circuit fault [8].

The major methods for analyzing the transient stability of the power system are time-domain simulation [9] (also known as step-by-step integration method) and direct method (also known as energy function method). Generally, the time-domain simulation is taken as the most dependable approach for studying power system transient stability problems because it simulates the behavior of the entire power system. The accuracy of simulation result usually only depends on the equivalent model of the power system components.

Online transient stability assessment [10, 11] inherits the same algorithm from offline assessment, using time-domain simulation for obtaining the power system operating state after the disturbance. The difference lies in that online assessment starts with the refreshed system snapshot, which is obtained from real-time EMS (Energy Management System).

The time-domain simulation method adopted by the online assessment system proposed in this chapter is briefly introduced as below,

11.2.1.1 Description of Simulation Process

Time-domain simulation is employed to capture the power grid's transient response and timing of some protection and control actions. To capture the transient response, a set of differential and algebraic equations (DAE) are numerically solved. These

11 Use of Online Transient Stability and Voltage Stability Assessment Tools...

DAEs, i.e., the mathematical model for time-domain simulation, include both mathematical description of the grid (network equation) and mathematical description of the dynamic characteristics of generator, load, HVDC/UHVDC, large-scale grid connected of renewable energy resources (Wind Turbines and PVs), together with controllers such as speed governor, power system stabilizer, relay protection, etc. The mathematical models can be divided into three parts:

1. Mathematical model of the power grid, i.e., the network equation, as shown in Eq. (11.1).

$$X = F(X, Y) \tag{11.1}$$

where $F = (f_1, f_2, \cdots, f_n)T$ are algebraic functions, and $X = (x_1, x_2, \cdots, x_n)$ are variables solved for network equations.

2. Mathematical model for generator and its regulators, load, HVDC, etc., i.e., differential equation, as shown in Eq. (11.2).

$$Y = G(X, Y) \tag{11.2}$$

where $G = (g_1, g_2, \cdots, g_n)$ are differential functions, and $Y = (y_1, y_2, \cdots, y_n)$ are state variables solved for differential equations.

3. Models for exerted disturbance and control measures, such as simple or complex faults, generator tripping, and load shedding, which will change the variables X and Y.

11.2.1.2 Simultaneous Solution of Differential Algebraic Equations

In the time-domain simulation, the differential equation is solved by the iterative method of ladder stability integral, the network equation is solved by the combination of direct triangulation decomposition and iteration, and the differential equation and the network equation are alternately iterative, until converge, and thus the solution for time period T is completed.

The trapezoidal implicit integral equation for transient stability is shown in Eq. (11.3),

$$Y^{(K+1)} = G\left(X, Y^{(K)}\right) \tag{11.3}$$

The iteration procedure for network equation is as below,

$$X^{(K+1)} = F\left(X^{(K)}, Y\right) \tag{11.4}$$

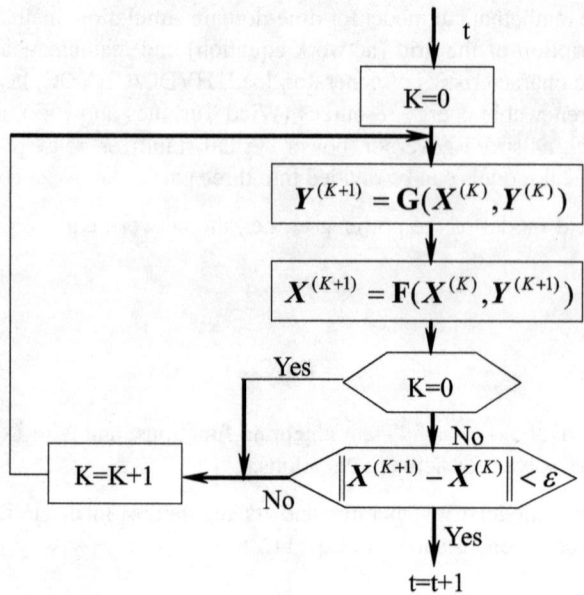

Fig. 11.8 Simultaneous solution of differential equations and network equations

The flowchart for simultaneous solution of differential equations and network equations is shown in Fig. 11.8.

11.2.2 Issues Concerned with Online Transient Stability Assessment

11.2.2.1 Computation Time

There is a contradiction between the limitation of time for online calculations and the need to scan for more contingencies. The use of parallel computing platforms and contingency screening can solve the problem to some extent, but these two methods are not perfect: the former requires more hardware investment, and the latter has the problem of the accuracy of the screening results.

11.2.2.2 Constraints on Modeling

There are two main sources of data for the online dynamic stability assessment system: the results of state estimation and the dynamic model of the grid. The former comes from the EMS, including the original power flow and models of static component. The latter refers to the model and parameters of dynamic components such as generators and HVDCs, which are obtained from the offline dynamic model library. The combination of these two types of data can achieve a comprehensive description of grid static and dynamic operation status, and carry out a variety of transient stability assessment.

EMS traditionally supports only static analysis, and there is often a gap between their grid models and the requirements of dynamic analysis. The gap mostly lies in the fineness of modeling such as:

1. Ignorance or keeping part of the distribution network. For the offline assessment, the load is typically connected to the busbar at 110 kV or lower voltage level, and thus preserving part of the 110 kV grid. But in the online assessment, due to fast convergence and other reasons, the 110 kV distribution network is often ignored during the online state estimation process for the large-scale interconnected power grid, and the load is simply processed as an equivalent on the 220 kV busbar.
2. Detailed or simplified model of dynamic reactive compensation. Controlled series compensation, static reactive compensator, controllable shunt compensation, etc., in each moment can be treated as a fixed impedance. During the traditional state estimation, these components are often replaced by a fixed compensation or condenser model, which has little effect on the accuracy of static analysis. However, for the calculation of power grid dynamics, the dynamic performance of these components may have a significant impact on power grid stability.
3. State of components at the terminal edge of the grid. The voltage of these terminal components of the power grid, such as the generator terminal voltage, has little effect on the accuracy of the static analysis, provided that voltage accuracy of the step-up transformer high-voltage side meets the requirement. Some online systems therefore do not pay enough attention to the direct measurement of information on the terminal component, while tending to derive the electrical variables from the high-voltage side under assumed ratio of transformer. However, in the transient stability assessment, the terminal voltage of the generator and load will determine the initial value of the state variable, and subsequent response, and thus accurate terminal voltage data are necessary.

The offline dynamic model library is built for setting the operational mode, and in some ways this may differ from the requirements of online assessment. For example, from the economic and safety point of view, in the dynamic model a lower limit may be set on the opening of the valve for some thermal power units, which will not generate any problem for the conservative operational mode analysis. But for the online system which may collect generator data during the process of starting up or shut down, the opening of the valve will be even lower than the limit, and thus brings trouble to the calculation.

Besides, there may be mismatches between offline dynamic models and online static models. If there are both a power plant and an industrial load connected at the same terminal end of a line, two components will be added in the offline dynamic model library, while only one equivalent load will be added in the online static model. How to configure the dynamic parameters for components that do not exist in the dynamic model library is one of the difficult problems that online stability assessment system has to solve.

11.2.2.3 Computation Accuracy

The computational accuracy and convergence of state estimation are mutually constrained. The order and complexity of the iterative equation matrix of transient stability assessment are greater than static analysis, and the accuracy requirement of the power flow distribution is also higher than that of the latter, and thus higher accuracy of the state estimation is required. Under some poor conditions where many errors exist in the measurements and parameters, the state estimation may only support the accuracy required for static analysis as the convergence criteria in order to ensure that system application can work. From higher accuracy point of view, this kind of processing may lead to small power imbalance on many busbars. How to deal with these power imbalances is a problem that must be tackled before further transient stability assessment.

In short, online transient stability assessment has put forward some new requirements on the traditional EMS, offline dynamic component library and the ability to handle large-scale computing, etc. These present a big challenge for the promotion of online transient stability assessment system.

11.2.3 Algorithm of Voltage Stability Assessment

As discussed in the first section, voltage stability now is a serious concern which must be examined carefully during planning and operational phase. Even the pre-contingency and post-contingency voltage levels are acceptable for the system states under study, the voltage instability still may occur when reactive reserves on specific generators reach certain values [12, 13].

First, calculate the P-V Curve [14] and static voltage stability margin for the base case. Voltage stability margin (VSM) is a measure of how close the system is to voltage instability [15]. For computation, the system load is increased step-by-step and the power flow is solved at each step, as follows.

Set the steady operating point of the system to be studied, i.e., (V_0, θ_0), to satisfy the power flow equation shown in Eq. (11.5),

$$\begin{cases} P_{G0} - P_{L0}(V_0) = f_p(V_0, \theta_0) \\ Q_{G0} - Q_{L0}(V_0) = f_q(V_0, \theta_0) \end{cases} \quad (11.5)$$

where P_{G0} and Q_{G0} are vectors consisting of active and reactive power of the generator at the current operating point, respectively; $P_{L0}(V_0)$ and $Q_{L0}(V_0)$ are vectors consisting of active and reactive power where static load characteristics are taken into account respectively; $f_P(V_0,\theta_0)$ and $f_Q(V_0,\theta_0)$ are the nodes that are determined by the network characteristics.

Considering the step increase of the system load, it can be represented by parameter k, as shown in Eq. (11.6),

$$\begin{cases} P_L(V,k) = P_{L0}(V) + kP_D(V) \\ Q_L(V,k) = Q_{L0}(V) + kQ_D(V) \end{cases} \quad (11.6)$$

where P_D and Q_D are vectors consisting of the step active and reactive power increase of the load, respectively.

The increased load active power is generally shared by multiple generators in a certain way, as shown in Eq. (11.7),

$$P_G(V,k) = P_{G0}(V) + kP_{DG}(V) \quad (11.7)$$

where $P_{DG}(V)$ is the step active power increase of the generator.

The voltage stability critical point is reached at the load level beyond which power flow solution does not exist. The increase in the system load from the initial operating point to the voltage stability critical point, i.e., $P_M - P_o$, is the voltage stability margin for the base case, as shown in Fig. 11.9.

Further, the VSMs for all the contingencies are calculated. At each load level, after solving the power flow for the base case, the contingencies are applied one by one and the power flows are solved. Here the loads are modeled as voltage-dependent. The last load level where the post-contingency power flow solution exists is the post-contingency critical point and the increase in the pre-contingency system load from the initial operating point to this point, i.e., $P_{CM} - P_o$ is the voltage stability margin for the given contingency, as shown in Fig. 11.9.

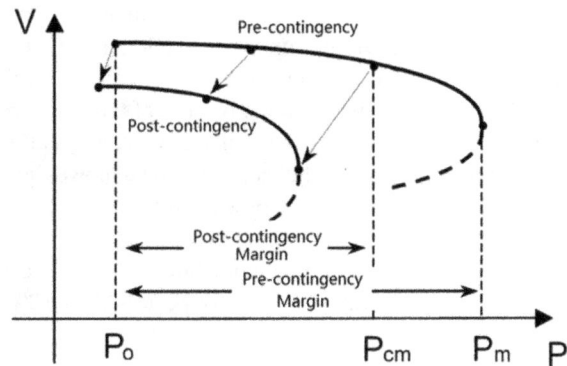

Fig. 11.9 P-V curves and VSMs

And for some selected critical contingency cases, time-domain simulation is used to validate the voltage instability problem will occur or not. Starting with the solved cases corresponding to the different load levels, the system is disturbed by applying the contingency, and the system dynamic response following this contingency is calculated. If the time-domain simulation shows that the system reaches its post-contingency steady-state equilibrium point after a finite time period, the system is stable. If the steady-state equilibrium of the post-contingency system does not exist, time-domain simulation will show that the bus voltages continue to decrease and therefore the system is voltage unstable.

11.2.4 Issues Concerned with Online Voltage Stability Assessment

In the offline voltage stability assessment environment, it is necessary to determine the VSM for all specified contingencies (such as single element outages, double outages of lines on the same tower, double elements lost due to breaker failure protection, etc.) for system conditions with all elements in service and for conditions with one or more elements out-of-service.

However, for the online voltage stability studies, the system state and topology are known through system measurements and state estimation. Therefore, it is necessary to study only the critical contingencies for all present elements in service. As a result, fewer scenarios need to be examined, and less VSM may be required than that of offline studies in which the system uncertainty is greater.

Voltage stability margin is an important output of online voltage stability assessment, and it can provide useful information, such as the inherent weakness of the network structure, impact of human operation, or grid fault on the current grid structure. The following aspects need to be considered during the application of the online assessment tool:

1. Load such as motor loads of auxiliary facilities in the power plant should be modeled as close as to the actual situation. Different facilities may exhibit different characteristics under different generation output of the power plant;
2. The generator needs to take into account the power limit so that the power margin will not exceed the rotational reserve of the whole power grid;
3. Performance of voltage stability may vary greatly in different seasons and in different regions. In the north regional power grid which is penetrated with a large amount of wind power, and has great demand for city heating, the voltage stability margin may exceed 100% during light load period on winter night. However, in the economically developed eastern provincial power grid, the voltage stability margin may be less than 10% during summer heavy load period.

11.3 Online Stability Assessment System Design

11.3.1 Framework and Functions of Online Stability Assessment System

The major functions of online stability assessment system include: online monitoring the operation of the power grid, analyzing the stability margin of power grid operation near real-time, discovering risks of instability problem, delivering early warning thus realizing online stability assessment, and early warning of the power grid operation status. Online stability assessment can be further divided into; online transient stability assessment, online voltage stability assessment, online small-signal stability assessment, online static security assessment, and online short-circuit current assessment. In this chapter, only the first two kinds of stability assessments are discussed in detail.

Based on the mature offline stability assessment algorithms, online stability assessment use the online power flow and grid model, together with the parallel computing platform, to achieve a comprehensive analysis and evaluation of online stability. According to the online analysis results, early warning about the status of the power grid may be issued to the operator directly through the Human–Machine Interface of EMS.

For online assessment system, the conventional offline calculation programs are upgraded to meet the requirements of online calculation, including:

1. Support the standard inputs and outputs: online system data in E file format for the D5000 platform and the calculation results also specified in D5000 E file format.
2. Adapt to online topology changes: For online data before and after the topology change, the calculation program is able to deal with the change of data, without any manually set conditions.
3. Support the Terminate-and-Stay Resident program mode for computation process, without any memory leakage or other duplicate call problems: for offline data, the computation process exits after each call, there is no memory leak or other duplicate call problems; however, for online computing, the computation process must be resident to memory and listen to calculation start signal, and thus put forward strict requirements for memory leaks and duplicate calls.
4. Support signal control mode based computation control: the platform calls the computation software by the system signal control means, where the platform starts the computation software by sending the Calculation Process Start signal to the computation software and begins to receive the calculation results after sending the Calculation Process Termination signal to it. For signal control mode, USR1 signal is specified as calculation result being stable with no need to upload the file, while USR2 signal as calculation result being unstable or being abnormal and the data file needs to be uploaded.

5. Support error handling: Online computation software requires continuous operation without human intervention, while offline software does not have this processing requirement. For all types of errors, including computational errors, data errors, and other errors, online computation software needs to be able to process and alarm automatically.

Besides the core function of online computation, the online stability assessment system needs to fulfill: online data integration, online stability analysis, assistant decision-making, stability margin assessment, and other functions. These functional modules together with the input parameters and settings constitute the framework of the online assessment system, as shown in Fig. 11.10. The online operation mode data is derived from the result of state estimation. Then the online operation mode data is integrated with the equipment models and parameters of the power grid to form a complete simulation dataset for online calculation. On this online refreshed simulation dataset, with applied contingency and specified constraint of operating limit, different application modules such as online stability analysis, assistant decision-making, and stability margin assessment module will be executed.

1. Online stability analysis and early warning module. This module mainly calculates the stability characteristics under the specific online operation mode and analyzes the ability to maintain or restore stable operation. During online stability analysis, different types of analysis, such as transient stability analysis, voltage stability analysis, small signal stability analysis, short-circuit current assessment, and static security analysis, are performed. If the system can keep stable, then the stability margin assessment module will be started to provide margin information for the operator. But, when the system stability level is found to be insufficient, early warning will be issued and corresponding assistant decision support module will be immediately started to tackle different kinds of instability problems.

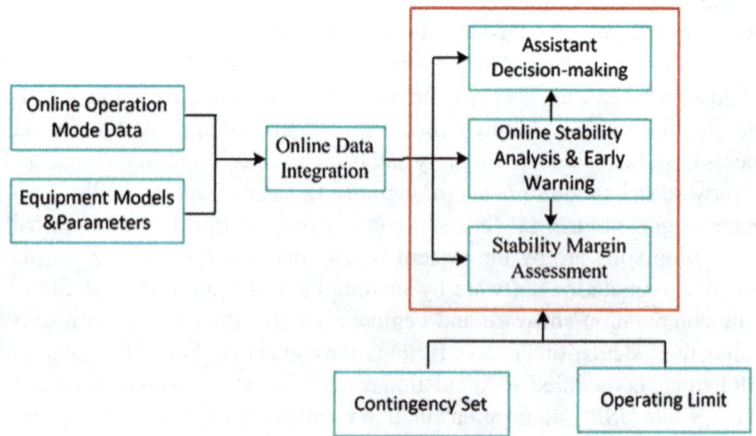

Fig. 11.10 Framework of the online stability assessment system

Finally, a feasible regulation scheme will be provided for the operator, so as to ensure the stable operation of the power grid.
2. Assistant decision-making module. According to the calculation conclusion of the online stability analysis and early warning module, to solve the instability problem, first the regulation object is selected by analysis, and then the regulation amount is derived by parallel computing. The regulation measures to eliminate the potential risk of the power grid operation are provided for the dispatcher. Regulation measures include changing the generator power output, adjusting transformer taps, closing or disconnecting the busbar switch or the line, adjusting reactive power compensation equipment, and shedding load.
3. Stability margin assessment module. The stability margin is used to aid the operator in evaluating alternatives and to stay away from situations which are extremely marginal. According to the specified or automatic power flow regulation scheme to adjust the transfer power of the transmission section, the available transmission capacity meeting the requirements of stable operation is derived online. Generally, this module includes two calculation steps, i.e., power flow regulation and verification of stability:

   (a) Power flow regulation. By means of generator power output regulation, switch operation of capacitor/reactor, switch of transformer tap, load regulation, etc., the transferred power on the specified transmission section is changed under the constraints of converged power flow calculation, and a new power system operation mode data is obtained.
   (b) Stability verification. The operation mode data obtained in the power flow regulation together with applied contingency set are input for stability analysis. Then the maximum transfer power on the transmission section which meets the stability requirement is selected, and the available transmission capacity is obtained.

11.3.2 Process of Online Assessment

Based on the high-speed computing power of parallel computing and open integration performance, the dynamic stability assessment system can fulfill all the stability assessment calculations based on the online data, and the whole calculation process can be completed in minutes. Comprehensive and rapid stability warning function changes the paradigm of traditional offline stability assessment (which is based on typically specified operation mode data), by offering more comprehensive and objective assessment results, and solving the problems such as, speed, comprehensive and credibility of stability assessment for the power system long-term cascading failure (or switch of key components).

Online stability assessment engineers need to closely monitor the analysis results of online operations on a daily basis, combined with comprehensive intelligent alarm, WAMS (Wide Area Measurement System) and other information, to analyze all kinds of alarm information.

The complete assessment process of the online computing platform starts with receiving online data and ends with assessment result display. The information flowchart of complete assessment process is illustrated in Fig. 11.11, which includes the following typical modules:

1. Data integration. Obtain online power flow snapshot from EMS and WAMS and prepare necessary models and parameters for stability assessment.
2. Distributed computing platform. After receiving the calculation data, the platform will broadcast the data to all the computation servers and historical servers to start a new round of assessment calculations. After the completion of various

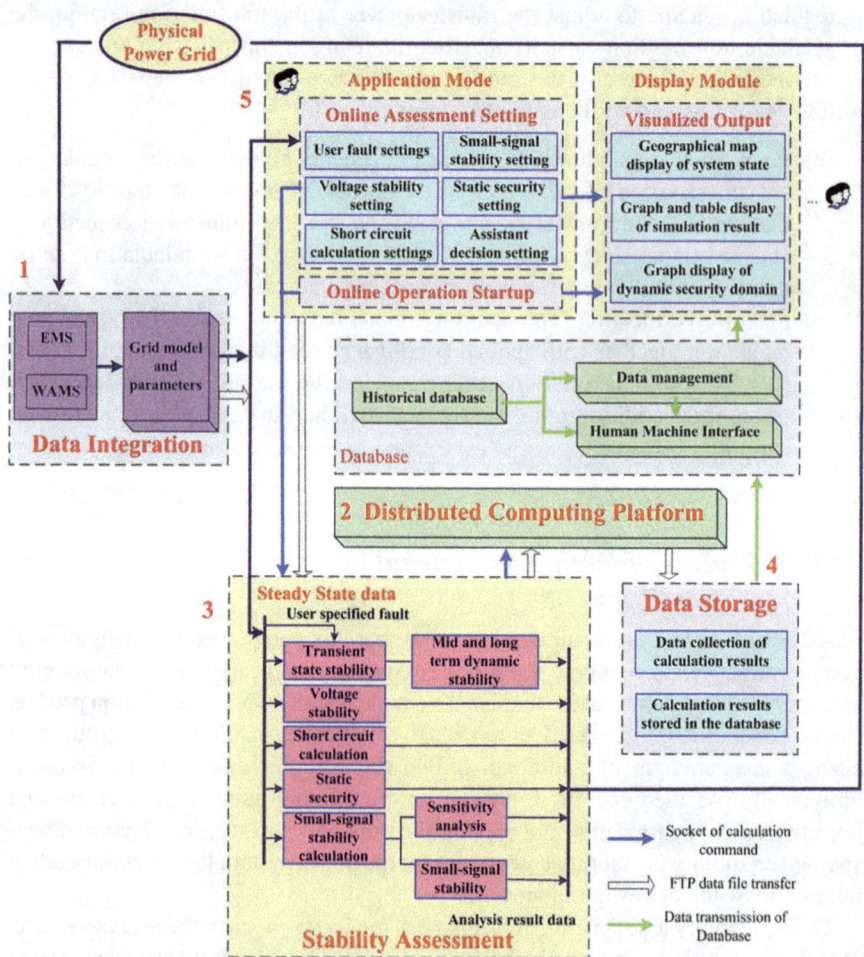

Fig. 11.11 Information flowchart of the online stability assessment process

types of simulation calculations, the platform will collect and summarize the assessment results.
3. Stability assessment. After receiving the calculation start-up command from the platform, the stability assessment module will be started based on the pre-allocated task, and the assessment result will be uploaded when the assessment process is completed.
4. Historical storage. Receive the assessment result forwarded by the platform, store to historical data server, and add the record of the storage database for archiving.
5. Human–machine information exchange. After the results of the stability assessment are collected and summarized, they will be displayed promptly, delivering the early warning information for various kinds of stability and security risks.

11.3.3 Software Architecture

Building an online dynamic stability assessment system is a very complex system of engineering. The online assessment system needs to integrate data and information from existing EMS, WAMS, offline stability assessment system, relay protection and fault information management system, special protection system, power market operation system, etc., and form the steady-state and dynamic data for simulation. The online assessment system can realize online dynamic monitoring and early warning, and provide assistant decision-making information for the grid operators.

The overall architecture of the online stability assessment system is shown in Fig. 11.12. The major functions consist of two kinds of platform functions, three kinds of computation functions, and two kinds of auxiliary functions. The two platform functions include data integration and parallel computing platform functions. The three computation functions include stability analysis and early warning, assistant decision-making, and stability margin assessment functions. The two auxiliary functions include historical data archiving and human–machine interfaces.

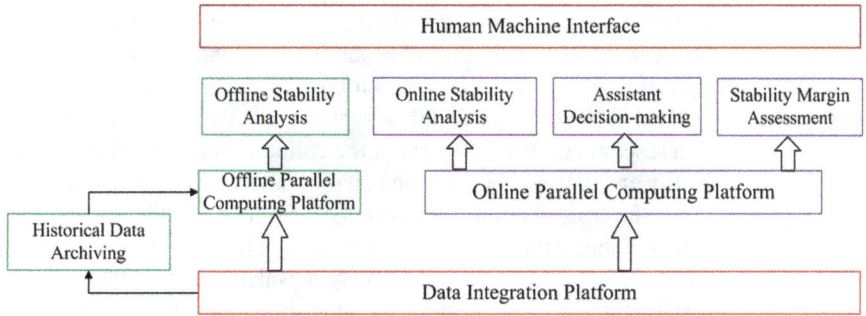

Fig. 11.12 Overall software architecture of the online stability assessment system

1. Data integration platform. By obtaining online power flow snapshots (including 1000 kv and some key 500 kV real-time grid data from national dispatching center, and 500 kV and 220 kV real-time grid data from regional dispatching center) from EMS and WAMS, together with necessary models and parameters, the online data integration platform forms a complete basic data for subsequent stability assessment. The data provided mainly include the online operation mode data (including online topology, online power flow snapshots) and component model parameters. The online operation mode data varies with the real-time operation state of the power system, and the component parameters also change with the development of the power system.
2. Parallel computing platform. By adopting cluster-based parallel computing technology, the parallel computing platform uses the computing task pre-allocation and parallel task computing, and achieves the calculation task scheduling and calculation result collection via UDP (User Datagram Protocol) multicast technique. Huge amount of computation jobs, such as stability analysis and early warning, assistant decision-making, and stability margin assessment, can be completed very quickly. Being the core part of the online stability assessment system, the parallel computing platform fulfills the processing of relevant grid data, the scheduling and execution of computing tasks, and the collection of the calculation results. The parallel computing platform can be further divided into online parallel computing platform and offline parallel computing platform according to the service objects.
3. Stability analysis and early warning. Based on the real-time data and dynamic information of the power grid, the online stability analysis is performed at a given time interval (e.g., 5–15 mins), including the evaluation of the stability for large-scale renewable energy resource connected to the power grid, and thus the secure and stable operation of the power grid is guaranteed. The most important job of the online stability analysis and early warning subsystem is to provide a comprehensive stability analysis and early warning by adopting a variety of analysis methods and integrating the offline stability analysis tool.
4. Assistant decision-making. For the unstable cases found in the stability analysis, this subsystem first selects the risk sets according to the online stability analysis conclusion, and then through the system linearization calculates the correlation coefficients between the system's adjustable variables and the system risk level. Finally, through the sorting of the correlation coefficients and calculation, the system components to be regulated and their regulation range are determined.
5. Stability margin assessment. With the help of the online system's powerful computation capability and data resources, based on the stability analysis results, this subsystem adjusts the cross-sectional power transfer by changing the distribution of power generation, under the premise of the overall balance of generation and load. The adjustment is based on the online margin evaluation algorithm, which increases power transfer on one or more sections simultaneously, while keeps the power transfer on the section under study control at the initial value or the specified value. The added power imbalance will be mitigated by configuring some slack buses. Finally, in the satisfaction of transient stability, voltage

stability, and N-1 thermal stability, the section power transfer limit is derived through the online stability margin search program. The results about power transfer limit are summarized together, and then through assessment to find out the cross-sectional power transfer margin. Transfer margin is a most direct index of the transmission capacity of the system, and can provide important guidance for practical grid operation, when some power grid component failure or unplanned shutdown occurs.
6. Human–machine interface. By using 2D and 3D visualization technology, this subsystem offers visual output, such as graphic and concise presentation about online data, online stability analysis results, early warning messages, assistant decision-making suggestions, and margin assessment conclusions. Besides, this subsystem also provides configuration and monitoring interface for the operation and maintenance of the whole online system.
7. Historical data archiving. This subsystem mainly fulfills the storage of power grid operation data (including grid model parameters) and some of the result data. It also can be accessed by offline stability assessment, human–machine interface, etc. Operations such as database storage, data file storage, data extraction, and data query are executed by this subsystem. It stores the system operation data and assessment conclusions of each time section, establishes the data index, and provides the corresponding stored system data to the offline parallel computing platform for further study.

11.3.4 Hardware Architecture

The major difficulties of hardware structure design for the power system online stability assessment and early warning system lie in the selection of parallel computing clusters and the determination of network types. First, the computing clusters should have enough CPUs or computing cores to meet the requirements of the computing power for a large amount of simulation tasks; Secondly, since the parallel computing of power system simulation relies on the network to transmit data and calculation results, the performance of the communication network needs to be evaluated in detail; the expansion scale and investment of the computing power should be considered from a comprehensive view.

According to the previous research on the task scheduling strategy for parallel computing, parallel computing of power system simulation is fit to take the form of parallel task between servers. The demand for communication network bandwidth during parallel computing is not very harsh, and the general Gigabit Ethernet can meet the requirements. Based on the above considerations, the cluster system consisting of multi-core servers and connected by Gigabit Ethernet is a reasonable choice for the major system hardware.

In order to make the system operation and maintenance simple and convenient, the hardware of the online dynamic security assessment system is chosen to consist of a suitable scale of parallel computing clusters, several human interactive

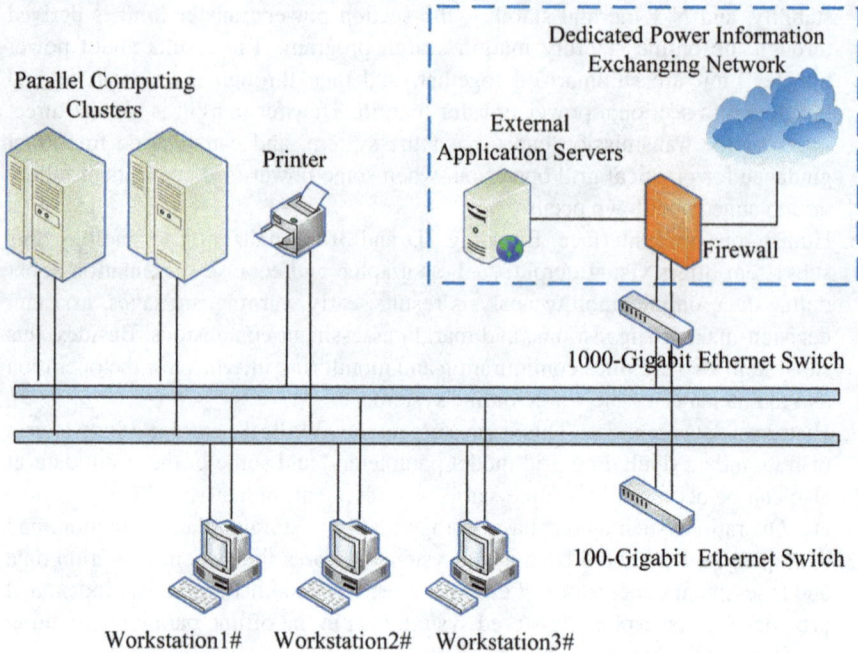

Fig. 11.13 Overall hardware architecture of online stability assessment system

workstations, and external communication network devices. The system should guarantee the required powerful computing power and high-speed data transmission channel. Thus, finally, the hardware architecture for the online dynamic security assessment system is determined, as shown in Fig. 11.13.

1. Parallel computing clusters. Parallel computing clusters consist of a number of application servers, node computers, and high-speed computing network.

    (a) Application servers are further divided into three categories: computation management server, data integration server, and data storage server, where the computation management server is used to manage parallel computing clusters, the data integration server is used for the exchange and integration of online data, and the data storage server is used for data storage and management. To prevent the poor effect of single component failure on the whole system's performance, all these application servers are generally taking the form of dual redundant configuration with standby.
    (b) Node computers have a certain scale of computing cores, forming a powerful computing power for all kinds of online assessment calculation. The lower limit for the number of node computers needs to meet the requirements of 15 min as online assessment interval.
    (c) High-speed computing network is formed by a suitable number of Gigabit Ethernet switches. Similarly, to prevent single component failure, the

high-speed computing network is also taking the form of dual redundant configuration with standby.
2. Human interactive workstations. Human interactive workstations are used for the operation and maintenance of the online system. Besides, offline stability assessment tasks are submitted via the workstations. It is also generally equipped with printer and other auxiliary devices.
3. External communication network devices. The external communication network devices enable communication to other power system applications outside the online assessment system, consisting of network switches and firewalls. For security reasons, external communication network devices generally take the form of hardware firewalls or direct access to information security protective networks.

11.4 Online Stability Assessment Implementation and Case Studies

11.4.1 Brief Introduction of Unified Dispatching and Control System D5000

In order to meet the great needs of the UHV power grid development, the State Grid Corporation launched the smart power dispatching and control system project in 2009, also named as unified D5000 system [7]. Based on unified specifications and standards, over 10 independent application systems of the dispatching center have been horizontally integrated into a new power grid dispatching and control system. Currently, the smart grid dispatching and control system, D5000, consists of one basic platform and four application parts, as shown in Fig. 11.14.

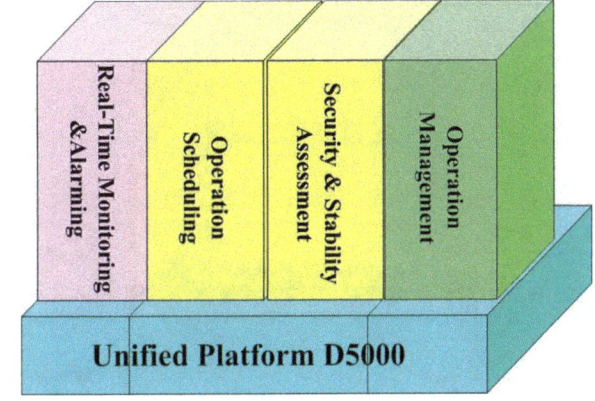

Fig. 11.14 Basic structure of unified dispatching and control system D5000

1. One platform: Unified D5000 Supporting Platform;
2. Four classes of applications:

   (a) Real-Time Monitoring and Alarming, including EMS/SCADA, AGC/AVC, WAMS, etc.;
   (b) Operation Scheduling, such as day-ahead scheduling, hydro/thermal generation schedule;
   (c) Security and Stability Assessment, including the transient stability assessment and voltage stability assessment discussed in this chapter;
   (d) Operation Management, such as operation and maintenance of the whole system.

Till now, the smart power dispatching and control system D5000 has been deployed on all the provincial level and above dispatching and control center, including one national dispatching and control center, six regional dispatching and control centers (North China, Northeast China, Northwest China, Central China, East China, Southwest China), and 27 provincial dispatching and control centers (Beijing, Tianjin, Shanghai, Jiangsu, Zhejiang, etc.).

The unified D5000 system has integrated the online stability assessment application, which exchanges the power grid data and model parameters with the unified D5000 platform. There are two schemes for the execution of online stability assessment, as shown in Fig. 11.15 (Schemes 11.1 and 11.2).

Scheme 11.1 Periodically running, in the fixed period of 15 mins (for a power grid with over 10,000 buses). As in Fig. 11.15, the assessment process will be executed at 9:00, 9:15, 9:30, etc

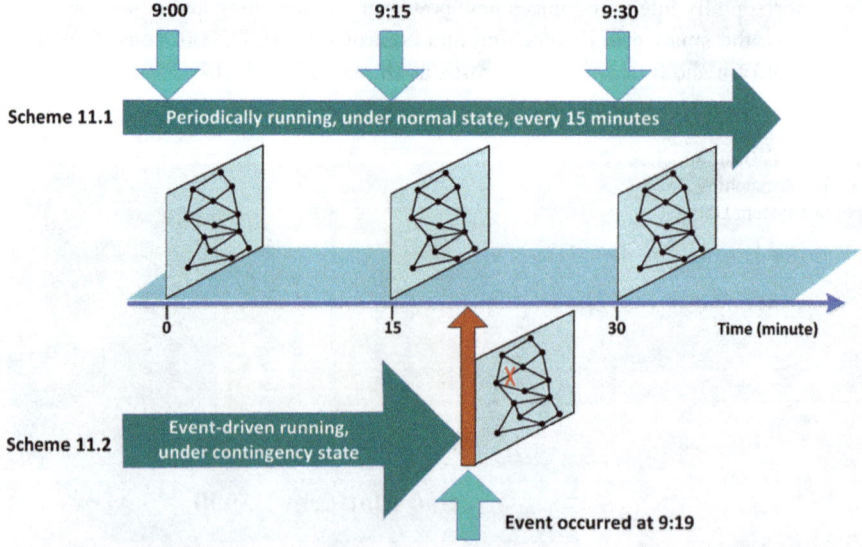

Fig. 11.15 Two schemes for execution of online stability assessment

11 Use of Online Transient Stability and Voltage Stability Assessment Tools... 239

Scheme 11.2 Event-driven running, under conditions when faults on the power grid lead to a real N-2 contingency or other kinds of contingencies. As in Fig. 11.15, when an event occurs at 9:19, the assessment process will be executed immediately

11.4.2 Demonstration of Online Transient Stability Assessment

Under the D5000 system, the online transient stability assessment interface is shown in Fig. 11.16, which is divided into four parts:

1. Description of triggering for the assessment process. Here the triggering type (periodically or event-driven), time, triggering cause, and data availability flag are displayed.
2. Link to transient stability assessment. Buttons displayed here provide links to the different stability analysis result interface, including transient stability assessment, small signal stability assessment, voltage stability assessment, short-circuit current assessment, static security assessment, etc.
3. Dynamic curves of the transient stability analysis result. Three curves, i.e., the maximum angle difference curve with the earliest loss of synchronism time, the voltage curve with voltage nadir, and the frequency curve with frequency nadir, are displayed. The table on the right side lists the summary about the

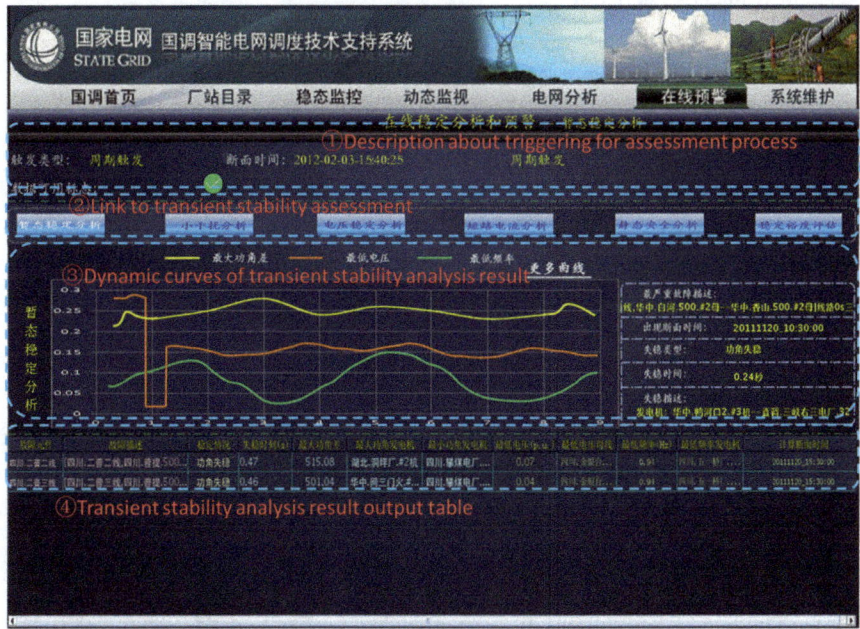

Fig. 11.16 Online transient stability assessment interface

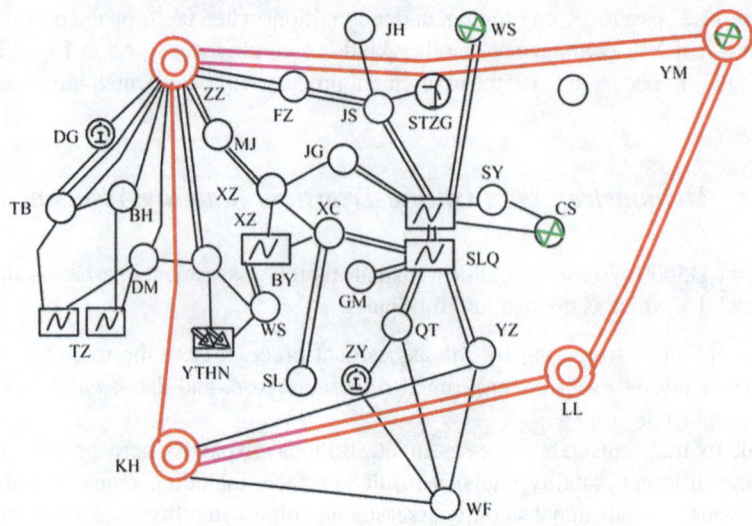

Fig. 11.17 Diagram of ZZ 220 kV area power grid in SD

corresponding transient assessment result. Click on the button "More Curves" to view more dynamic curves.
4. Transient stability analysis result output table. This table lists the output of the stability analysis in detail.

Take the ZZ 220 kV area power grid, part of SD provincial power grid, as an example, with the commission of 500 kV KH substation, the connection between ZZ area power grid and rest of the SD power grid will be strengthened by one 500 kV transformer in KH substation and two 500 kV transformers in ZZ substation, as shown in Fig. 11.17, where YM 500 kV substation and LL 500 kV substation belong to the rest of SD power grid.

On May 20, 2015, at 15:30, according to the maintenance schedule, the 500 kV busbar 1# of ZZ substation would be out-of-service, while at the same time, the generator 8# of SLQ power plant in this area power grid had to be started to support the generation export. At this time, if fault leads to the tripping of 500 kV busbar 2# of ZZ substation, the two 500 kV transformers in ZZ substation would be disconnected from the rest of SD power grid. As the result, the ZZ area power grid would be connected to the rest of SD grid only through the 500 kV transformer 4# in the KH substation, and the generators in SLQ power plant and TZ thermal power plant would have even weaker connections with SD grid. Large risk of transient instability existed in this area power grid under this case.

The deployed online transient assessment system in SD dispatching and control center detected this risk in time. The online system found that the generators in SLQ power plant would lose synchronism with the generators in TZ thermal power plant, and the calculated angle difference curve between generators of these two plants is shown in Fig. 11.18.

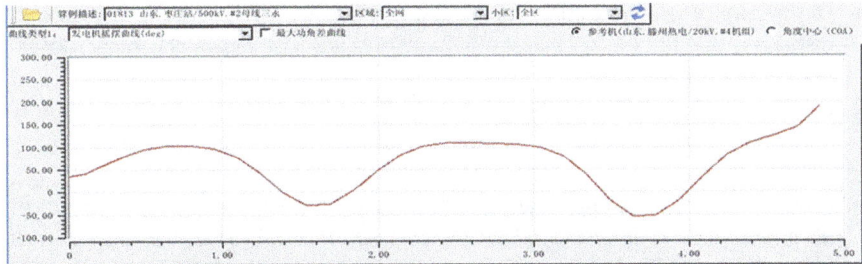

Fig. 11.18 Angle difference curve between SLQ #8

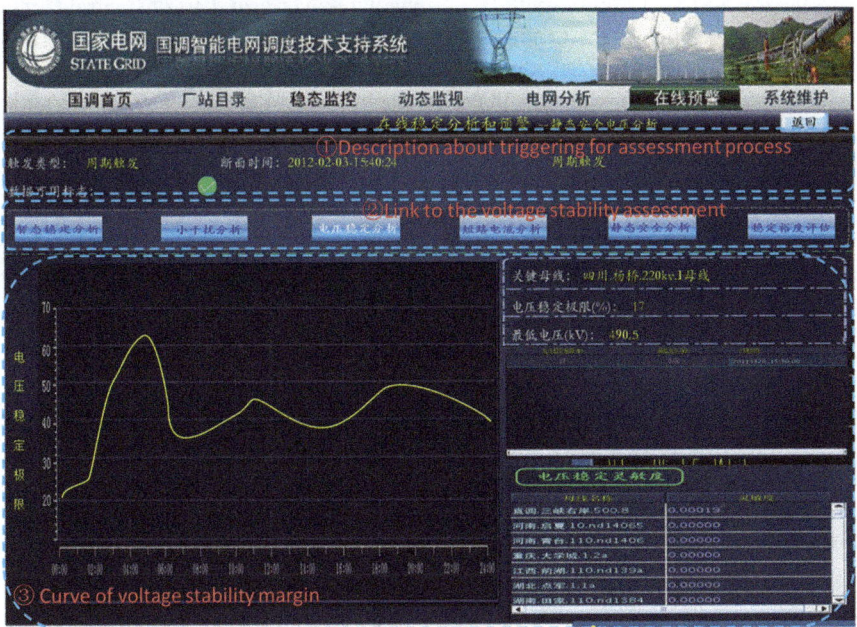

Fig. 11.19 Online voltage stability assessment interface

The online assessment system further issued a warning message and presented the regulation suggestions for the operators. After taking some measures such as decreasing the local generation, the risk of transient instability was eliminated in time.

11.4.3 Demonstration of Online Voltage Stability Assessment

Under the D5000 system, the online voltage stability assessment interface is shown in Fig. 11.19, which is divided into three parts:

1. Description about triggering for assessment process. This part is the same as displayed in Fig. 11.16 since both functions are realized in the same D5000 system.
2. Link to voltage stability assessment. Among the listed buttons, "voltage stability assessment" can provide the complete assessment for the power grid voltage stability.
3. Curve of voltage stability margin. The voltage stability margin in the form of 96 points per day is displayed in this part. On the right side, the most critical bus, voltage stability margin, minimum voltage, and voltage sensitivity result are listed.

Take the 500 kV Central/South HN provincial power grid, part of Central China regional power grid, as the example, and the grid diagram is shown in Fig. 11.20, where ±800 kV Q-S UHVDC is connected at SS converter station in this area power grid. The UHVDC line transmits the electric power from the renewable generation located in the Northwest China regional power grid of SGCC. The transmission section between the central HN and the south HN consists of three 500 kV lines, i.e., line CYP-MF, line CS-SS, and line CS-GT.

On October 19, 2017, at 13:50, line inspectors found there were some problems with the insulators of the 500 kV CS-GT line, and a temporary outage of this line for maintenance was needed. After the outage of line CS-GT, only two 500 kV lines are in operation between the central HN and the south HN, which indicates a weak connection between these two areas.

Based on the conclusion from offline stability assessment, there existed the risk of voltage instability for this area power grid, when a three-phase permanent fault

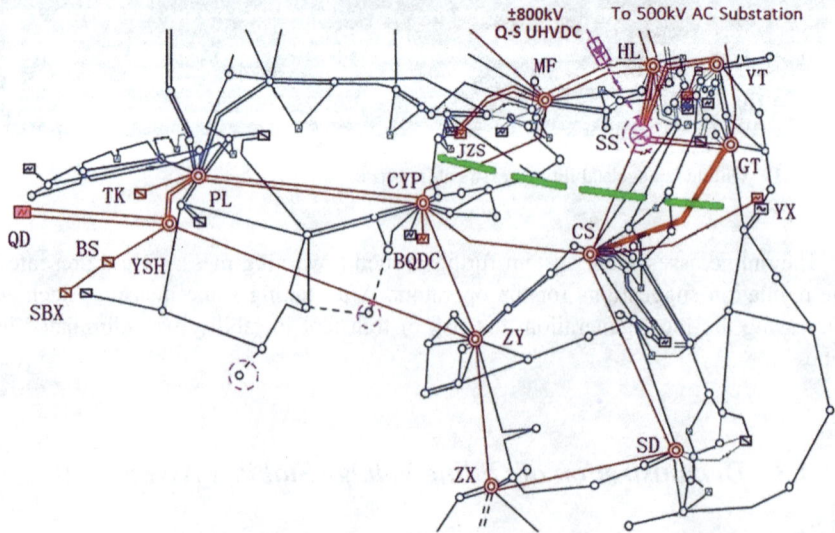

Fig. 11.20 Diagram of Central and South HN area power grid

happened on the HL side of the 500 kV AC-HL line (the 500 kV AC substation is located in the north of HN power grid). As a result, online voltage stability assessment was needed to check the voltage stability under the same contingency and current operating mode with line CS-GT outage.

Since in SGCC, the 500 kV lines are dispatched by the regional dispatching and control center, under this case, the deployed online transient assessment system in Central China dispatching and control center performed the online assessment task. Under the 500 kV line CS-GT outage and three-phase permanent fault on the HL side of the 500 kV AC-HL line, two scenarios for different transmission power over the Q-S UHVDC line, i.e., 1400 MW and 800 MW DC power, were online simulated and analyzed. The transient voltage dynamics of 500 kV buses in the South HN area power grid were output for comparison.

1. Scenario 1: Q-S UHVDC transmission power being 1400 MW

    Given the online power flow and specified UHVDC transmission power, calculate the voltage dynamics of the south HN area power grid, with the applied three-phase permanent fault on the HL side of the 500 kV AC-HL line. The voltages at the 500 kV busbars of ZY substation and PL substation were illustrated in Fig. 11.21. Voltages at other 500 kV busbars of the south HN area power grid were similar to Fig. 11.21.

2. Scenario 2: Q-S UHVDC transmission power being 800 MW

    Given specified UHVDC transmission power, regulate the local generation on the basis of the online power flow and reach the new balance, then calculate the voltage dynamics of south HN area power grid, with the same applied three-phase permanent fault. The voltages at the 500 kV busbars of ZY substation and PL substation are illustrated in Fig. 11.22. Voltages at other 500 kV busbars of the south HN area power grid were similar to Fig. 11.22.

By checking the calculation results for these two scenarios, the voltages at 500 kV busbars of the south HN area power grid would have a large drop with the

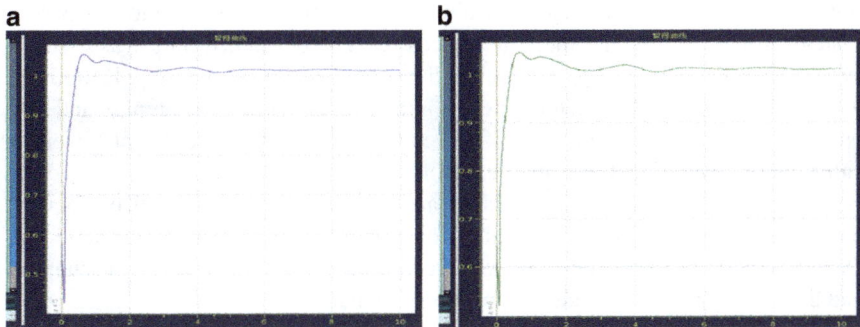

Fig. 11.21 Voltage dynamics of HN area power grid under scenario 1. (**a**) Voltage at 500 kV ZY busbar. (**b**) Voltage at 500 kV PL busbar

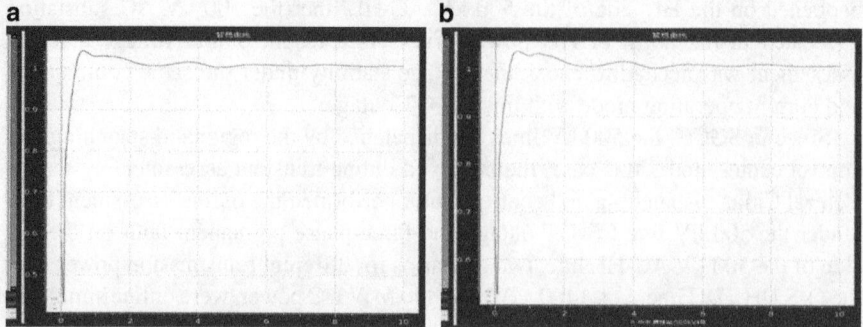

Fig. 11.22 Voltage dynamics of HN area power grid under scenario 2. (**a**) Voltage at 500 kV ZY busbar. (**b**) Voltage at 500 kV PL busbar

applied three-phase permanent fault. When the transmission power on the UHVDC decreased from 1400 to 800 MW, the nadirs of the dynamic voltage curve will increase about 1% upto 2%.

But for both scenarios, the voltages could recover to 1.02 p.u. within 2 s, which indicated that no voltage instability risk for the current operating condition existed. This conclusion was quite different from the offline stability assessment, which used more severe operating mode data to derive even more conservative conclusion.

With the help of an online assessment system, the operator can guarantee the current system operation without reducing power exchange on the tie lines or other control measures as required by the offline assessment suggestion.

11.5 Conclusions and Future Work

In this chapter, a novel online transient stability and voltage stability assessment system is proposed and deployed in SGCC for enhancing situational awareness and stable operation. Principles and algorithms of implementing the online assessment system in real-time environment are discussed. The developed online assessment system has been deployed in the unified dispatching and control system D5000 and has achieved satisfactory performance in improving situational awareness and safe operation of the power grid. The effectiveness of the online assessment system is validated by several field operation examples.

For future work, additional research and development efforts will be focused on developing more advanced applications such as hybrid electromagnetic and electromechanical simulations, modeling complete protection and control schemes, and developing more high-fidelity power grid dynamic models.

References

1. Z. Liu, *Electric power and energy in China*, 1st edn. (Wiley Press, Hoboken, 2013)
2. Z. Liu, *Ultra-high voltage AC/DC grids*, 1st edn. (Academic Press, Cambridge, 2014)
3. Y. Shu, G. Chen, Z. Yu, et al., Characteristic analysis of UHVAC/DC hybrid power grids and construction of power system protection. CSEE J. Power Energy Syst. **3**(4), 325–333 (2017)
4. S. Yinbiao, Z. Zhigang, G. Jianbo, et al., Study on key factors and solution of renewable energy accommodation. Proceedings CSEE **37**(1), 1–8 (2017)
5. L. Shengfu, Z. Ailing, L. Shaohua, et al., Study on transient stability control for wind-thermal-bundled power transmitted by AC/DC system. Power Syst. Protect. Control **43**(1), 108–114 (2015)
6. Z. Chao, M. Shiying, S. Canhui, et al., Transient voltage stability control based on the HVDC Inverter Station acting as dynamic reactive source. Proceedings CSEE **36**(34), 6141–6149 (2014)
7. X. Yaozhong, S. Junjie, Z. Jingyang, et al., Technology development power flows of smart grid dispatching and control system. Autom. Electric Power Syst. **39**(1), 2–8 (2015)
8. P.M. Anderson, A.A. Fouad, *Power System Control and Stability*, 2nd edn. (Wiley-IEEE Press, Hoboken, 2002)
9. J. Arrillaga, C.P. Arnold, *Computer Analysis of Power Systems*, 1st edn. (Wiley Press, Hoboken, 1990)
10. R. Schainker, P. Miller, W. Dubbelday, et al., Real time dynamic security assessment: Fast simulation and modeling applied to emergency outage security of the electric grid. IEEE Power Energy Magazine **4**(1), 51–58 (2006)
11. K.W. Chan, D.P. Brook, R.W. Dunn, A.R. Daniels. Time domain simulation based on-line dynamic stability constraint assessment. International Conference on Electric Utility Deregulation and Restructuring and Power Technologies (2000), pp. 384–389
12. Y. Mansour, Voltage stability of power systems: Concepts, analytical tools and industry experience, IEEE Publication 90TH0358-2-PWR, (1990)
13. CIGRE Task Force 38.02.10, Modeling of voltage collapse including dynamic phenomena. Electra (147):71–77, (1993)
14. W. Ma, J. Thorp, An efficient algorithm to locate all the load flow solutions. IEEE Trans. PWRS **8**(3), 1077–1083 (1993)
15. P. Lof, T. Smed, G. Andersson, et al., Fast calculation of a voltage stability index. IEEE Trans. PWRS **7**(1), 54–64 (1992)

Chapter 12
Online Voltage and Transient Stability Implementation at ISO New England

Omar A. Sanchez, Yuan Li, Xiaochuan Luo, Slava Maslennikov, and Song Zhang

12.1 Introduction of New England Power System

New England's power system is part of the Eastern Interconnection with more than 350 generating plants and over 8000 miles of transmission lines, serving 14 million people across six states in the New England region. It has 12 interconnections to the neighboring systems, including seven AC tie lines and one HVDC to New York, two HVDC ties to Hydro Quebec in Canada, and two AC tie lines to New Brunswick. Figure 12.1 shows the geographic transmission system of the New England power system. The all-time summer peak demand was 28,130 MW on August 2, 2006; the all-time winter peak demand was 22,818 MW on January 15, 2014. ISO New England is responsible for the regional power system planning, reliable operation of the power system, and the efficient operation of the electricity markets.

The New England grid has gone through significant changes in the past decade ranging from new transmission infrastructure, change of resource mixes, and deployment of new technologies for wide area monitoring and situational awareness to support system operations. This trend is expected to continue in years ahead. As of 2019, there are about 31,000 MW of generating capacity for summer and 33,000 MW for winter, of which about 47% use just-in-time natural gas as primary fuel (roughly 15,000 MW). The region's large dependence on natural gas has created energy security issues during the winter season when there is a supply constraint of natural gas for power plants. About 21,000 MW of new generating capacity, mostly wind (65%) are in the interconnection queue and proposed to be built. Since

O. A. Sanchez (✉) · Y. Li · X. Luo · S. Maslennikov · S. Zhang
Real-Time Support, EMS Applications, Business Architecture and Technology, ISO New England, Holyoke, MA, USA
e-mail: osanchez@iso-ne.com; yli@iso-ne.com; xluo@iso-ne.com; smaslennikov@iso-ne.com; sozhang@iso-ne.com

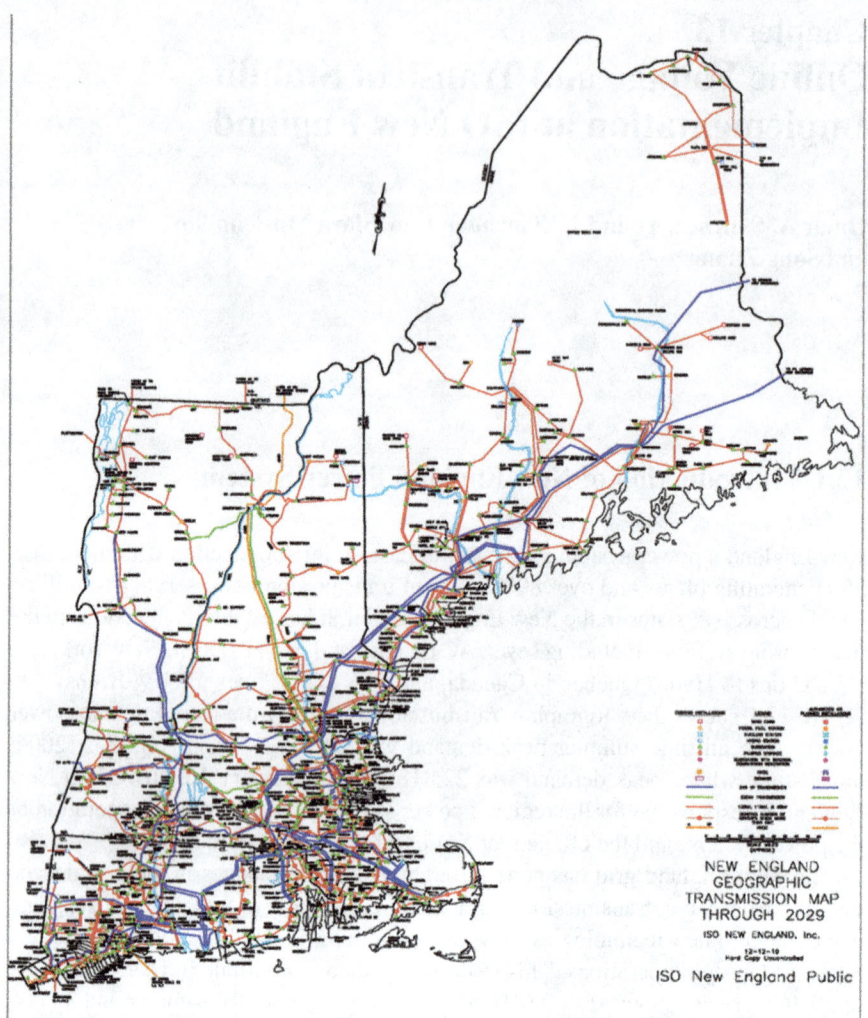

Fig. 12.1 New England geographic transmission system

2013, roughly 7000 MW of generating resources have retired or will retire in the next few years, with another 5000 MW from coal- and oil-fired plants are at risk of retirement in the coming years. In addition, there are about 3100 MW of active demand response (DR) and energy efficiency resources registered in New England, and over 150,000 solar installations totaling about 3400 MW, with most connected "Behind the Meter" (BTM). The total BTM solar will continue to grow to 6700 MW by 2028.

Besides the significant changes in resource mix, the transmission system has also gone through substantial changes as well. From 2003 to 2019, there were about $11 billion invested in the New England transmission system. Additional investment in transmission infrastructure is also planned to meet both reliability and the states' policy directives for renewable energy and decarbonization. FERC Order 1000 also has driven competitive transmission investment to solve longer term reliability needs. The first Order 1000 Request for Proposal (RFP) to solve Boston 2028 reliability needs was issued successfully by ISO-NE at the end of 2019. New transmission technologies such as VSC HVDC, SVC, STACOM, D-VAR, and advanced energy storage devices are deployed all over the New England grid to enhance system reliability.

All these changes to the New England transmission system make maintaining the grid reliability imperative and challenging. Since renewable energy resources are weather-dependent with high temporal and spatial variability, additional transmission investment will help balance the generation mix, relieve transmission constraints, and enhance system reliability even more. With the explosion of transmission enhancements, a large number of planned transmission outages can be expected including tighter maintenance windows for existing transmission elements and new topology that will require conducting numerous operation studies and developing new operating guides.

The large transfer of power across New England and with neighbors continues to stress the system and push the operating point closer to the system limits. Additionally, the combination of resource shifts and new system topology have changed the landscape of the New England power system, and require operating the system in such a way that is less familiar to the system operators. As a result, we have implemented various technologies to help our control room operators and operation engineers assess system reliability by providing them with better tools including online management of system operating limits.

In this chapter, we focus our discussion on the technology deployment of advanced online voltage and transient stability tools for operations support, key issues, and recommendations to improve future installations. The authors envision that the deployment and application of these tools will continue to evolve and include more functionality to assess the state of the power system.

12.2 Online Voltage Stability Assessment

12.2.1 Introduction

Prior to the integration of online voltage stability assessment, ISO New England (ISO-NE) required performing scores of off-line studies using PSS/E to determine transfer voltage limits for huge amount of scenarios to support outage coordination, generation commitment decisions for area reliability, adequacy assessment of local area reserves as well as for the online assessment of area voltage limits.

While this methodology served the needs of ISO operations for many years, the rapid expansion of the transmission system required constant updates to these limits to keep pace with new transmission topology. The whole process became more complex; therefore, ISO-NE decided to find an engineering solution that would meet at least the following objectives:

1. Integrate easily with the GE Energy Management System (EMS)
2. Present an accurate analysis of the real-time system condition
3. Present voltage limits in a way so that operators could design countermeasures

ISO-NE employs currently a combination of homegrown tools and a cutting-edge power system Voltage Stability Assessment Tool (VSAT) interfaced with the ISO EMS. These homegrown tools were developed using off-line parametric studies of the system after collecting large sets of data for many scenarios and then applying mathematical models to estimate voltage limits for a myriad of system conditions. These homegrown tools covered a wide range of outage conditions; however, they are less predictable or accurate when there are multiple layers of forced or planned transmission outages or when new transmission elements are energized before the tool is updated.

In addition to the less sophisticated tools, ISO-NE also employs an online VSAT with a real-time network and solution. The VSAT solution is used by outage coordinators in the off-line mode to evaluate interface voltage constraints and limits caused by existing and expected outages of the transmission system, and in the online mode by operators to identify transmission reliability concerns related to voltage. In New England, many subregions can naturally withstand N-1 and N-1-1 contingencies without additional actions taken. N-1-1 assumes a 30-min interval between contingencies as a planning criterion. For exporting areas, 30-min recovery is inherently achievable since it involves the reduction of generation resources within the area. Many importing-constrained areas, on the other hand, do not inherently have enough resources to provide acceptable recovery; thus, ISO-NE must act to ensure that following the loss of the first major transmission element and within 30 min, the area can withstand the loss of another transmission element. The actions taken by ISO-NE preparing for second contingency requires managing transfers, operating reserves, and transactions with neighboring systems.

Managing the impact of external transactions and making sure sufficient local reserves exist to recover the area from the worst contingency pair is paramount for ISO-NE operators. Operators at the ISO use two-dimensional analysis as a way to understand the impact of power transfers across the system. The analysis also provides actions that would maximize transfer capability to mitigate voltage issues in the system and have sufficient reserves to recover the system following the loss of the worst two contingencies.

12.2.2 2D Voltage Stability Assessment

A two-dimensional or "2D" transfer analysis in VSAT is a type of analysis designed to identify a stability boundary by stressing the system in the space of

Fig. 12.2 Example of nomogram between two interfaces in New England

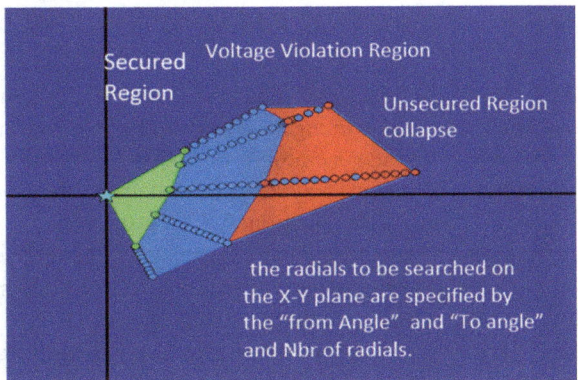

two coordinates [1]. Each coordinate is a scalar along a vector reflecting multiple injection points (generators and loads) in the system stressing. Stability boundary estimation in VSAT consists of the system stressing by using predefined number of directions in 2D transfer space. Each point along the stress direction represents a system state, which is evaluated using contingency analysis. The last secured point from each stress direction forms a boundary point of the secured operating region. Stability boundary is approximated by linear interpolation between calculated points on the boundary.

Figure 12.2 illustrates the 2D stability area/nomogram in coordinates of two interface flows. The green area inside the nomogram represents the secured operating region meaning none of the tested contingencies inside the green region showed any voltage violations; the blue area represents the area wherein the tested contingencies showed voltage violations; and the red area is the area where contingencies showed voltage instability.

In ISO-NE's installation, one of the 2D coordinates is a reportable interface and the other coordinate corresponds to the power flow of the external transaction. External transaction is treated as an independent interface, which could be adjusted to maximize the limit of the reportable interface.

Business requirements for online voltage analysis required the processing, analysis, reporting, and display of voltage limit including 2D results every 6 min. That is challenging for the application performance. Calculation of 2D stability boundary for 100+ contingencies is the most computationally intensive scenario. Computational time for a given set of contingencies depends mainly on the number of stress directions.

Minimization of computational time was achieved as a compromise between the appropriate number of stressing directions and the accuracy of the estimated stability boundary. The used assumption here is that the stability boundary between calculated points is convex (at least not significantly concave) and the linear interpolation between calculated points at the boundary can be used as a reasonable approximation of the stability boundary.

12.2.3 Process Setup and Integration with EMS

As discussed before, some subareas of ISO-NE system can be limited by voltage conditions post-N-2 contingency. Depending on operating conditions, the limitation can happen after the generator + line (G + L) or line + line (L + L) contingency. The subarea is protected by keeping the reportable interface flow below the limit, which is calculated via the online VSAT software.

To meet all the operational requirements, the online voltage stability tool was setup with five scenarios each of them looking to determine the reportable interface limit for the following contingency categories: N-1, G + L, L + L, 2D G + L, and 2D L + L. The reportable interface limits correspond to the steady-state post-contingency conditions after automatic control actions of tap changers, phase shifters, and switched shunts.

Numerical robustness of the process and the efficiency of calculated limit critically depend on the scenario setup including the selection of source and sink, monitoring conditions, system stressing, available MW capacity to reach the limit, and realistically the available capacity to compensate for a generation lost due to a contingency. Available MW capacity from generating resources in real-time case could be insufficient to reach the interface limit; therefore, a work-around to this issue is to use a combination of generation and load in the source/sink definitions to increase or decrease transfers to avoid pushing the solution beyond the boundaries of the power flow convergence region.

The generation lost in contingency is compensated through AGC mechanism in VSAT. We extended the dispatchable range of assigned AGC generators to avoid numerical problems. Small generators and particularly having very limited reactive power capability for voltage control were excluded from the source. Elimination of those resources from the source improved the numerical robustness of the process by avoiding earlier termination of the transfer analysis due to local voltage collapse stemming from the lack of voltage support by small generators.

Excluding radially connected network elements not related to transfers from monitoring is also important to the robustness of the process. The radial connection can exist in the base case or be created as the result of contingencies.

12.2.4 Online VSAT Architecture

1. Architecture Design

    Figure 12.3 illustrates the system configuration and data flows between the real-time applications on the basis of GE *eterra*-platform EMS system. The closed data flow loop is driven by RTVSA, which takes the valid network topology data from state estimation solution (RTNET) and contingency definition (RTCA). RTVSA then generates the input files and saves them in the shared drive for the DSA Manager to retrieve and trigger the VSAT engine on the

Windows server. Once the VSAT computation is completed, the DSA Manager consolidates the results from all computation servers. RTVSA retrieves the result file and conducts post-processing to feed the data for display and any downstream applications such as Double Contingency Analysis (DOUBLC) used by operations for calculating double contingency local reserve requirements. For the study mode, the data flow loop is the same as the real-time application, except that Study Network (STNET) application is the driver and previously developed study saved cases are used.

On VSAT server side, DSA Manager of Powertech Labs is chosen over standard VSAT release version even though the core engine is identical. DSA Manager allows more flexibility to include ISO-NE customizations. In addition, it does not require extra modeling from the EMS side. Generally speaking, the update of VSAT software package and EMS system upgrade are independent of each other. To take advantage of the powerful hardware of the VSAT server, the DSA Manager RT (Real-time) and ST (study) are configured to process the computation requests from RTVSA and STNET, respectively.

2. Input

The input data file set as seen in Fig. 12.3 consists of predefined modeling data and real-time system data generated from real-time EMS application RTVSA. The predefined modeling data is generated based on the existing network model and configured on the VSAT server to support each defined scenario. The scenarios define the study contingency set, resources participating as source and sink, contingency compensation definition, power flow solution and bus monitoring, and voltage criteria.

The real-time system data includes a raw data file in bus-branch format generated from GE's NETMOM EMS database platform and a node-breaker data

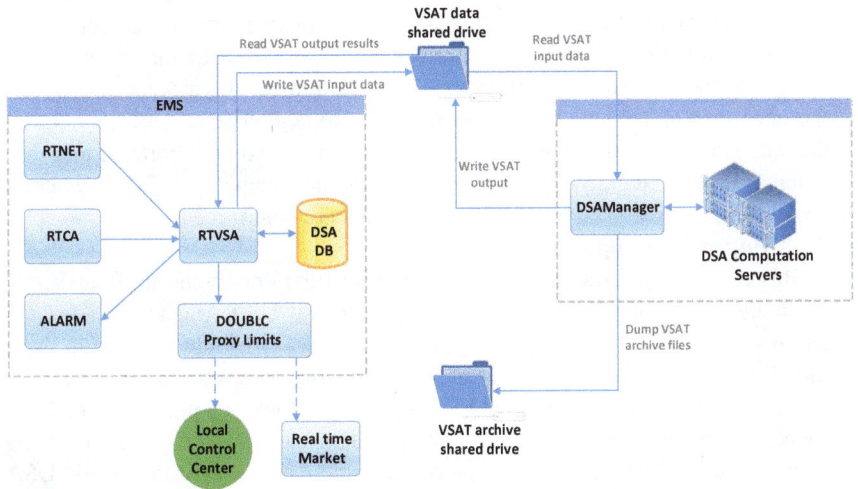

Fig. 12.3 System diagram for RTVSA (real-time voltage stability assessment)

file in XML format containing the mapping of all the breakers at each station. These two files when combined provide all the details about the network as if we were processing the real-time system network. This novel approach to process real-time network data was first developed and implemented in ISO-NE as a way to handle the contingency breaker definition in the EMS model and solution of the power flow using the bus-branch model-based VSAT engine.

Since the study assumptions are driven by engineering requirements, some standard input files from the EMS for the VSAT solution are replaced by model modification files that contain these requirements. These engineering model modification files are applied in DSA Manager and include the redefinition of voltage monitoring zone and the control status of shunt capacitors.

The implemented architecture of the online voltage stability tool was designed to protect the VSAT input files by leaving them outside the common user interface from the EMS. This way data changes can be managed and controlled. Since the input files reside outside the EMS and in the DSA Manager, any changes to the contingency list driven by topology are handled by a customized feature provided by Powertech Labs that allow contingencies in the contingency list to be deactivated when the study user or operator deems it necessary. The activation control from the EMS Contingency Analysis tool is able to disable any N-1 contingency and that in turn results in any N-2 contingency containing this N-1 contingency to be deactivated in VSAT.

In order to minimize and improve maintenance of N-2 contingency definition for VSAT configuration files, a new approach was designed at ISO-NE that allows the creation of N-2 contingency definition automatically based on N-1 definitions from EMS and a list of N-2 names consisting of pairs of N-1 names.

3. Output and Post-processing

As it is explained in the previous section, in addition to the conventional limit result file for the 1D scenario, the results from the 2D scenario are parsed and interpolated by the ISO EMS to display the relationship between the reportable interface limit and independent interface flow. Armed with this information, operators are able to select from the interpolation results a value that meets the operating objective without violating any voltage criteria. Once a value of the independent interface flow is selected, the corresponding reportable interface voltage limit is taken by a downstream application named DOUBLC where a proxy limit and the local reserved requirements are determined.

4. Archiving and Backup System

Based on the strict archiving requirements of the operations staff, all VSAT result files (zip format) generated from production DSA Manager servers are retained for up to 3 years. EMS script is triggered to periodically backup these files for troubleshooting and for future engineering postmortem investigation. Moreover, as the model data and real-time data files are processed in DSA Manager independently, the parallel DSA study system is configured, i.e., using EMS script to transmit another copy of real-time data file set to the study DSA server. Then with different settings or scenario definitions on the study DSA server, operation's study users are able to tune the cases with different solution

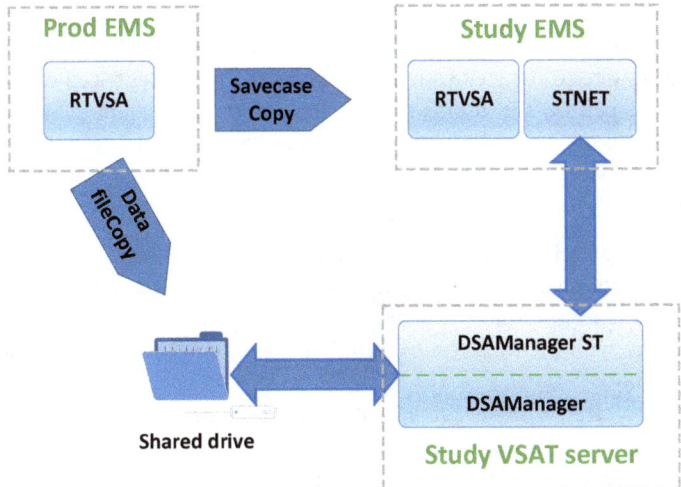

Fig. 12.4 Data archiving and parallel setup

parameter settings or assess other possible transfer scenarios without touching the existing production systems. This process is illustrated in Fig. 12.4.

5. Alarming

   Considering the complexity of the system architecture in addition to the purpose of situational awareness, different categories of alarms were created to monitor the healthy status of RTVSA application in EMS system and DSA Manager application on VSAT servers. For system integration abnormalities such as network outage or server out of service, EMS support engineers are notified. While for solution related issues, there are corresponding alarms issued to help identify the causes so that operators and support engineers take timely corrective actions.

   In the event of a VSAT failed solution or if the calculation process hangs up, the RTVSA application is not affected because it is triggered periodically outside the EMS real-time sequence.

6. Performance Enhancement

   Powerful CPUs having multiple cores and hyper-threading are being used as production servers to handle the computing needs of VSAT due to the multiple stressing directions in the 2D scenarios. It takes quite some time for all the RTVSA scenarios to complete. Through some scalability studies suggested by Powertech Labs, the optimal number of servers allocated by DSA Manager is a function of nature of transfer analysis, number of contingencies, communication cost, memory usage, etc. Testing with actual implemented software and hardware showed that the optimal number of cores for the number of defined real-time scenarios was between 10 and 15.

   For study applications, the ISO EMS Applications Staff setup four study DSA Study (ST) servers to serve more than 20 study users. To take advantage of the hardware resources and save maintenance effort, the study users are grouped according to the nature of their case studies. Each group of study users is then assigned to one of the studies DSA ST servers. Therefore, based on their EMS application permit, the users' request for VSAT calculation is automatically dispatched to the corresponding servers to shorten the waiting time period and avoid unnecessary waste of resources.

7. Future Work

- By parsing each result files archived from the continuous VSAT runs, we are able to obtain voltage stability trends for all operating conditions throughout the year. In the future, we may take advantage of the interface in the DSA Manager with the PI server to further analyze limit trends and conduct more research.
- Currently, RTVSA is triggered either manually or periodically with a predefined cycle. To make it more intelligent and efficient, a flexible triggering scheme depending on various operation conditions will be more desirable such as the system load level or the flow direction along the transfer path, or a combination of criteria.

12.2.5 How are the Results Used by ISO-NE Operations

ISO-NE employs the voltage stability results from the online voltage stability tool to determine a proxy limit. A proxy limit is a VSAT calculated N-2 limit plus area's 30 min response between contingencies including fast response units and spinning reserves. This approach enables ISO-NE to ensure that enough resources exist within importing areas to successfully recover the system from the first contingency and guarantee appropriate recovery after the second contingency.

The online voltage stability tool is also critical in the assessment of how scheduled transactions across the system's inter-ties could affect the interface operating limits. By anticipating changes in the interface operating limit due to wheeling of power across the system, operators are able to determine appropriate corrective actions to maintain appropriate level of reliability and satisfy the area's voltage requirements.

12.2.6 Recommendation for Future Improvement

Significant statistics of VSAT results from various power system operating conditions have confirmed the critical importance of careful scenario setup in order to get feasible and realistic results all the time. Sudden and significant jumps of interface limit over time without any reasonable physical explanation may happen due to numerical issues and unexpected monitoring conditions. These jumps in the transfer limit must be investigated and dealt with in order to keep improving the numerical solution.

Selecting realistic source/sink systems for stressing will avoid numerical solution issues, improve robustness, and ensure that sufficient MW capacity exists to reach interface limits during stressing. In addition to this, a radial search algorithm needs to consider typical as well as atypical system conditions in which a bus becomes radial due to either planned outage or contingency.

Since a single limit value could not be used to address all business requirements, we could not use a standard VSAT-EMS integration architecture provided by the vendor. We had to implement a custom solution where results of all five scenarios are sent back to EMS and a GUI, supporting a final operator's decision, was implemented as a custom addition to EMS. Such a customized solution works well for one interface, but unfortunately cannot be scaled for other interfaces without additional work of EMS vendor. It would be desirable to implement such treatment of results of multiple scenarios outside of EMS to avoid customization in EMS.

Current VSAT solution cannot account for remedial actions between N-1-1. ISO-NE uses so-called N-2 proxy limit in dispatch, which is an approximation of N-1-1 limit. Both contingencies in "N-2 proxy" limit are applied simultaneously and the impact of expected/feasible remedial actions between contingencies in the assumed 30-min period are accounted as MW adder to the limit. More efficient solution can be obtained by adding a capability of N-1-1 calculation into VSAT. Another option for accounting the impact of remedial actions would be to engage SPS functionality in VSAT to model the remedial actions such as generation re-dispatch, bringing online available fast start resources, load shedding, and cutting wheel-through transactions.

12.3 Developing Stability Transmission Operating Guides (TOGS)

ISO New England establishes stability limits from off-line analysis for selected defined interfaces and captures the limits in written instructions for operators. These stability limits are represented in a base plus adders limit structure in which the adders are listed as related to specific transmission elements, dynamic devices, or generating units known to help stability performance if they are online.

For some interfaces, the stability limits are only required during "Facility Out" conditions. These "Facility Out" stability limits are applied during a single planned or forced outage condition. Since it is not practical to anticipate every outage condition or network configuration and then develop a stability guide, combinations of multiple planned or forced outages are not addressed by these facilities out stability guides. When an outage condition arises that is not addressed by an existing stability guide, operations engineering staff is notified to conduct an off-line stability analysis.

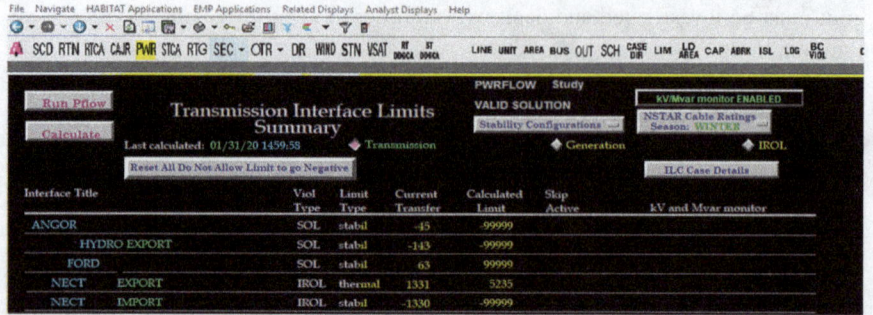

Fig. 12.5 Transmission interface limits in ILC

The stability guides are stored and managed by the Operations Document Management System (ODMS). The ODMS contains both confidential and nonconfidential information of the power system and the guides are searchable by the facility name.

12.3.1 How Stability Guides are Used

Stability Guides at ISO New England are used to limit interface transfers to prevent the potential violation of a System Operating Limit (SOL) or an interconnected Operating Limit (IROL). These stability limits are programmed into a customized application in the ISO EMS named Interface Limit Calculator (ILC) application, as shown in Fig. 12.5.

The application monitors every defined interface flow in ILC and compares it against the calculated limit from either thermal, voltage, or stability.

Alternatively, operators may use the ODMS where all the transmission operating guides are accessible to see a full paper version of a stability guide to identify the critical factors that affect a particular stability limit. Depending on the status of surrounding facilities, operators may opt to change the configuration to gain modest increases in the stability limit. For instance, if the limiting contingency is a stuck breaker contingency, operators may decide to open the breaker to eliminate the most limiting contingency. The stability guide will then indicate the next limiting contingency and the new stability limit.

12.3.2 Future Work to Formalize Transmission Operating Guides (TOGs)

ISO-NE creates Transmission Operating Guides (TOGs) in the form of text documents designed for operators; however, they do not follow a standardized EMS naming convention and sometimes require additional clarifications to avoid ambiguous interpretation. As a result, existing TOGs are not suitable for automated processes; thus, it requires manual and tedious procedures in order to make TOGs usable for other down-streaming applications such as ILC, TARA, and PowerWorld.

For example, manual programming of TOGs into ILC is required for the utilization of TOGs in EMS. Any use of TOGs outside of EMS, such as outage coordination study, requires manual efforts to select interface limits corresponding to specific operating and topological conditions. With the total number of TOGs close to 300, manual selection of interface limits is a tedious and prone to human error process. Searching ODMS to find the right guides for specific operating conditions require some efforts even for a well-trained and experienced person.

The business needs in TOGs will continue in the future before they could be completely replaced by online stability assessment as described in Sect. 4. Additionally, with the increased penetration of renewables and higher uncertainty and variability in operating conditions, we may need to update TOGs more often in the future, which will aggravate the issues discussed above. TOGs are very useful instruments; however, the nonstructured nature of text documents is the main limiting factor for the efficient use of TOGs.

ISO-NE has developed a methodology for a formalized representation of TOGs and a pilot infrastructure for the use of digitized TOGs. Digital TOG is a standardized structure, which contains a formalized description of all TOG related information including logical conditions, interface limit values, and metadata for version tracking and automated search of information. A standardized TOG structure enables the following benefits:

- Automated and unique interpretation and selection of interface limit value for any given power system operating state. Prevention of ambiguous interpretation or misinterpretation of the logical conditions and the interface limit selection procedure.
- Automated use of TOGs in all downstreaming processes.
- Reduction of new TOG deployment from days to minutes and automated use of TOGs in EMS and for external to EMS applications.
- A user-friendly process to create a digital TOG including automated testing, preserving integrity of the entire TOG repository, automated tracking changes related to the modification of EMS model.

Upon testing the pilot infrastructure, ISO-NE will develop a production grade infrastructure for the use of formalized TOGs.

12.4 Online Transient Stability Assessment

12.4.1 Introduction

Unlike other advanced real-time applications such as N-1 thermal, N-1, and 2D voltage stability assessment, which are tightly integrated with EMS and have created a synergy with the latter, the online TSA has not been widely implemented by utilities as of now. Some published reports are extensively focused on a few sporadic exemplary online TSA projects [2–5], ISO-NE's implementation is one of them.

As opposed to off-line dynamic study that uses bus-branch network model and steady power flow solutions, online TSA is based on node-breaker model with dynamic network topology and periodically updated solutions from state estimator in EMS. As such, online TSA can inherit real-time situational awareness from up-to-date EMS network topology at both the transmission and sub-transmission level. The downside of relying on EMS network model is that adequate details are absent for dynamic security studies [4]. For example, in EMS modeling, it is very common to simplify an entire substation network as an aggregated generator and load that directly connects to the high voltage side of a step-up transformer. Furthermore, some sophisticated components, such as High Voltage Direct Current (HVDC) and Flexible Alternating Current Transmission System (FACTS) devices, are usually represented by an equivalent generator for power injection in EMS. These modeling approaches may be sufficient for State Estimation solution; however, if we simply use the EMS network model without appropriately modifying it, the results produced by online DSA will be inaccurate and cannot be directly used by real-time operations.

To perform real-time dynamic security assessment, ISO New England initiated an effort to implement an online TSA about 10 years ago. This homegrown solution is a combination of tools including PowerWorld Simulator for topology processing and external model merge, Powertech Labs Transient Stability Analysis Tool (TSAT) for nonlinear time domain simulation, and DSA Manager for job queueing and scheduling. Details of our implementation are described in the following sections.

12.4.2 Network Modeling

12.4.2.1 EMS Network Model vs. Planning Network Model

Node and breaker are two fundamental components of EMS network model. Different combination of breaker statuses results in various network topologies and power flow distributions. As a result of the variation in system topology during grid operation, the bus numbers are assigned dynamically to nodes and therefore individual components are usually identified by a unique equipment name. In

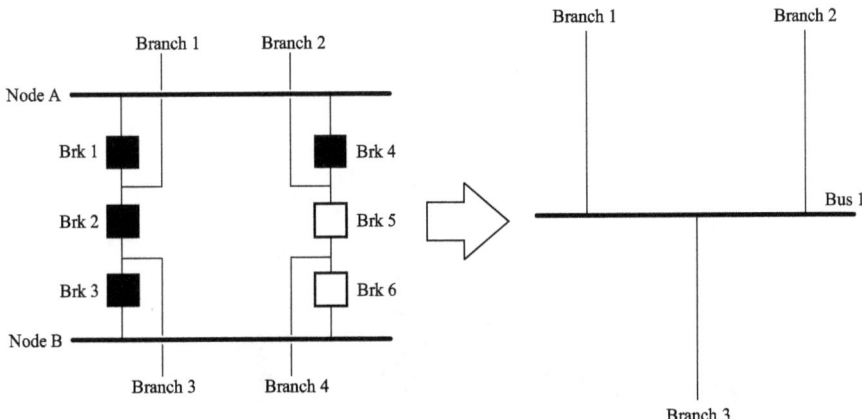

Fig. 12.6 Node-breaker model and bus-branch model

contrast to node-breaker EMS network model, the bus-branch model used in offline planning studies is comprised of buses and branches with breakers consolidated. Fixed bus numbers are usually tied to buses for equipment identification. Figure 12.6 has shown an example of these two types of models and how a node-breaker model is mapped to a bus-branch model. Consolidating a full node-breaker topology into a bus-branch model will improve not only the computational efficiency but also the numerical stability of the solution algorithms.

12.4.2.2 Network Model Mapping and Modification

As mentioned earlier, EMS node-breaker model is not directly usable for TSA without proper modifications. Since the dynamic data is based on planning model, EMS model needs to be modified to represent the system in a similar way as the planning model. The modifications to the EMS network model are primarily fourfold: (1) consolidating the breakers that are not involved in any dynamic contingency definition; (2) mapping individual generator and adding Generator Step-Up (GSU) transformer if not modeled in EMS; (3) splitting the aggregated generator and detailing the electrical connections between the individual unit and the remainder of the network; and (4) replacing equivalent generators that represent HVDC and FACTS with detailed models. Among these four tasks, breaker consolidation requires no human intervention and has been taken care of by a commercial software program, while the remaining modification jobs can be classified into four categories of mapping tasks: One-to-One Generator Mapping, One-to-N Generator Mapping, Special Generator Mapping, and HVDC/FACTS Mapping.

- *One-to-One Generator Mapping*
    This type of mapping is applicable to situations when both the EMS and planning network case have only one individual generator modeled at corre-

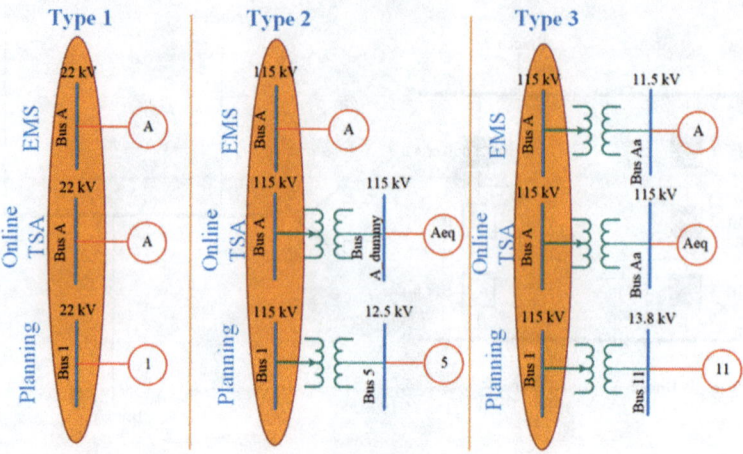

Fig. 12.7 Scenarios of One-to-One Generator Mapping

sponding buses. Depending on where the generator is connected and whether a GSU transformer is modeled, a corresponding GSU transformer is added or modified for online TSA to make the generator connections consistent with that in the planning case. Figure 12.7 has illustrated several One-to-One Mapping scenarios.

- *One-to-N Generator Mapping*

In this type of mapping, a generator in EMS corresponds to multiple generators that are tied to the same substation or in the vicinity of the substation in the planning case. As shown in Fig. 12.8, a "common bus" with connections to all nearby substations is identified first. Then the equivalent GSU transformers and feeders of corresponding generators are added to the EMS network model and attached to this common bus.

The equivalent GSU transformer and feeder impedances are calculated using the following formulas.

$$\text{TR}_{ei} = \sum_j \text{TR}_{ij} \frac{\text{MVA}_i}{100}, \text{TX}_{ei} = \sum_j \text{TX}_{ij} \frac{\text{MVA}_i}{100} \quad (12.1)$$

$$\text{LR}_{ei} = \sum_j \text{LR}_{ij} \frac{\text{MVA}_i}{100}, \text{LX}_{ei} = \sum_j \text{LX}_{ij} \frac{\text{MVA}_i}{100} \quad (12.2)$$

where TR_{ij} and TX_{ij} are the GSU transformer parameters on system MVA base, and LR_{ij} and LX_{ij} are the feeder parameters on system MVA base. Likewise, TR_{ei} and TX_{ei}, LR_{ei} and LX_{ei} are equivalent GSU transformer parameters

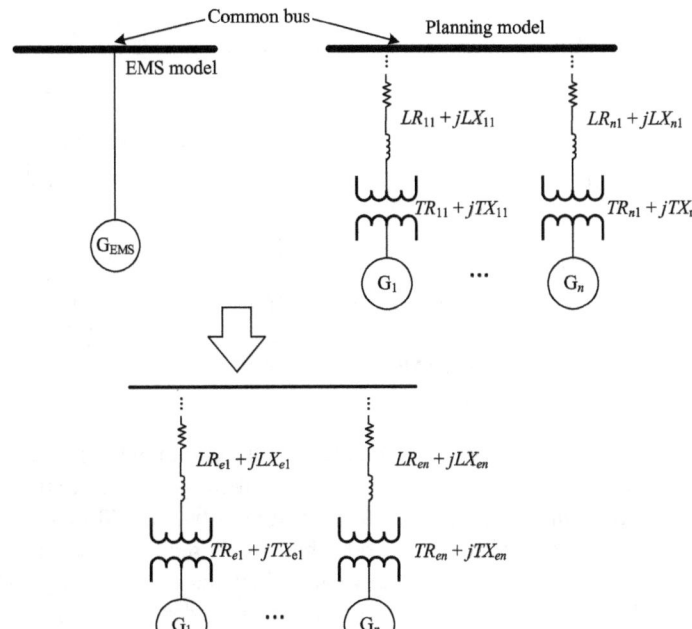

Fig. 12.8 Schematic Diagram of One-to-N Mapping

and feeder parameters used in the online TSA model, respectively. MVA_i is the MVA base of ith generator in planning case.

- *Special Generator Mapping*

A more complicated case in planning model is a group of generators that may be connected to a common bus through a complex meshed network, making it extremely difficult to model them using the One-to-N Mapping method for online TSA case. Retaining the entire substation network in the planning case for online TSA use is impractical because the substations of our particular interest might have very complicated connections. Sometimes they may connect to their neighboring substations via lower voltage networks. Fortunately, as more and more details in network connectivity are modeled in EMS, the number of generators, which need a special mapping approach is greatly reduced. Most of the generators that need this type of mapping are small units.

In special mapping, as shown in Fig. 12.9, the EMS generator will be replaced by the exact cluster of planning generators, while the network between the mapped bus and generator terminal buses are ignored. If there are more than one generator in EMS case, their output is equally distributed to the mapped generators.

As illustrated in Fig. 12.10, a workflow has been developed at ISO New England to determine the mapping type of a given generator. Basically, the approach is using the Breadth-First-Search (BFS) method to loop through all

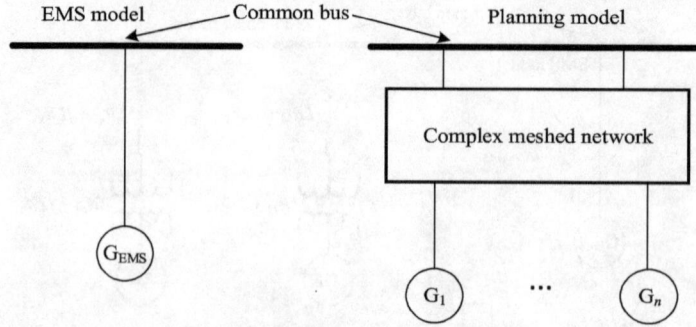

Fig. 12.9 Schematic Diagram of Special Mapping

branches from the terminal bus of a generator to the common bus in the planning case. The common bus, if not specified, is automatically assigned to the first bus identified at the same voltage level as the common bus in EMS case. Starting from the common bus, all searched branches are stored. The special mapping will be used to modify EMS case, if any pair of generators in planning case share a common branch, i.e., any GSU transformer or feeder. Additionally, special mapping is also used if the number of search steps exceeds a cutoff threshold.
- *HVDC/FACTS Mapping*

In addition to mapping for the generators, we also need to modify the equivalent generators in EMS that represent HVDC and FACTS with a certain level of modeling details, such as converters, inverters, AC/DC transformers, switched shunts, and filter banks. These modeling details are necessary to simulate the contingency of HVDC and FACTS, and represent accurately their dynamic behaviors and impacts on the system stability.

12.4.3 Dynamic Modeling

12.4.3.1 Dynamic Model Integration

With appropriate EMS network model modification, we are able to bridge majority of dynamic data in planning case to the online TSA case. Since the dynamic models in planning case are saved in PSS/E .dyr format and most of them are PSS/E standard library models (e.g., GENSAL, GENROU, GENCLS for generators, PSS1A, STAB1 for stabilizers, and EXAC1, EXAC4 for exciters), these models can be recognized by TSAT automatically. The remainder of the dynamic models in planning case are user-written models (USRMDL), which come from equipment manufactures. These user-written models are like black-boxes in TSAT because very little information related to the control system structure or parameters is available. Even though ISO New England will no longer approve any nonstandard library

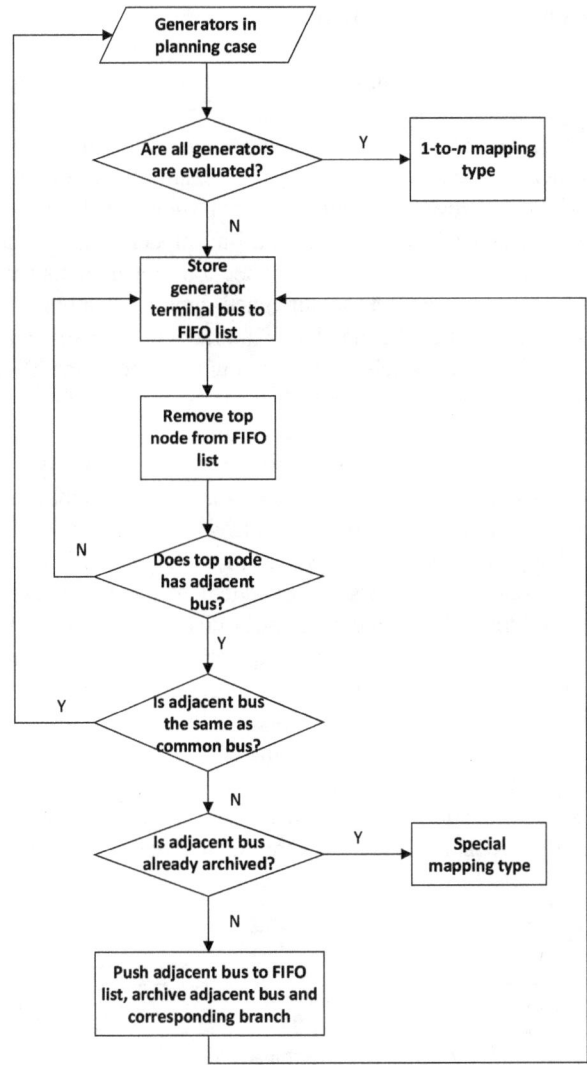

Fig. 12.10 Flow chart of mapping type determination using BFS algorithm

models, there are still dozens of grandfather legacy models currently being used in the planning case. ISO-NE contracted Powertech Labs to convert these nonstandard models into User-Defined Model (UDM) in TSAT and integrate them into the online TSA study. As part of the NERC MOD-26 and MOD-27 process, most of these legacy nonstandard models will eventually be converted to standard models by plant owners through field testing and verification.

12.4.3.2 Dynamic Equivalent of External Systems

EMS snapshot reflects the system conditions in real-time, including most recent network topology, generator outputs, load level, and distributions within its own territory. However, the external system is either simplified or not modeled at all due to limited measurements and visibility. For example, in our EMS we model details in New England, New York, and New Brunswick, but stops at the New York boundary with PJM and IESO with each tie line modeled as a generator. Since New England is part of the Eastern Interconnection, not modeling the remaining interconnected system could affect the simulated dynamic response, especially for inter-area oscillations. On the other hand, it is impractical and inefficient to retain all the external areas' dynamic models as in planning case for online TSA. Therefore, it is necessary to reduce the dynamic models for the external systems while keeping essential properties that may influence the inter-area modes.

To reduce the external dynamic models, we have explored two types of methods. Coherency-based [6] and measurement-based dynamic equivalent [7]. Considering the limited access to interconnection-wide measurements, we have adopted the coherency-based approach built into Powertech Labs Dynamic Reduction Program (DYNRED) to aggregate generators as equivalent machines for our online TSA study. A schematic illustration of the dynamic equivalent is shown in Fig. 12.11.

The details of coherency-based dynamic equivalent can be found in [8, 9].

In addition to the dynamic equivalent, another crucial task is to match the network variables at the boundary buses. Specifically, the bus voltage, phase angle, tie-line MW, and MVAR flows need to be matched before the external equivalent models are merged with the EMS snapshot. To solve this challenging problem, we developed an optimization approach, which re-dispatch the equivalent external generators from DYNRED in such a way that the boundary conditions are matched.

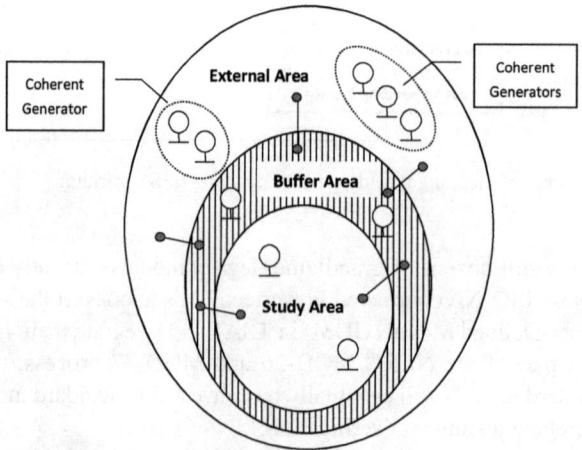

Fig. 12.11 A schematic illustration of dynamic equivalent

12 Online Voltage and Transient Stability Implementation at ISO New England

The formulation is based on a modified DC-OPF as given below.

$$\min w_i \, |P_i - P_{i0}| - \text{minimize MW output change}$$

$$s.t. B\theta = P - \text{network constraint}$$

$$P_i^{\min} \leq P_m - P_{mi}^{\text{ref}} \leq P_i^{\max} - \text{dispatchable units}$$

$$\theta_m^{\text{tol}} \leq \theta_m - \theta_m^{\text{ref}} \leq \theta_m^{\max} - \text{boundary buses}$$

where θ_m^{ref} and P_m^{ref} are the real-time voltage angle at boundary bus and MW tie-line flow in EMS snapshot, respectively. P_{i0} is the initial external generator MW output. w_i is the dispatching weight.

The entire process is executed in PowerWorld through its APIs. At the final step, pseudo phase shifter and shunt devices will be added at the boundary to compensate for any mismatch in terms of voltage phase angle and tie-line MVAR flows. A simplified diagram as shown in Fig. 12.12 has illustrated how the output of the equivalent generators and boundary buses are determined for a specific network snapshot.

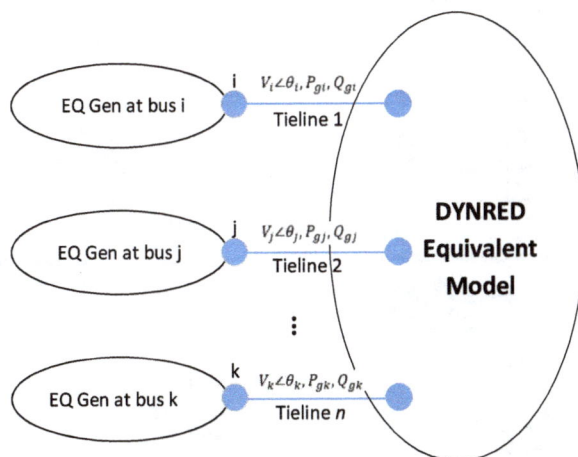

Fig. 12.12 Output of dynamic equivalent generators at the boundary buses

12.4.3.3 PMU-Based Power Plant Model Verification

Thanks to the PMU deployment across the New England region, we are able to leverage synchrophasor data to validate some of our generators' dynamic models per NERC MOD-26/27 standards. We have developed a tool named Automatic Power Plant Model Verification (APPMV), which can automatically perform the model verification once a disturbance is detected online by our synchrophasor system.

Basically, the model verification process can be summarized into three steps [10], as shown in Fig. 12.13: (1) detect online if there are any system disturbances such as generator trip, line trip, or oscillation which has excited generators' dynamics, and then retrieve PMU data from the database; (2) use PMU playback function to inject the PMU measurements of voltage magnitude and voltage angle at the Point Of Interconnection (POI), and then acquire the simulated active power and reactive power outputs from the generator model; and (3) compare the simulated model output, i.e., P and Q with the corresponding PMU measurements to verify the dynamic models. The entire process described above is fully automated. If a generator model shows unacceptable results across multiple events, the ISO will follow-up with the plant owners for further investigation.

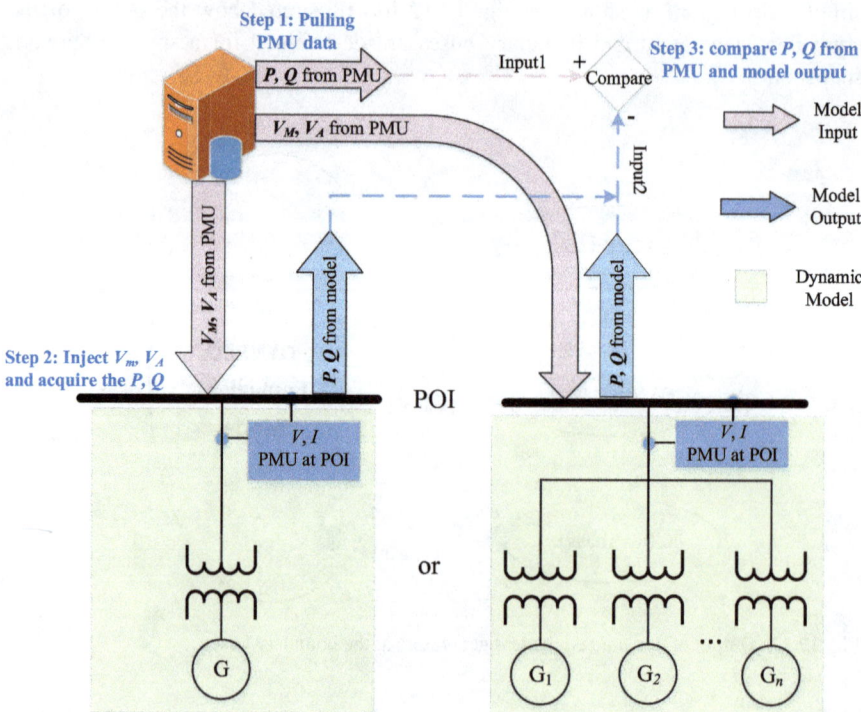

Fig. 12.13 Process of PMU-based power plant model validation

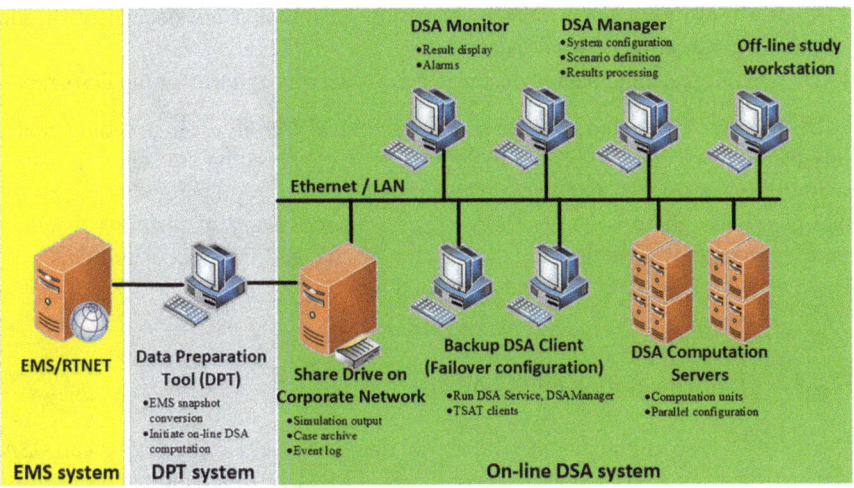

Fig. 12.14 System configuration for online TSA

12.4.4 Online TSA Program

12.4.4.1 System Architecture

Figure 12.14 has illustrated the system configuration for online TSA program at ISO New England.

When a cycle of online TSA is initiated, predefined TSA input files such as contingency, interface, security criteria, and monitoring, along with system real-time conditions from the state estimator are sent to the online TSA program to determine system stability and interface transfer limits. These limits are based on angle instability, transient voltage dip, and undamped oscillations. The real-time TSA cases are also archived and can be retrieved by engineers for post-event analysis. On-call engineers can also simulate the impact of a forced outage or develop remedial actions if there are any stability concerns.

12.4.4.2 Data Preparation and Model Update

The Data Preparation Tool (DPT) was developed in-house by the ISO-NE. It has four major functions.

1. Translate the node-breaker network model in EMS to bus-branch network model used by the TSA
2. Call PowerWorld APIs to solve the OPF problem described before and then merge the external equivalent with the bus-branch EMS case

3. Modify those parameters in wind dynamic models that are dependent on the operating condition of wind machines
4. Transfer all real-time data to designated DSA servers and trigger the DSA run

DPT can run in two modes, *real-time mode* and *off-line mode*. In real-time mode, only the network data is updated in each study cycle based on the state estimation real-time system status. Off-line mode is typically used to prepare other necessary data files when there is a new EMS release, a new library of planning dynamic models, change of generator mappings, new or changed interface definitions, etc. These files are only updated on an as-needed basis.

12.4.4.3 Implementation and Configuration

The online TSA was implemented on virtual machines including a hot/backup DSA Manager, five DSA computation servers, and several off-line study workstations. The interval between each study cycle is currently set at 20 mins considering that system stability will not change in a relatively short period of time. The DSA process can also be triggered by users on-demand.

Our current DSA scenarios include one base case and three transfer analyses. Each transfer analysis is used to compute stability-based interface limits. The computation process takes about 3 min with most of the computing time spent on transfer analysis. We are in the process to add more scenarios of transfer analysis so that all stability interface limits in New England can be computed online by TSA. As shown in Fig. 12.9, the DSA architecture is very flexible. We can easily add more computation servers if time of completion is a concern in the future.

References

1. M. Papic, M.Y Vaiman, M.M Vaiman, *A New Approach to Constructing Seasonal Nomograms in Planning and Operations Environments* (Idaho Power Co.)
2. L. Wang, K. Morison, Implementation of online security assessment. IEEE Power Energy Magazine **3**(5), 46–59 (2006)
3. J. Viikinsalo, A. Martin, K. Morison, L. Wang, F. Howell, Transient security assessment in real-time at southern company, in *IEEE PES Power Systems Conference and Exposition (PSCE' 06)*, October 29–November 1, 2006, pp. 13–17
4. J. Jardim, N. Carlos, M.G. Santos, Brazilian system operator online security assessment system, in *IEEE PES Power Systems Conference and Exposition (PSCE' 06)*, October 29–November 1, 2006, pp. 7–12
5. S.J. Boroczky, G. Ellis, Real-time transient security assessment in Australia at NEMMCO, in *IEEE PES Power Systems Conference and Exposition (PSCE'09)*, 15–18 March 2009, pp. 1–6
6. R. Podmore, A comprehensive program for computing coherency-based dynamic equivalents, in *IEEE Power Industry Computer Applications Conference*, Cleveland, OH, May 1979
7. Z. Jiang, X. Zhang, Y. Xue, A.G. Tarditi, Y. Liu, Measurement-based power system dynamic model reduction using ARX equivalents, in *IEEE PES General Meeting*, Atlanta, GA, August 2019

8. L. Wang, G. Zhang, *DYNRED Enhancement Project Final Report*, EPRI, April 2010
9. L. Wang, G. Zhang, *DYNRED Software Manual—Dynamic Reduction*, EPRI, April 2010
10. M. Wu, W. Huang, F.Q. Zhang, X. Luo, S. Maslennikov, E. Litvinov, Power plant model verification at ISO New England, in *2017 IEEE Power and Energy Society General Meeting, PESGM 2017* (Vol. 2018-January, pp. 1–5). IEEE Computer Society (2018). https://doi.org/10.1109/PESGM.2017.8273867

Chapter 13
Stability Assessment at CAISO

Aftab Alam, Ruili Zhao, and Ran Xu

13.1 California Independent System Operator (CAISO)

In 1996, the Federal Energy Regulatory Commission issued Orders Nos. 888 and 889. These orders required utilities that own transmission to provide nondiscriminatory access to all transmission customers. One way for a utility to comply with this requirement was to allow an independent system operator or "ISO" to operate its transmission system. California Independent System Operator (CAISO) [1] is one of the nine independent system operators in North America shown in Fig. 13.1. Collectively, the independent system operators deliver over 2.2 million gigawatt-hours of electricity each year and oversee more than 26,000 miles of high-voltage power lines. Two-thirds of the United States is served by these independent grid operators. ISOs do not own the electricity transmitted over the grid, and they allow market participants to buy, sell, and transmit electricity at the best available price. In 1998, as a result of Order 888 and state legislation (AB 1890), the California ISO was incorporated as a nonprofit public benefit corporation to fulfill this mission.

CAISO is also the largest of the 38 balancing authorities (BA) in the Western Interconnection. As a balancing area, it handles approximately 35% of the electric load in the West and manages about 80% of California and a small part of Nevada, which encompasses all of the investor-owned utility territories and some municipal utility service areas. There are some pockets where local public power companies manage their own transmission systems. A balancing authority is responsible for operating a transmission control area. It matches generation with load and maintains consistent electric frequency of the grid, even during extreme weather conditions or

---

A. Alam (✉) · R. Zhao · R. Xu
Operations Planning, California Independent System Operator, Sacramento, CA, USA
e-mail: aalam@caiso.com; rzhao@caiso.com; rxu@caiso.com

Fig. 13.1 Independent system operators in North America [2]

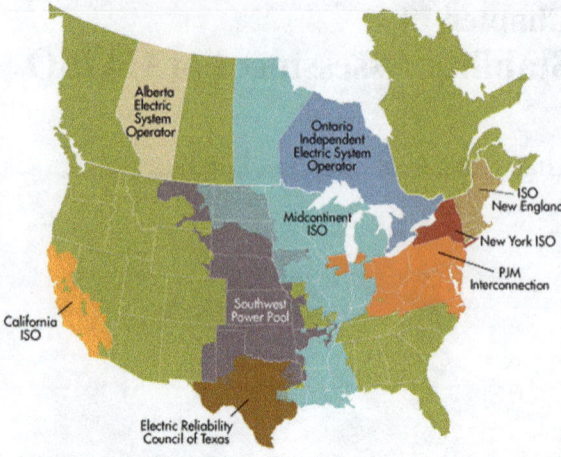

Fig. 13.2 CAISO balancing area footprint [3]

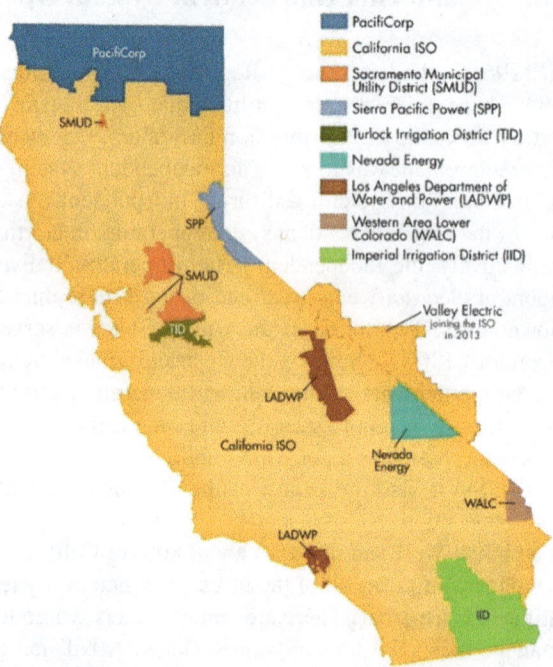

natural disasters. In addition to balancing area responsibilities, CAISO also serves as the Transmission Operator (TOP) for a majority of the same footprint (Fig. 13.2).

In addition to BA and TOPs responsibilities, the ISO's reliability coordinator RC West [3] is the reliability coordinator (RC) of 42 balancing authorities and transmission operators in the western United States. As a reliability coordinator, the ISO provides core reliability coordinator services as required by NERC standards, including outage coordination, day-ahead operational planning analysis, real-time

Fig. 13.3 RC West entities [3]

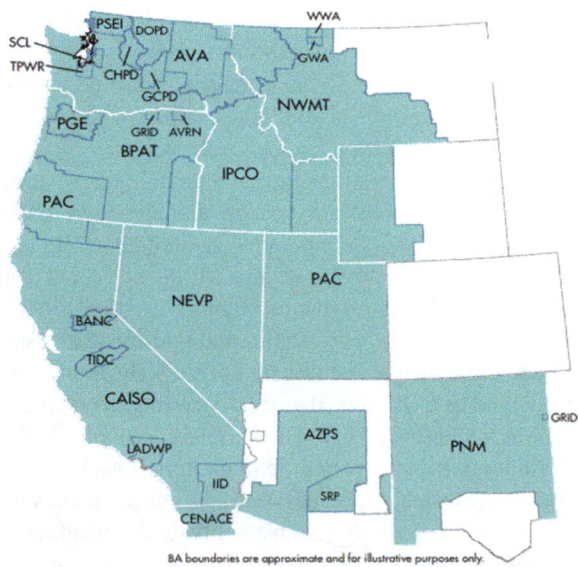

assessment, real-time monitoring and analysis, and system restoration coordination (Fig. 13.3).

13.2 System Operating Limits (SOL) Methodology

As a transmission operator and reliability coordinator, the ISO is required by NERC standards to continually assess and evaluate projected system conditions within the operations horizon with the objective of ensuring acceptable system performance in real-time. These assessments are performed in an iterative fashion, typically as part of the seasonal studies followed by assessments as part of the outage coordination process followed by day-ahead operational analysis and ultimately concluding with real-time assessments. The methodology utilized by a transmission operator or reliability coordinator in establishing the various system operating limits is commonly known as the system operating limits methodology.

The main purpose of the establishment of system operating limits is to allow operation of the bulk electric system so that acceptable performance is achieved in both the pre- and post-contingency states. Pre-contingency conditions are those that exist normally prior to the unforeseen loss of a transmission line or generator or other equipment. Post-contingency conditions are those that exist following the unexpected loss of a transmission line or generator or other equipment. The unexpected loss of transmission equipment is generally referred to over here as contingencies. Typically the scope of contingencies for the operations horizon is limited to credible contingencies. These are a smaller subset of contingencies compared to those evaluated in the transmission planning horizons. The selection of these contingencies is also described in the system operating limit methodologies. Generally, the loss of a single transmission line or generator is treated as always credible and the loss of multiple equipment depends on their proximity to each other, frequency of tripping, and other factors.

The system operating limits methodology [4] requires that the bulk electric system shall demonstrate voltage and dynamic stability in the pre- and post-contingency states among other requirements. In addition, it is required that cascading or uncontrolled separation do not occur following the loss of contingencies. Voltage and dynamic stability are typically ensured by operating within limits calculated through studies and real-time assessments. Operating outside the voltage and dynamic stability limits established through these assessments would be considered as SOL exceedance as operating parameters indicate that a contingency could result in instability. The various forms of stability have a wide spectrum of reliability impacts—from little to no impact such as losing a unit due to instability, all the way to major and devastating impact, such as losing a major portion of the bulk electric system due to instability. Some of these voltage or transient stability concerns that cannot be confined to a localized contained area of the bulk electric system and have a critical impact on the operation of the interconnection warrant the establishment of interconnection reliability operating limits. Any exceedance of interconnection reliability operating limits is required to be mitigated within 30 mins. The RC must ensure that SOLs and Interconnection Reliability Operating Limits (IROLs) for its RC Area are established and that the SOLs and IROLs are consistent with its SOL Methodology. RC West performs real-time monitoring and real-time assessments to determine SOL exceedances and to determine if the system has unexpectedly entered into a single Contingency or credible MC insecure state. It also performs analysis in the operations planning and day-ahead time frames for any revision of the established stability limits due to maintenance outages.

13.3 Real-Time Voltage and Transient Stability Analysis

Real-time and day-ahead voltage and real-time transient stability analyses are conducted to ensure sufficient margins exist between the actual flows and determined limits on transfer interfaces typically associated with voltage or transient stability concerns [5]. Figure 13.4 shows the real-time setup of the various processes involved to enable the calculation of real-time voltage and transient stability analyses. After state estimation is complete, dynamic CIM XML [6] payloads are sent to both Market and real-time voltage stability (RT-VSA) applications. RT-VSA is able to utilize the dynamic CIM XML along with the static CIM XML to run the base powerflow and the subsequent transfer analysis using various scenario definitions for each of the monitored interfaces. The static CIM data for the network model is updated on a monthly basis. One of the outputs of each iteration of RT-VSA is a PSS/E raw file that represents a hybrid bus-branch model equivalent of the full node-

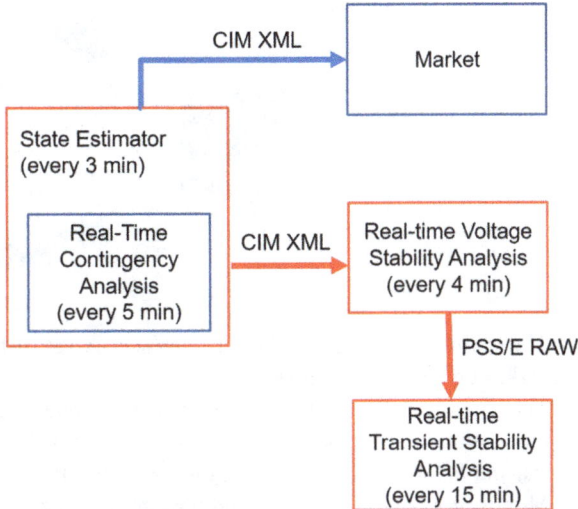

Fig. 13.4 Real-time analyses setup

breaker format of the full network model with some node-breakers representations saved to support implementation of the needed scenarios. This PSS/E raw file is the starting point for performing transient stability analysis to evaluate the impact of critical contingencies on the base powerflow and also calculate transfer limits for critical interfaces.

Multicore parallel processing is utilized to simultaneously process contingencies to allow these analyses to complete in a reasonable amount of time. CAISO's voltage stability analysis [7] is set to process all monitored scenarios in approximately 5 min. The analyses for all interfaces involve the detection of path flow levels at which thermal, absolute voltage, and voltage stability violations are observed. Real-time transient stability is more computationally intensive due to the time required to complete dynamic simulations. Parallel processing on faster servers allows the simultaneous simulation of contingencies and helps to complete CAISO's dynamic stability assessment [8] within 15 min or less. Real-time engineers and operators monitor the results of these analyses. Appropriate actions are taken for monitored interfaces that are associated with operating guidelines.

13.4 Real-Time Voltage Stability Monitoring

Voltage stability monitoring is performed for various interfaces within the CAISO balancing area and also for IROLs within the RC West footprint.

The setup of each monitored interface involves:

1. Defining the sink and source area: The sink and sources are usually defined by a combination of generators and loads

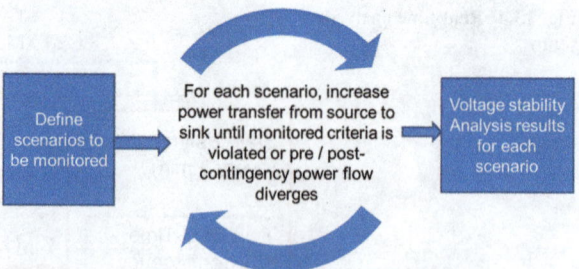

Fig. 13.5 Real-time voltage stability monitoring

2. The definition of the monitored interface: A set of lines/transformers typically makes up a monitored interface
3. The contingencies to be evaluated: Contingencies deemed to be credible for the operations horizon that is known to cause voltage stability concerns as the transfer is increased across the monitored interface are included here
4. Monitoring criteria: The pre- and post-contingency steady-state voltage and thermal concerns are set to be monitored allowing RT-VSA to calculate additional transfer limits in addition to the voltage collapse point. The list of monitored equipment for steady-state voltage and thermal concerns are usually the buses, lines, and transformers.

For each of the monitored interfaces, the source and sink definitions are utilized to increase power transfer across the monitored interfaces and at each level of the transfer, all contingencies are evaluated and assessed for violation of any of the monitored steady-state voltage and thermal criteria and also for voltage collapse concerns (Fig. 13.5).

At the end of the analyses, the following are provided:

1. Base powerflow on the monitored interface: This is the powerflow on the monitored interface prior to any transfer. This typically matches the powerflow on the monitored interface in the input data from state estimation.
2. Maximum power transfer on the monitored interface: Different maximum powerflow transfers are provided that were achieved based on the steady-state monitored criteria and the voltage collapse point. In addition, the maximum power transfer based on the voltage collapse point is also provided.
3. Margin: The difference between the base powerflow conditions on the monitored interface (powerflow in state estimation) and the maximum power transfer on the same interface represents the available margin on the monitored interface
4. Limiting contingency: If post-contingency power conditions limit the transfer, the respective contingency is listed as the limiting contingency
5. Limiting equipment: Depending on where the steady-state pre- or post-contingency voltage or thermal violations might be occurring or where the voltage collapse is occurring, the respective equipment is reported as the limiting equipment.

13.4.1 Utilization of Real-Time Voltage Stability Monitoring Application for Grid Operations

13.4.1.1 Voltage Stability Limit Monitoring

As mentioned the primary purpose of the real-time voltage stability monitoring is to ensure that system reliability is always maintained by ensuring that a positive margin exists across the monitored interfaces with voltage stability concerns. When flows do start approaching the real-time limits or when margins start diminishing, operators follow established protocols in operating procedures to coordinate between various TOPs and BAs to reduce flows and increase margins.

Figure 13.6 shows an example of the variation of a real-time limit provided per unit with reference to the seasonal limit of the interface. It is observed that the limits change gradually through the course of the day but can have sudden drops that can highlight the challenges with the use of state estimation inputs.

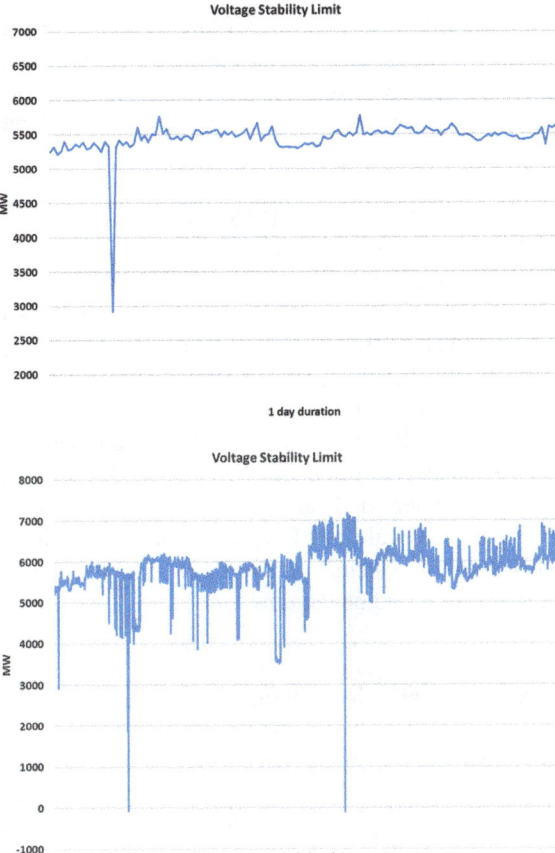

Fig. 13.6 Sample voltage limit plots from real-time voltage stability analysis

Fig. 13.7 Sample voltage limit plots from real-time voltage stability analysis

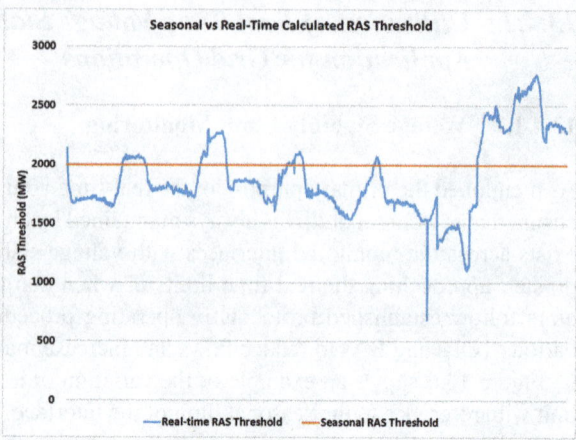

13.4.1.2 Real-Time RAS Threshold Determination

Remedial action schemes (RAS) are utilized for increased power transfer capabilities. One example of a remedial action scheme is where the RAS is utilized to increase power transfers across northern California through the 500 kV backbone. RAS schemes exist to trip generation and load based on the loss of multiple contingencies. The RAS schemes are set to operate after the flows across the major path involving the contingent lines exceed a certain threshold referred to here as the RAS thresholds. These RAS thresholds are typically set after numerous off-line studies using operational planning cases. However, RAS thresholds can vary based on operating conditions. This can lead to RAS arming either too early or even too late, which in turn can lead to thermal concerns appearing in real-time contingency analysis (RTCA) [9]. The heavy influx of solar and wind generation leading to high volatility in generation and flow patterns can also warrant more frequent tuning of such RAS thresholds.

The capability to conduct transfer limit calculations based on the thermal concerns allows to calculate RAS thresholds based on real-time conditions leading to more efficient and reliable utilization of RAS actions. Figure 13.7 shows an example of a real-time RAS threshold calculation compared with the seasonal RAS threshold calculation for a duration of 2 weeks. Significant differences can be observed based on the time of the day.

13.4.1.3 Congestion Management

There may be situations that warrant having situational awareness to know when to bring on long start units in order to be able to mitigate any post-contingency voltage or thermal concerns in areas with limited generation mitigation options. In addition, depending on load conditions, there can be areas where post-contingency concerns frequently fluctuate between low voltage concerns and thermal concerns. Operators may want to bring on units early in order to be ready to mitigate any post-contingency low voltage concerns that need to be mitigated immediately or any thermal concerns that require a long start unit to come on and require proactive monitoring. This can be established by creating an interface between the sink area consisting of the load and long start units and the rest of the system. The transfer analysis can provide the margin available and operator experience can be used to determine at what point the resources should be started.

13.4.1.4 Total Transfer Capability Calculations

The application is designed to determine limits beyond the traditional voltage stability limit based on point of collapse. In addition, the application is designed to provide transfer limits based on thermal and steady-state voltage limit criteria. These criteria typically fall into the following groups:

(a) Pre-contingency thermal loading: This criterion allows all pre-contingency loading on the monitored transmission lines and equipment to be lower than the normal rating
(b) Post-contingency thermal loading: This criterion allows all post-contingency loading on the monitored transmission lines and equipment to be lower than the emergency rating.
(c) Pre-contingency and Post-contingency voltage limit criteria: This criterion allows all pre- and post-contingency voltages to be within the normal and emergency bus voltage limits, respectively.

The ability to limit transfers based on thermal and voltage limit criterion allows the determination of total transfer capability (TTC) calculations for interfaces that are typically limited by pre- or post-contingency thermal concerns. CAISO establishes TTCs for paths operated by CAISO on a seasonal basis. In addition, the TTCs can be impacted by maintenance or forced outages on transmission equipment. Generally, these TTCs are established through off-line studies performed on operational cases from WECC that are tuned for the season for the appropriate load and generation profile. However, real-time conditions are always changing. This can lead to the TTC being over-estimated or under-estimated based on the study assumptions. An over-estimated TTC can lead to thermal violations appearing in the RTCA which is a sign that the transfers over the path effective in mitigating the overload may need to be curtailed. The setup at CAISO allows the tracking of a real-time TTC against the seasonal fixed TTC values to see if an adjustment is necessary based on real-time conditions.

TTCs are typically established using contingencies that are determined to be always credible for normal operations. However, adverse operating conditions sometimes warrant the establishment of TTCs based on contingencies that would normally not be credible during normal operations. For example, the loss of two lines in a common corridor may or may not be considered as a credible multiple

Fig. 13.8 Sample total transfer capability calculations considering normal and adverse conditions

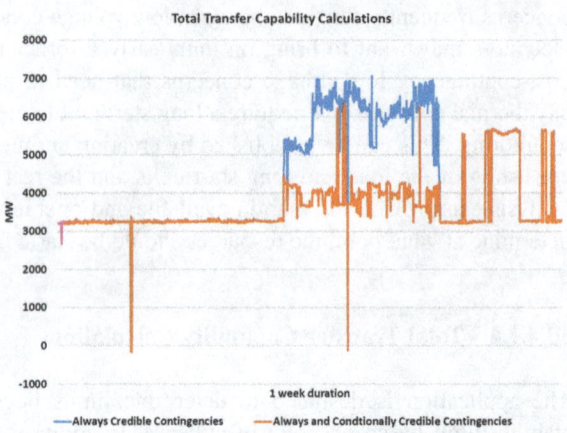

contingency depending on criterion and risk assessment. Real-time adverse conditions such as fire near the common corridor or high windy or storm conditions can warrant the RC or TOP declaring a conditionally credible multiple contingency to be credible for real-time operations. This in turn could impact the TTC calculations that were based on the always credible contingencies. Defining multiple scenarios with different sets of contingencies can help to quickly determine the impact of contingency credibility on transfer capability.

Figure 13.8 shows an example of TTCs for imports from Oregon to California. The TTC calculations are conducted for two scenarios. One that includes only the always credible contingencies and another that includes the always credible and the conditionally credible contingencies. During planned outage conditions on a 500 kV line that is part of one of the conditionally credible contingencies, we can see that the TTC results merge into one, as the loss of the conditionally credible multiple contingency across the common corridor has now effectively become the same as the loss of a single line which is an always credible contingency. When both the lines are in service, it is observed that the TTC results are significantly different reflecting the impact of conditionally credible multiple contingencies.

13.4.1.5 Supplementing Seasonal Studies

Seasonal IROL or TTC numbers are typically established based on off-line operational cases tuned to expected load/generation patterns. However, based on changing generation and load profiles, in some cases the real-time flows may never reach the established limits. Setting up the respective scenarios in real-time allows to collect data on limit calculations based on real-time conditions. This data can provide input on actual margins in real-time which can further lead to either confirming the need to continue to establish the seasonal limit calculation for operational purposes or open up an opportunity to remove the need to calculate limits for the respective path as part of operator guidelines.

13.4.2 Challenges with Real-Time Voltage Stability Monitoring

Some of the major challenges introduced with real-time voltage stability application and monitoring are discussed next.

One of the most significant challenges comes from the quality of state estimation and the network model utilized for the analyses. As state estimation is dependent on good quality data inputs, bad data or lack of data can often introduce various challenges in the determination of the transfer limits. The impacts could lead to local pockets in the system with bad voltage profile creating weak areas prone to voltage collapse.

Another occasional concern is the modeling of the reactive capability of aggregated resources. Multiple units at a common location may often be aggregated into a single resource in the EMS model. However, the reactive capability would need to be adjusted, as the aggregated resource would need to have dynamic reactive capability depending on how many underlying resources are online.

Sources and sinks are generally set up based on off-line studies. These are typically far away from the area of study. When studying the impact of transmission or generation or other equipment outages on the respective interface limits, these outages do not typically create local bottlenecks in the source or sink areas. They do however create degraded conditions in the area of study. One of the challenges with online analysis is that local outages in the areas of the sources and the sinks can lead to local congested transmission corridors which in turn lead to either local voltage collapse zones as the sources or sink generation or load is ramped up or down or also lead to other pre- or post-contingency concerns. There is a need therefore for such tools to have a smarter approach on how the sources or sinks can be adjusted on a real-time basis.

Another challenge, in general with all online real-time applications is the constant maintenance of the application setup and configuration due to updates to the network model as different projects come into service and system updates are made. In most online applications a common identifier such as the name of the equipment in the EMS network model is utilized for inclusion in the setup of the scenarios. However, when network updates are made, equipment names may change and this introduces a regular task to determine the impact of network model updates on the setup of online applications.

13.5 Real-Time Transient Stability Monitoring

CAISO runs online stability analysis for certain interfaces for which corresponding operator guidelines have also been provided. Most known stability issues have very high stability limits as compared to the normal flows on the respective interfaces, i.e., there is usually a very high margin. In addition, stability issues only occur after a multitude of outage conditions. These interfaces were previously operated with fixed seasonal limits based on extensive off-line studies using operational planning cases. However, real-time transient stability assessment allows to utilize real-time limits and the fixed seasonal limits established from off-line studies are utilized as a back-up when the real-time application or results may not be available.

Figure 13.9 shows an overview of the end-to-end processing and data flow of the real-time dynamic security assessment (DSA) implementation. Also, indicated are some other uses of the DSA application which are explained further in the later sections.

As shown in Fig. 13.9, one of the outputs of the real-time voltage stability application is a bus-branch equivalent of the full node-breaker CIM model. This bus-branch model is utilized along with the dynamic data associated with the off-line operations planning case for the season to perform the following two analyses:

1. Basecase analysis: The state estimation solution is assessed with a set of credible contingencies deemed to be significant from a transient stability perspective to make sure the contingencies are all secure and do not cause instability issues.
2. Transfer limit analysis: For interfaces associated with known stability concerns, maximum transfer limits are calculated based on an associated scenario definition

The setup of each monitored interface involves:

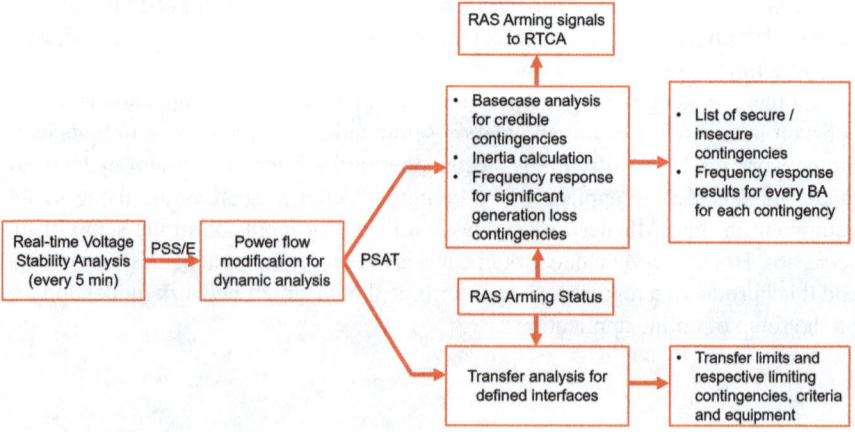

Fig. 13.9 Overview of online transient stability setup

1. Dynamic Data: The dynamic models and data to be used for each generator, line, load, and relays
2. Remedial Action Schemes: The list of remedial actions that can be triggered for any of the contingencies during the transfer analysis
3. Defining the sink and source area: The sink and sources are usually defined by a combination of generators and loads
4. The definition of the monitored interface: A set of lines typically makes up a monitored interface
5. The contingencies to be evaluated: Contingencies deemed to be credible for the operations horizon that are known to cause transient stability concerns as the transfer is increased across the monitored interface are included here.
6. The list of monitored equipment for reviewing transient stability results: These are usually the generator MW and MVAR outputs and other generator parameters like speed, field voltage, mechanical speed, rotor angles, and other transmission equipment parameters such as line flows and bus voltages.
7. Monitoring criteria: The transient stability criterion such as angular stability, transient voltage and frequency, damping and relay margins are used to limit the transfer analysis to ensure transient stability.

For each of the monitored interfaces, the source and sink definitions are utilized to increase power transfer across the monitored interfaces and at each level of the transfer, all contingencies are evaluated and assessed for violation of any of the monitored transient stability criterion.

At the end of the analyses, the following are provided:

Basecase Analysis

(a) List of any insecure contingencies
(b) Frequency response results for contingencies monitored for frequency response. These results include the minimum frequency (Nadir point) observed during the simulation and the various metrics from the NERC BAL-003 standard [10] for every Balancing Authority.
(c) In addition, the inertia levels of each BA are also provided for the no-fault run

Transfer Analysis

1. Base powerflow on the monitored interface: This is the powerflow on the monitored interface prior to any transfer. This typically matches the powerflow on the monitored interface in the input data from state estimation.
2. Maximum power transfer on the monitored interface: The maximum powerflow transfer is provided that was achieved based on the monitored transient stability criterion.
3. Limiting factor: The type of limitation observed at the maximum transfer level is provided. These could be the transient stability margin, the voltage or frequency drop, or insufficient generation or load reserves in the source or sink to increase the powerflow across the monitored interface

4. Limiting contingency: The respective contingency leading to transient stability concerns is provided as the limiting contingency
5. Limiting equipment: Depending on where the pre- or post-contingency transient voltage or frequency violations might be occurring or where the transient stability margin reductions are occurring, the respective equipment is reported as the limiting equipment.

13.5.1 Utilization of Real-Time Transient Stability Monitoring Application for Grid Operations

13.5.1.1 Transient Stability Limit Monitoring

As mentioned, the primary purpose of the real-time transient stability monitoring is to ensure that system reliability is always maintained by ensuring that there are no instability issues for significant contingencies in the current operating state and that a positive margin exists across the monitored interfaces with transient stability concerns. Depending on system topology conditions, when flows do start approaching the real-time limits or when margins start diminishing, operators follow established protocols in operating procedures to coordinate between various TOPs and BAs to reduce flows and increase margins.

Figure 13.10 shows an example of the variation of real-time stability limits provided by the online application. Also shown are the basecase flows on the interface being monitored in the state estimation output and the actual flow on the same interface as observed in EMS. Additionally, the fixed stability limit calculated from seasonal studies using the operations planning powerflow cases for various seasons is also shown for comparison. It is observed that the limits change gradually through the course of the day but can have sudden drops that can highlight the challenges with the use of state estimation inputs. Additionally in general, a healthy margin is observed between the flows and stability limits. Also, it is observed that stability limits based on actual conditions can be higher or lower than the fixed limits determined from seasonal studies. This highlights one of the major advantages of using online applications where the calculation of limits based on real-time conditions provides a more accurate situational awareness to the operators.

13.5.1.2 Frequency Response Estimation

NERC BAL-003 standard requires every balancing area to carry sufficient amount of frequency response capability by making sure that median frequency response measure (FRM) for at least 20 events selected by NERC exceeds the frequency response obligation (FRO) of the respective BA. Every BA has a FRO which is a share of the Interconnection frequency response obligation (IFRO) which is

Fig. 13.10 Comparison of stability limits calculated in real-time versus real-time flows on the monitored interface and limits calculated from seasonal cases

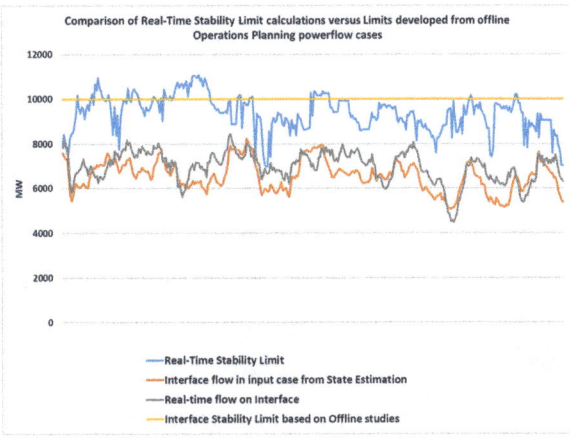

Fig. 13.11 Primary frequency response illustration

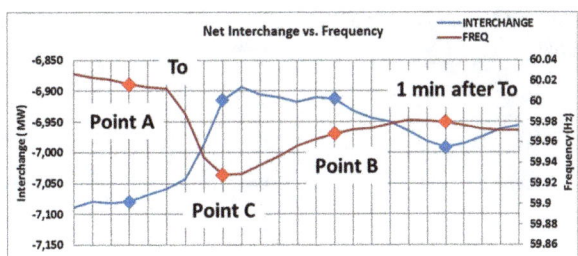

indicative of the interconnection's minimum frequency response to avoid under frequency load shedding.

Figure 13.11 shows how to calculate the FRM for an event. Point A is the average frequency and interchange from the time that that event occurred T_0 up to $T_0 - 16$ s. Point B is the average frequency and interchange after primary frequency response has responded to recover the decline in frequency from $T_0 + 20$ to $T_0 + 52$ s. The ratio of the difference of interchange over the difference in frequency in between these two key points and is used to establish the FRM in MW/0.1 Hz. Point C is the nadir point of the frequency during the frequency event.

One of the outputs of the basecase analysis is the frequency response measure calculations for a defined set of contingencies that lead to a significant generation drop. Various contingencies in different geographical regions are defined as generation loss in different areas that can lead to different frequency response measures for the same amount of generation loss. So tracking contingencies from different geographical regions helps to track the worst-case scenarios. Figure 13.12 shows an example of the real-time frequency response measures calculated over a period of time in comparison to the frequency response obligation.

Fig. 13.12 Primary frequency response illustration

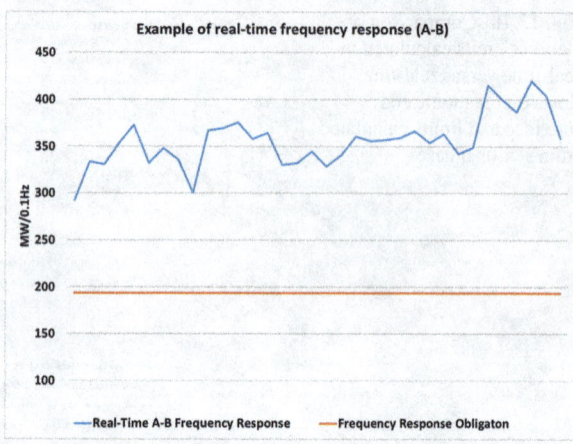

Fig. 13.13 Sample inertia trend

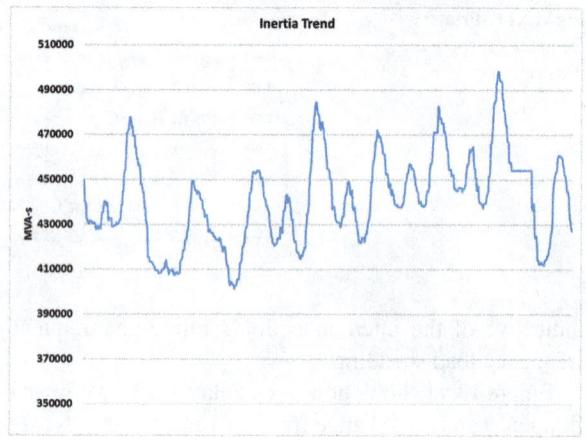

13.5.1.3 Inertia Tracking

Another use of the basecase analysis is tracking the inertia of the system. The powerflow basecase from state estimation provides the status of every unit. The H and MVA values are picked up from the dynamic data of the units that are online and are utilized to calculate the total inertia of the various areas. Figure 13.13 shows an example of the inertia trend for the RC West area.

13.5.1.4 Utilization of Transient Stability Results in Real-Time Contingency Analysis

Certain contingencies trigger RAS actions that are simulated in the real-time contingency analysis. Many of these RAS actions actually occur based on the dynamic response for contingencies such as:

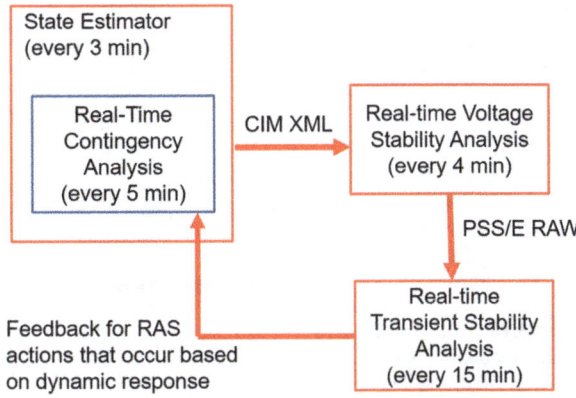

Fig. 13.14 Feedback of RAS action between real-time transient stability and contingency analysis applications

(a) Reactive insertion based on time duration and level of voltage dip
(b) Generator tripping based on angular acceleration

When implemented in RTCA, they can be assumed to be getting triggered or not getting triggered based on worst-case scenarios. A more accurate way to determine the assumptions of these RAS actions getting triggered is to run dynamic simulations to check if the conditions to trigger such RAS actions are met and then run the steady-state contingency analysis with or without the corresponding RAS actions.

The online transient stability application at CAISO is interfaced with the real-time contingency application to provide feedback of such RAS actions as shown in Fig. 13.14.

13.5.1.5 Model Validation

The availability of real-time snapshots along with the applicable dynamic models makes it possible to validate the model by simulating the actual events using a snapshot closest to the occurrence of the event and then comparing the dynamic responses to high-speed data such as synchrophasor data from phasor measurement units (PMU). Figure 13.15 shows an example of the comparison of the simulated and actual frequency response for a generator trip event.

It is observed that the frequency plots have the same trends but do not match closely. However, if the governor baseload assumptions are changed in the powerflow cases, then the frequency response shows a closer match as seen in Fig. 13.16.

Figures 13.17 and 13.18 shows a comparison of powerflow and voltages, respectively.

The model validation process allows improvement of network, dynamic model parameters, and the modeling of remedial action schemes.

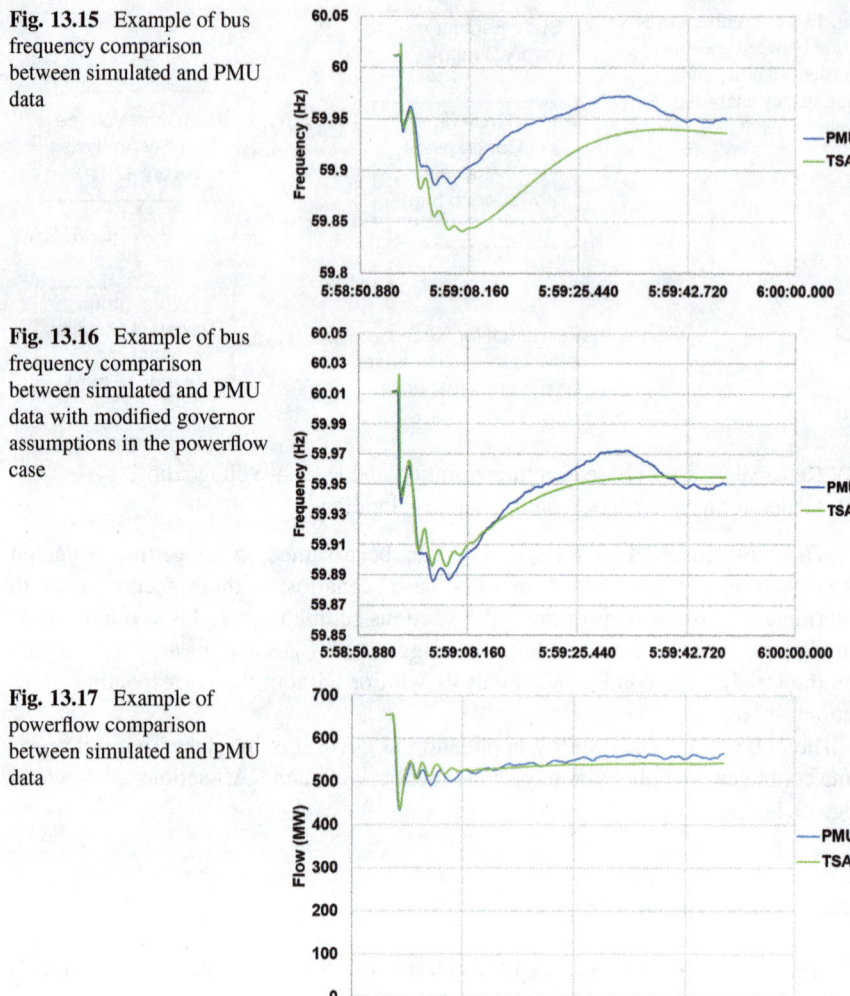

Fig. 13.15 Example of bus frequency comparison between simulated and PMU data

Fig. 13.16 Example of bus frequency comparison between simulated and PMU data with modified governor assumptions in the powerflow case

Fig. 13.17 Example of powerflow comparison between simulated and PMU data

13.5.1.6 Small Signal Stability Analysis

Another extension of the transient stability application is to perform small signal stability analysis to determine the natural oscillation modes [11] of relevance in the system and also to determine mitigation actions such as reducing inter-area transfers to increase the damping of a mode. The same exact powerflow and dynamic data utilized for the transient stability simulation are utilized here. One significant advantage of this approach is that mitigation actions can be validated for real-time system operating conditions using real-time cases as opposed to completely relying on mitigation actions developed.

Fig. 13.18 Example of bus voltage comparison between simulated and PMU data

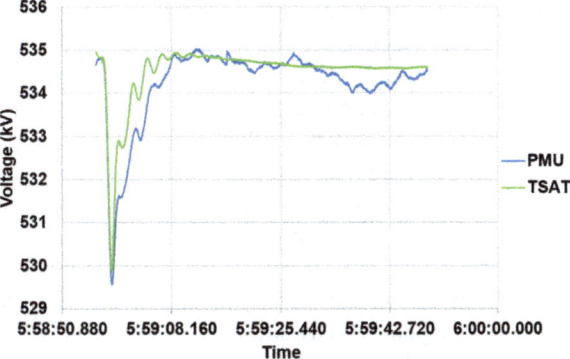

Fig. 13.19 Example of mode-shapes for an inter-area mode with a real-time case

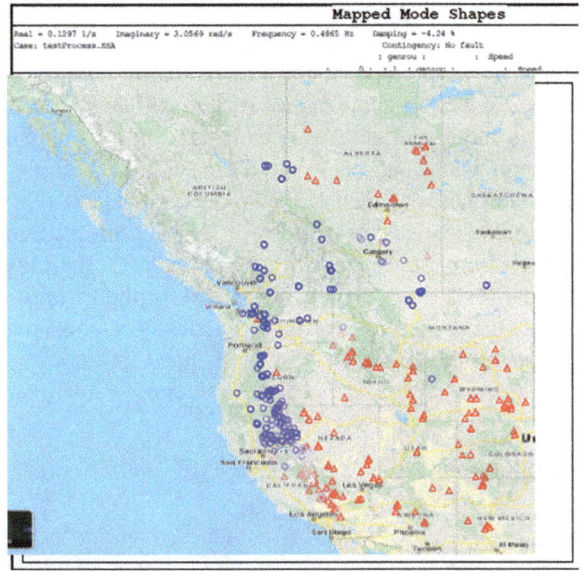

Figure 13.19 shows an example of mode-shapes for the North-South Mode observed in a real-time snapshot. One advantage of viewing mode-shapes with real-time cases allows to view the participating generators in the mode based on the system operating conditions.

13.5.2 Challenges with Online Real-Time Transient Stability Analysis

Similar to online real-time voltage stability analysis, there are various challenges with real-time transient stability analysis. One of the most significant challenges come from the results of state estimation. As state estimation is dependent on good quality data inputs, bad data or lack of data can often introduce various challenges in the basecase analysis or the determination of the transfer limits. The impacts of state estimation could lead to generators with negative real power outputs. When used along with dynamic models, such negative real power outputs can cause units to go unstable in a dynamic solution or respond oddly. This is more common with models used for inverter-based renewable resources. Addressing these issues may require different approaches such as improving the powerflow solution utilized for the dynamic analyses and potentially making dynamic models more robust.

Another occasional concern is the dynamic modeling of aggregated resources. Multiple units at a common location may often be aggregated into a single resource in the EMS model. The dynamic model of each of the individual units would need to be aggregated so that real-time transient studies can use the same set of dynamic data with aggregated resource models.

Load modeling is another significant challenge for online applications. Current composite load models for utilization in real-time transient stability analysis introduce various complexities such as a significant number of parameters for each load in the model and the need to modify load parameters in real-time based on different factors such as location and season. These are expected to be incorporated over time as the composite load models are adapted for real-time use.

Similar to real-time voltage stability analysis, another challenge, in general, with real-time transient stability application is the constant maintenance needed to address updates to the network model for the various system upgrades, which can impact the setup of the various scenarios, RAS modeling, and dynamic data utilized for the stability analysis.

References

1. http://www.caiso.com/
2. ISO/RTO Council, https://isorto.org/
3. http://www.caiso.com/informed/Pages/RCWest/Default.aspx
4. RC West System Operating Limit Methodology for the Operations Horizon, http://www.caiso.com/Documents/RC0610.pdf
5. A. Alam, G. Gopinathan, B. Shrestha, R. Zhao, J. Wu, R. Xu, High performance computing for operations and transmission planning at CAISO, in *2018 IEEE/PES Transmission and Distribution Conference and Exposition (T&D)*, Denver, CO, 2018, pp. 1–9
6. Core International Electrotechnical Commission Standards, https://www.iec.ch/smartgrid/standards/
7. BSI, Bigwood systems Inc, http://www.bigwood-systems.com/
8. DSA Tools, http://www.dsatools.com/
9. G. Gopinathan, A. Alam, Real-time contingency analysis with RAS modeling at CAISO, in *2018 IEEE/PES Transmission and Distribution Conference and Exposition (T&D)*, Denver, CO, 2018, pp. 1–9
10. NERC BAL-003-2, Frequency response and frequency bias setting, https://www.nerc.com/pa/Stand/Reliability%20Standards/BAL-003-2.pdf
11. NERC Interconnection Oscillation Analysis—Reliability Assessment, July 2019, https://www.nerc.com/comm/PC/SMSResourcesDocuments/Interconnection_Oscillation_Analysis.pdf

Chapter 14
Use of Stability Assessment Tools at Midwest ISO

Raja Thappetaobula

14.1 Introduction

Midcontinent Independent System Operator (MISO) is an independent, not for profit organization that delivers safe, cost-effective electric power across 15 US states and the Canadian province of Manitoba. MISO operates one of the world's largest markets with more than $29 billion in annual gross market energy transactions [1]. Figure 14.1 provides the details of the foot print for MISO.

With the increased amount of renewable energy resources in the footprint and energy transfers that result in high flows across multiple interfaces results in potential for transient/voltage stability issues in MISO system.

To better prepare and maintain stability of the system MISO utilizes stability assessment tools in both real-time horizon and as well as operational planning horizon. MISO utilized Powertech DSA tools to develop both, the operational planning as well as real-time stability assessment process.

14.2 Operatinal Planning Stability Assessment Tool

MISO has developed operational planning stability assessment tools to help outage coordination engineers and operation engineers run transient stability and voltage stability analysis for any planned or forced outage studies. This approach ensures stability of the grid for planned outage and ensures more accurate interface limits.

R. Thappetaobula (✉)
Midcontinent Independent System Operator (MISO), Carmel, IN, USA
e-mail: rthappetaobula@misoenergy.org

© Springer Nature Switzerland AG 2021
S. (NDR) Nuthalapati, *Use of Voltage Stability Assessment and Transient Stability Assessment Tools in Grid Operations*, Power Electronics and Power Systems,
https://doi.org/10.1007/978-3-030-67482-3_14

Fig. 14.1 MISO footprint

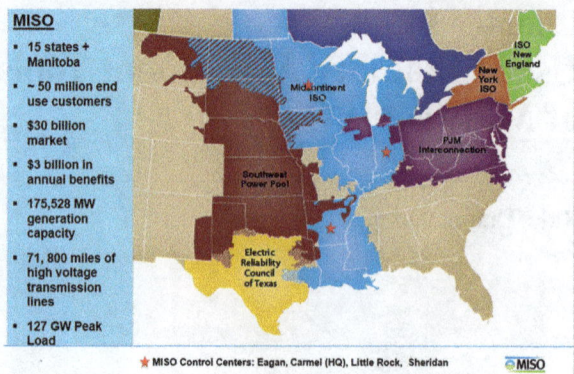

This package was developed to be compatible with the daily power flow base cases that outage coordination engineers develop for their normal thermal studies. This helps outage coordination engineers to utilize daily power flow base cases and perform detailed stability studies on known interfaces for planned outages.

Without the availability of these tools in the operational planning horizon, historically engineers selected worst-case scenarios and simulated specifically chosen operating points to determine a safe loading limit for various critical lines and interfaces. During real-time operations, these pre-calculated limits were then used to make decisions on re-dispatching generation in order to restrict the flows to within limits.

Traditionally, stability assessment involved many manual processes. Consequently, studies were rather laborious considering the number of simulations required to be run for different scenarios, especially in the planning horizon. Moreover, the performance criteria specified by different transmission operators (TOP) varies which complicates studies and determination of accurate limits.

When MISO shifted to using DSATools for performing dynamics studies, several of the manual processes were automated. MISO's DSA Tools based operational planning stability assessment tool utilizes standard PSS®E dynamic models. The PSS®E user-defined models were also converted to corresponding TSAT format. MISO has developed automation scripts to standardize the disturbance files to allow the Transmission Owners (TO) to provide their disturbances in any legacy format which can be quickly converted to TSAT format.

Currently, MISO performs transient stability and voltage stability assessment for several constrained interfaces within the northern portion of its footprint, such as the Minnesota Wisconsin Export (MWEX) interface, the North Dakota Export (NDEX) interface, and Manitoba Hydro Export Interface (MHEX) limit.

The MISO northern portion footprint is shown in Fig. 14.2. Conceptually, increasing levels of power transfers are simulated across these interfaces by varying the generation/load levels across the interfaces and the reliability of the system is assessed against a set of severe, yet credible disturbances. The transfer studies and

Fig. 14.2 The MISO footprint (northern portion)

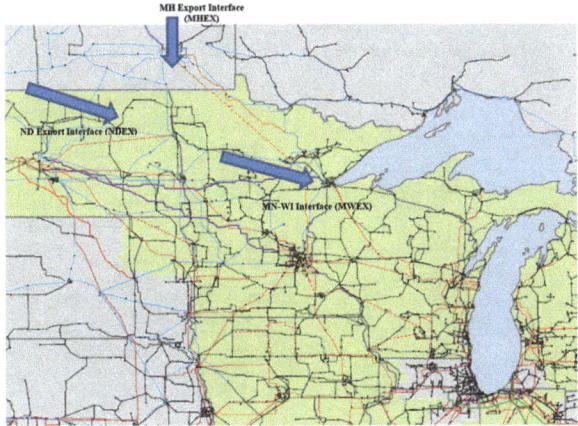

the post-processing of the solutions are automated by applying the individual TO criteria, which saves time and effort.

Some of the basic building blocks of MISO operational planning stability process (Fig. 14.3) are as follows:

- Seasonal peak, off-peak or high transfer or custom-biased base cases from MISO Outage Coordination Case-Builder process.
- Dynamic Data from the Planning Models.
- HVDC, SVC, SPS, Fast Switch Caps, Relay, Reductions, and other parameters from the real-time TSAT/VSAT process.
- Outage Coordination daily cases are generated from these base cases to analyze planned outages that may cause voltage stability and transient stability limitations.

14.2.1 Operational Planning Stability Assessment Tool Process Flow

The below flow chart describes stability assessment process flow.

14.2.1.1 Base Power Flow Model Development Process

MISO Model on Demand (MOD) process is used to develop the base power flow models for the stability studies. MISO creates a monthly base case, with the daily case being derived from the monthly base case.

MISO and the regional entities utilize power flow models for seasonal transfer analysis studies, outage analysis studies, and various other study applications. It is

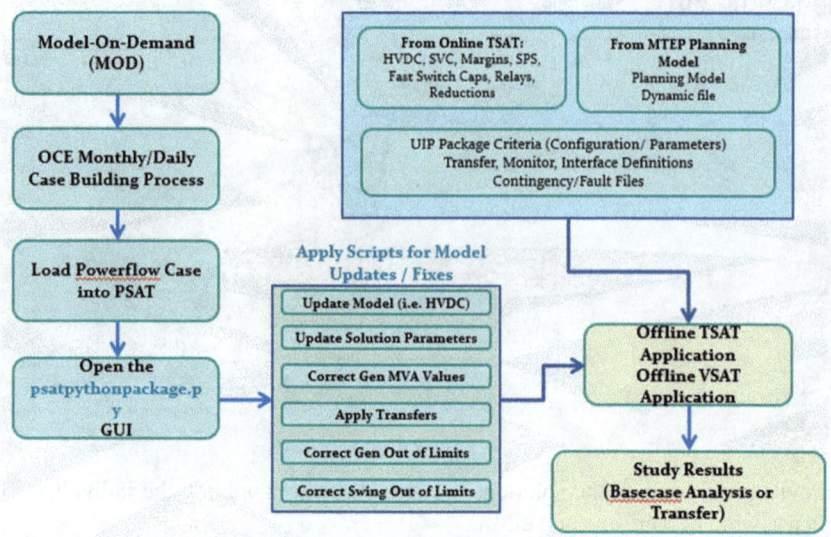

Fig. 14.3 Operational planning stability process overview flowchart

beneficial to have all the regional entities in the MISO area utilize the same base power flow model when performing studies.

14.2.1.2 Operational Planning Stability Assessment Tool Benefits to MISO

With the development of the operational planning stability process, MISO achieved enhanced reliability as well as market efficiency benefits.

These benefits can be characterized as below:

- Process Synchronization and Utilization between RT/Forward Ops.
- Common methodology and Tool for Transient Stability studies for multiple departments.
- Study Accuracy Improvement with More options.

  - TSAT/VSAT Transfer Simulation utilizes a single case reducing possibilities of user-introduced errors.
  - TSAT transfer simulations are automated, resulting in an increase in the study process efficiency.
  - Enhanced Study Efficiency/Flexibility.
  - Ability to run studies that were previously impossible due to resource/tool restrictions for our external stakeholders.

- Tremendous value creation for the TO's.
- TSAT/VSAT simulations improve study efficiency (67% of process improvement; 264 h/year).
- Increased Transmission Utilization, improving efficient Market dispatch in this region as this was a heavily constrained path during certain prior outage conditions.
- New Standing guide(s) are being developed using this tool in line with increased interface limit by as much as 300 MW in MWEX.
- Reductions in Day-Ahead/Real-Time Market binding with Higher Transmission Interface limits.

14.3 MISO Real-Time Stability Assessment Tool

MISO is one of the first Reliability Coordinators (RC) in the US to implement such a sophisticated real-time transient and voltage stability assessment tool. Real-Time capability is especially useful as NERC and the industry is working to clarify stability related IROLs and corresponding operating obligations. MISO implemented the real-time TSAT/VSAT (DSATools™) package in May 2012 to perform calculations in real-time and determine the stability limits. Currently, the tool iteratively performs several voltage stability studies and transient stability studies that are used in real-time operations. The voltage stability tool repeats the study every 6 min while the transient stability studies are performed every 15 min, respectively. MISO monitors eight different transient stability interfaces and more than 30 voltage stability interfaces in real-time.

The real-time TSAT/VSAT tool enables the calculation of actual transfer limits in near real-time using the most recent system operating state. This is a very important enhancement which essentially allows the operator to maximize the system usage while maintaining reliability.

When a forced outage on an element that has the potential to cause stability concerns occurs in real-time, the operational planning calculated stability limits are still used, but only until the real-time TSAT/VSAT tool captures the system states and returns the new limits. The operational planning stability assessment tool is used to calculate the stability interface limits at the day-ahead and operational planning (outage coordination) timeframes, while the results from the real-time tool are used in real-time operations.

Figure 14.4 provides the details on how MISO real-time dynamic security assessment works.

Analysis is performed for real-time system conditions captured by SCADA and solved by state estimator. The state estimator case is converted to a PSSE format power flow and that model is sent through a power flow modification process. Power flow modification specifies the necessary modifications need to be made to the real-time power flow data to make the power flow suitable for real-time Dynamic Security Assessment. This process creates a base power flow that is solved

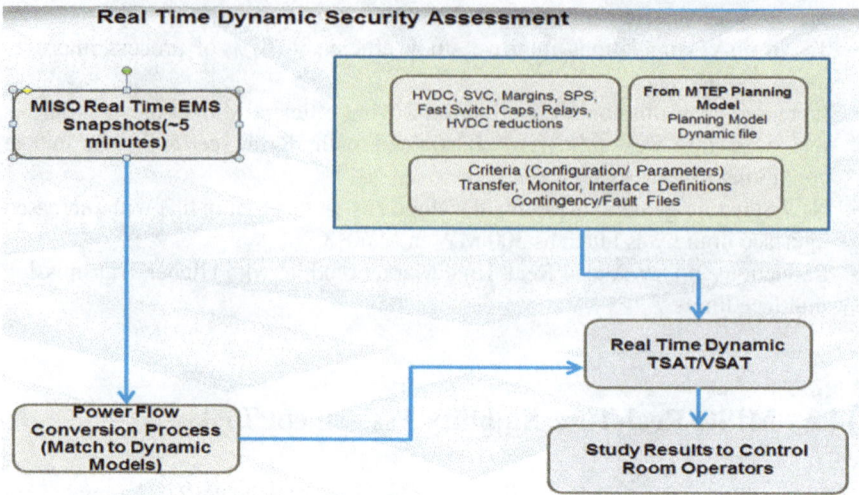

Fig. 14.4 MISO real-time dynamic security assessment

and ready for transient and voltage stability analysis. Both base case and transfer analysis are performed on all monitored stability limited interfaces. After the base case analysis is secure the safe transfer limits are then calculated by using a binary search approach to determine the maximum transfer limit at the interfaces and the corresponding limiting constraint. The tool automatically performs the stability analysis based on the criteria specified and cycles through calculation process continuously and providing results for operators to manage reliability.

Other key inputs to the real-time dynamic security assessment as shown in Fig. 14.4

- Dynamic Data from the Planning Models.
- HVDC, SVC, SPS, Fast Switch Caps, Relay, Reductions, and other parameters from the real-time TSAT/VSAT process.
- MISO TO's/TOP's transient stability criteria are implemented.
- MISO TO's/TOP's critical contingencies are applied.

14.4 MISO Real-Time Dynamic Security Assessment: Real-Time Operations

When MISO started working on implementing real-time dynamic security assessment one of the main challenges was how we would present results from the complex stability assessment to system operators and enable quick, accurate decision-making. This challenge provided us an opportunity to develop criteria to visualize the results and a new display environment where operators can view the results and operate the system.

For voltage stability limited interfaces to ensure a sufficiently conservative assessment, 95% of calculated voltage stability limits are posted as interconnection reliability operating limits (IROL) and 95% of the IROL is posted as the system operating limit (SOL), respectively. MISO operators monitor the results and make sure the flows on the stability interfaces are below the conservative operating limits. They utilize MISO congestion management procedures to manage the stability interface limits.

For transient stability limited interfaces, we utilize below criteria to monitor and manage system reliability on the interfaces.

- If angular stability issues or damping issues at a generator are identified in the real-time dynamic security assessment MISO system operators will implement generation reductions if the real-time unit output is greater than the generator stability limits.
- If voltage violations are identified at interfaces/Flowgate MISO system operators utilize MISO congestion management procedures to manage the stability interface limits.

MISO developed a process to read the results from the real-time dynamic security assessment and display all the relevant results in visual format for the control room operators. This process is developed with the help of MISO IT groups.

14.5 Illustration of Transient Stability Results

Transient Stability results are shown on a display as shown in Fig. 14.5. This stability study MISO performs basecase and transfer analysis studies and depending on type of stability issues identified the display gets populated with limits on flowgates that system operators need to manage in the real time. If there are no stability issues identified the display shows a message as shown in Fig. 14.5, to let system operators know that there are no stability issues on the interface and system is secure. Voltage stability and transient Stability results are displayed as shown in Fig. 14.6.

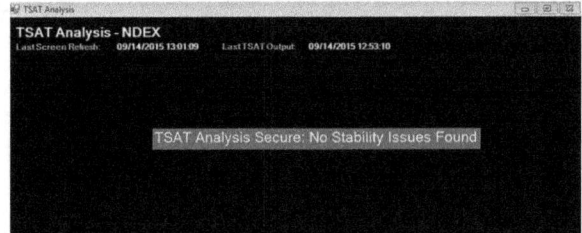

Fig. 14.5 Results of transient stability analysis when there are no issues

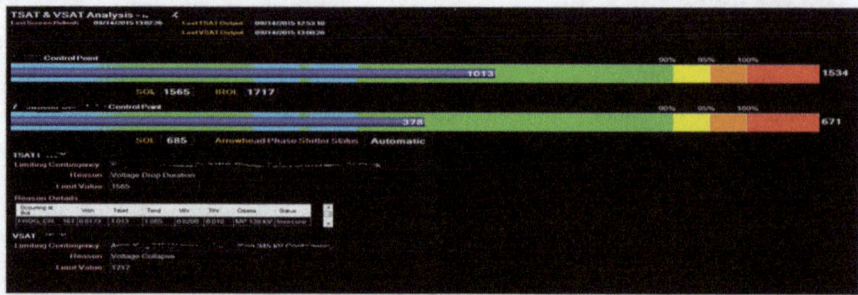

Fig. 14.6 Display of results for TSAT and VSAT

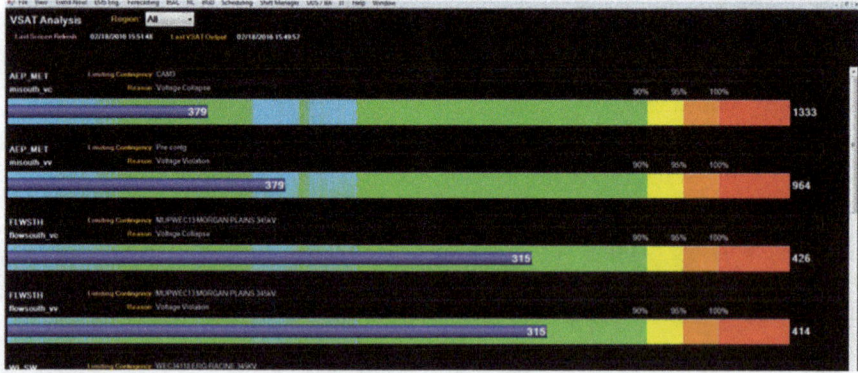

Fig. 14.7 Voltage stability interface display

MISO performs both voltage stability studies and transient stability studies on an interface as shown in Fig. 14.7. Based on the limits calculated from both studies MISO identifies interconnection reliability operating limits (IROL) and system operating limit (SOL) respectively and display the results to operators. The studies identified 1717 as the IROL limit and 1565 as the SOL limit on the interface. The display also shows real-time flow on the interface with the limit identified in bar chart format which is easy for operators to understand and make a quick decision as needed and utilize MISO congestion management procedure to resolve the stability issues.

This display shows voltage stability interface limits and real-time flows on those interfaces in a bar chart format which is easy for operators to understand and make a quick decision as needed.

Fig. 14.8 MISO real-time DSA architecture

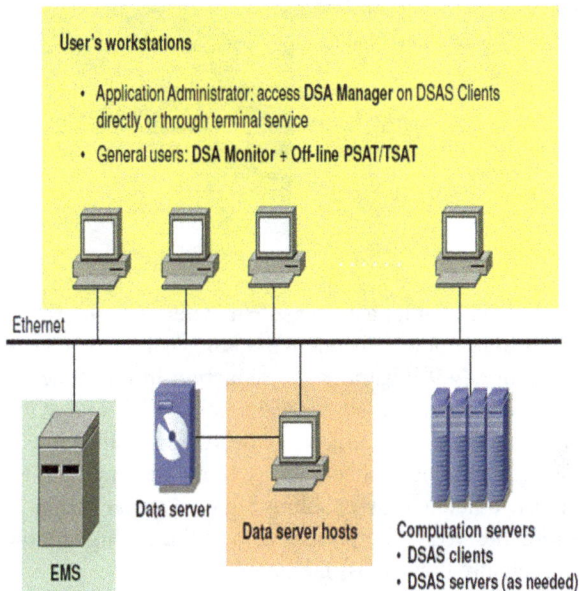

14.6 MISO Real-Time DSA Architecture

MISO DSA tool is set up on three environments, namely the production, failover, and the staging environment. Each environment has 8, HP DL380 G8 servers and each server consists of 16 cores. The simulations are run with parallel processing for fast performance (Fig. 14.8).

MISO routinely performs business continuity tests and make sure production and failover systems are working as designed.

14.7 Benefits of MISO Real-Time Dynamic Security Assessment

With the implementation of dynamic security assessment, MISO saw both enhanced reliability benefits as well as market efficiency benefits. MISO can calculate real-time stability limits on real-time system conditions which enhanced system reliability. Limits are calculated for all system conditions rather than one peak condition operating guide limits are calculated for. Real-time Stability limits are also being calculated for multiple generators that encounter local stability limits.

This process also helped reduction in real-time market binding with higher transmission interface limits from real-time TSAT/VSAT instead of operating guide

Fig. 14.9 Illustration of reduction in real-time binding

Year	Interface limit calculated	# Real-time Binding Hours	Total real-time congestion costs (Million $)
2014	Offline	502	31
2015	Online	7	0.7
2016	Online	3	0.1

limits. This enables a higher safe transfer of power from one region to another region and enables more efficient market dispatch.

Figure 14.9 illustrates the reduction in real-time binding on a stability limited interface. Since we started calculating real-time stability limits on this interface congestion costs interface has decreased tremendously with the real-time assessment calculating higher stability limits than operating guide limits. This increased market efficiency by enabling higher transfers from one region to another region within the MISO footprint. It also enabled a higher amount of wind generation that can reliably flow across the interface.

Reference

1. MISO Independent Market Monitor (2013/2014) Reports. https://www.misoenergy.org/markets-and-operations/independent-market-monitor2/#nt=%2Fimmtype%3AQuarterly%20State%20of%20the%20Market%20Report&t=10&p=1&s=Updated&sd=desc

Chapter 15
Reactive Power Management in Real-Time at Tennessee Valley Authority (TVA)

Tim Fritch, Ulyana Pugina Elliott, Josh Shultz, Patrick Causgrove, and Gilburt Chiang

15.1 Introduction

Reactive power is essential to the flow of real power through the bulk transmission system. Sources of reactive power come mostly from generation units and shunt devices in the network. However, the control of reactive power is incredibly challenging for utility operators of large transmission systems. Unlike real power, reactive power cannot be moved far across long-distance transmission power lines to accommodate the highly variant loads. Yet, reactive power is needed and necessary to maintain proper voltage levels and move active power through transmission lines. Reactive power can cause the voltage to rise or fall. Therefore, not having sufficient components to produce or absorb reactive power can cause voltage instability issues. Still, reactive power is often overlooked in system reliability studies, which can lead to voltage instability and potential collapse if not properly monitored. Again, the amount of reactive reserves at generating stations is a measure of the degree of voltage stability.

T. Fritch · U. P. Elliott · J. Shultz
Tennessee Valley Authority (TVA), Chattanooga, TN, USA
e-mail: tnfritch@tva.gov; uvpugina@tva.gov; jshultz@tva.gov

P. Causgrove · G. Chiang (✉)
Bigwood Systems, Inc., Ithaca, NY, USA
e-mail: Pat@bigwood-systems.com; Gilburt@bigwood-systems.com

15.2 Background

Tennessee Valley Authority (TVA) is a federal agency that was created by Congress in 1933 and tasked with a unique mission of service: to make life better for the people of the Tennessee Valley through the integrated management of the region's resources. TVA carries out the mission by focusing on the three key areas of energy, environment, and economic development. As one of the largest transmission systems in North America, TVA spans 16,200 miles of the transmission line. TVA partners with 154 local power companies to serve ten million people and 700,000 businesses across 80,000 square miles in parts of seven southeastern states (North Carolina, Georgia, Tennessee, Kentucky, Virginia, Alabama, and Mississippi). TVA's transmission system achieved 99.999% reliability for the 20th year in a row. In addition to operating the electric power system, TVA also manages the Tennessee River system through land management, navigation management, and flood control. TVA serves as the NREC Reliability Coordinator for the TVA area and eight (8) nearby Member Balancing Authorities and Transmission Operators.

To combat concerns with reactive power, TVA worked with software vendor Bigwood Systems, Inc. (BSI) to implement a real-time VAR Management System (VMS) for regional reactive reserves, which has become a key tool in the TVA control center for maintaining and enhancing voltage security while ensuring proper management of localized VAR resources in the TVA electric transmission system. The VMS tool, which is designed for large-scale interconnected power systems, produces results that allow system operators to anticipate voltage issues before they occur and take action to prevent possible reliability constraints in real-time operations, thereby improving situational voltage awareness. Also, as demonstrated in the Implementation and Use Cases section below, there were several regions that were identified for voltage issues with certain system conditions present at the time. These discoveries aid in identifying potential areas on transmission systems that need to be evaluated further. If the voltage issues arose due to inadequate local area shunt sources such as capacitor banks, then this could aid in timely maintenance or prioritization of various shunt resources to improve local area voltage capability in the future.

The VMS implementation evolved from TVAs need to replace an obsolete built-in-house system for management and reporting on reactive reserve regions. This effort helped TVA move past a stalled effort to use a voltage stability system that was supplied with the then current EMS. At the time, the Voltage Stability Analysis and Enhancement (VSA&E) package tool supplied by BSI was in use at TVA and working for off-line studies by the reliability engineering group that supports real-time operators. TVA asked BSI to integrate and leverage their online voltage stability assessment capabilities, who then successfully implemented a pilot online Var Management System (VMS) for a selected Reactive Region. This early success spurred an effort with the TVA reliability engineers and system operators to develop a full Online Var Management System for Reactive Reserve Monitoring and Control. VMS has been running continuously online in the TVA control center since 2013.

During a recent Transmission Operations audit at TVA, the BSI VMS was cited by the South East Reliability Corporation (SERC) as a *best practice solution for electric utility reactive power management.* SERC is one of eight regional electric reliability councils under North American Electric Reliability Corporation (NERC) authority and is responsible for ensuring reliable and secure electric grids across 16 states in the southeastern and central United States. Having VMS helps strengthen compliance posture. Currently, there are several NERC reliability standards related to voltage and reactive control. These standards help mandate utilities to ensure that voltage levels, reactive flows, reactive resources, and system operating limits are identified and properly monitored within limits in real-time operations to protect equipment and reliable operation of the electric grid. Having VMS helps fulfill the requirements of these standards successfully.

In the past few years, TVA's power portfolio has been changing in the face of various demands and regulations. The emphasis has moved away from traditional coal-based units and toward cleaner forms of power generation. This led to decommissioning some of TVA's oldest coal-fired units that were historically placed around load pockets to make way for the new forms of power production. Additionally, large new businesses have been locating in the Tennessee Valley, which places variously sized loads throughout the system and may be spaced out further from generation sources. This combination of retiring generation and adding new load throughout the system, which may not be placed close together electrically, may lead to potential voltage instability scenarios. Hence, it is imperative to have reactive power monitoring capabilities such as VMS.

As previously mentioned, TVA is uniquely placed in the central United States. This geographical position offers a set of advantages as well as disadvantages. One of the disadvantages may be daily power transfers through TVA's transmission system by other major utilities in the Eastern Interconnection (EI). These power transfers happen hourly, and they may reach large amounts of power transfer at a time. The exceedance of surge impedance loading of transmission lines due to these large power system transfers can result in voltage instability and therefore it is important to monitor these scenarios.

15.3 Implementation and Use Cases

15.3.1 The Reactive Reserve Zone Definition

The first step in monitoring and managing system reactive reserves on a regionalized basis is to properly identify and define reactive reserve "zones" such that stations within the same zone are electrically coherent to each other. Bigwood Systems began by performing this engineering study for TVA's service territory before VMS was implemented.

Both the electrical and physical distances between the buses were considered in determining the zones, ensuring that buses within the same zone will have similar electrical properties and will be physically close to each other.

15.3.2 Reactive Power Management

The TVA's online VAR Management System (VMS) software platform leverages aspects of Bigwood Systems' flagship voltage stability analysis and control tools for real-time visualization, monitoring, and maintenance of reactive power reserves. Operators and engineers use VMS to determine load demand and the ability to meet the demand for each zone. The same is determined for the capability of reactive power imports from neighboring zones while identifying the weakest buses and critical generators. For mitigation, operators can also use the tool to generate control switching suggestions to maintain adequate reactive power. These capabilities facilitate thorough voltage analysis for power systems with renewable energy under large numbers of contingencies, all while ensuring sufficient reactive power in each zone.

15.3.2.1 Online Assessment

TVA engineers and operators use VMS to assess and compute the reactive reserves of reactive regions using online state estimator case snapshots. The system is applied to conduct operation studies for regional VAR management and voltage security of the power system under a range of online operating conditions. Each case run performs analysis for a large set of credible contingencies ("disturbances" or "what if" events), over 2000 contingencies, using screening and ranking functions for fast identification of insecure contingencies with zero or negative load margins, as well as critical contingencies with small load margins. Load margins are the margin between the current operating point and the point of voltage violation or voltage collapse. Exact voltage security load margins are calculated for the top-ranked contingencies. The exact load margin to voltage collapse for the base case and each of the top-ranked contingencies is provided to the operators. This allows for the real-time study of voltage security and VAR reserves of each reactive region under the list of user-designated contingencies for the network.

Engineers use two modes of operation: (1) online real-time mode for real-time monitoring and analysis of reactive power and voltage security, 24 h per day, 7 days per week; and (2) study mode, using archived real-time actual or simulated cases, which allows users to easily execute shared planning activities and cooperative problem studies leading to solutions that mitigate potential operating shortfalls in voltage security and handle significant penetration of intermittent renewable energy.

15 Reactive Power Management in Real-Time at Tennessee Valley Authority (TVA)

Users can visualize the overall current operating state of the system and are alerted when potential issues might occur. The main display presents a base case summary side by side with the limiting contingency case summary, with each zone represented in a schematic geographic image for the area monitored for VAR reserve margins.

Current Status					
Current MVAR Reserve :	1212 MVAR		Current MVAR Load Margin :		163 MVAR
Regional Load Demand			Regional Power Capability		
	MW	MVAR		MW	MVAR
Total	832	37	Total	2586	1249
Regional source	417	54	Regional source	583	562
Import	415	-17	Import	2003	687

Comprehensive status reports are generated for each region in the system and are updated during each real-time processing cycle. The real-time summary displays current MVAR reserves and the current load margin in MVAR as well as total demand in the region for MVARs and MWs, broken down by the amount of demand supplied by local regional sources as well as the amount from sources outside of the region. In addition, the total capability for the region or "zone" is computed and shown, broken down by the amount available in regional sources and the amount imported from external sources.

MW/MVAR Import					Voltage at Sample Buses	
Branch/Inf. Name	MW	MVAR	Max MW	Max MVAR	Bus Name	Voltage (pu)
5LOWNDES -- 8LOWNDES 11	284	8	829	369	5SEBASTO	1.02
5FIVEPNT -- 5HOMEWOD 1	-106	8	110	59	5W_POINT	1.02
5LOWNDES -- 5CALDNIA 1	41	6	140	29	5PHILADE	1.03
5CALHOUN -- 5DERMA 1	0	9	115	29		
5W_POINT -- 8W_POINT 12	145	-44	541	139		
5STATELN -- 5SMTH_MS 1	23	3	88	15		
5OKOLONA -- 5EGYPT_P 1	27	-7	182	47	Average Voltage	1.02
RR-5	415	-17	2003	687	Comment:	

Additional information shows the details of the MW/MVAR flowing in and out of the reactive reserve region by displaying the amount and direction of the flow for each regional interface branch. Details for each regional (local) VAR source (generators, capacitors) in the reactive reserve region are also shown.

15.3.2.2 Online Mitigation

Operators are supplied with rapid, effective, and economically sound recommendations for control actions that improve the reactive reserve capability and ensure sufficient reactive power in each zone.

For insecure contingencies, the Preventive Control engines recommend feasible control actions to mitigate the need for load shedding. Additionally, the Enhancement Control engines recommend actions to enable an increase in regional reactive reserves where current reserves are judged to be insufficient.

A module for handling Special Protection Systems (SPSs) and Remedial Action Schemes (RASs) monitors post-contingency system conditions for certain criteria violations and, if there is a violation, performs adjustments to simulate stable post-contingency system conditions.

In addition, this system will meet the compliance requirements for the Basic Operating Policies that are monitored by the North American Electric Reliability Corporation (NERC). Regional entities are charged with maintaining adequate reactive reserve regions throughout their control areas. The operation of a deregulated power market has pushed power systems ever closer to their security boundaries. As such, a global and effective methodology and its implementation for reactive power reserve monitoring and management in power system operations and operational planning environments are needed for those entities responsible for scheduling and carrying out power transfers, such as the TVA's role as a NERC Reliability Coordinator and Balancing Authority.

15.3.2.3 Testing to Production

At first in 2011, the VMS was used as an additional online reliability tool that runs in parallel with other tools and determines reactive reserves within local areas of the TVA system. However, after a period covering 18 months of testing, the demonstrated system required functional capabilities and produced accurate results and operational benefits on production data in a test system configuration. During this time when VMS was running locally in the Reliability Engineering office on the real-time data feed outside of the production systems at TVA, several events were experienced where the BSI VAR Management System identified a pending power network problem and was used to remediate the issue prior to an unstable contingency occurring.

These events are presented in the following scenarios.

Scenario Case 1: Hiwassee/Murphy 161 kV Looped System
VMS Successful Result

- A temperate summer day, and the area loads were moderate but not heavy.
- The 161 kV looped system was open for maintenance outage on a 161 kV line.
- The only local generation was hydro, and it was low due to lack of water.
- The system was stable in the base case; VMS identified an issue for the next contingency with load increase.
- Engineers noted that VMS identified an issue and backed-up the study with an off-line tool.
- The line was returned to service before the issue became real-time.

- Would have been in post-contingency load shed if not identified in time; approximately 75–100 MW were at risk.
- No other online tools in production or development identified this issue.

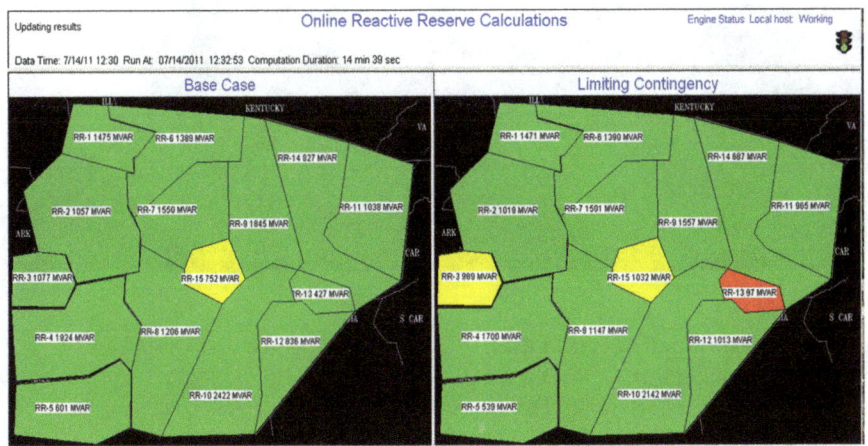

Scenario Case 2: Lower Mississippi 161 kV System
VMS Successful Result

- 9/22/11—Early fall day, area loads were moderate but not heavy, outage season was starting.
- One 161-kV source line (from W. Point) was open for maintenance outage.
- The only local generation was Red Hills and Kemper; Red Hills was in outage and Kemper was not running.
- The system was stable in the base case; VMS identified an issue for the next contingency with load increase.
- Engineers noted that VMS identified an issue and backed-up the study with an off-line tool.
- The line was not available to immediately return to service before the issue became real-time.
- Two Kemper CTs were brought online to mitigate the next contingency.
- Would have been in post-contingency load shed if not identified in time; approximately 50–75 MW were at risk.
- *No other online tools in production or development identified this issue.*

Scenario Case 3: Johnsonville Bus Outage—Potential Collapse
VMS Successful Result

- The TVA system peak load forecast for 4/20/2015 was ~16,600.
- Johnsonville Fossil Plant Bus 1–3 was scheduled for a one-day outage on 4/20/2015 to perform bus differential testing.
- The following elements in the area were outage:

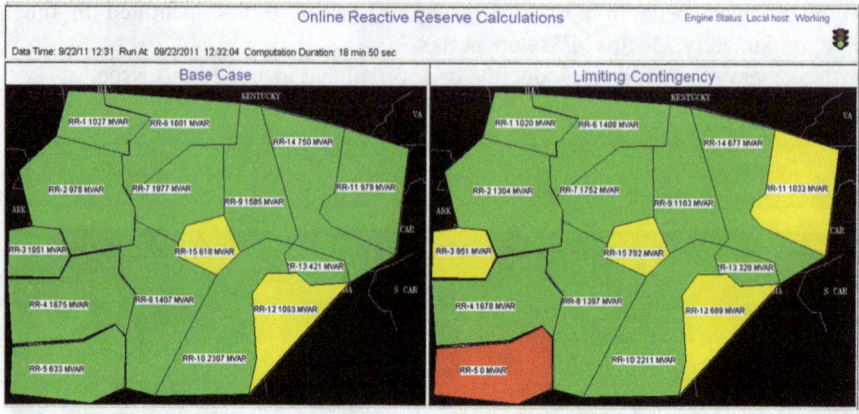

- Marshall-Mayfield (Oak Level)—Out for line upgrade (3/23–5/08).
- Johnsonville Bus 1–3—Out for Bus Differential testing (4/20).
- Monsanto-Johnsonville (Hilltop)—Out for pole replacement (4/13–4/29).
- Colbert Fossil Plant-Lawrenceburg (Waynesboro)—Out for SOL mitigation for the Oakland-Wilson line during the Colbert bus 2–1 outage (4/17–4/20).

- *No other online tools in production or development identified this issue.*

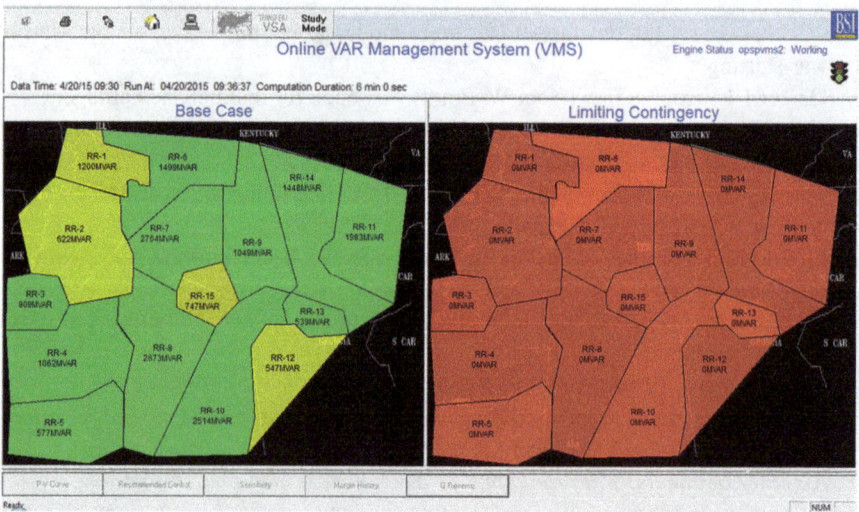

Scenario Case 4: Member's Bus A Station
VMS Successful Result

- The VMS tool identified a potential collapse in a TVA member's area on the morning of 2/13/2017.

- Breaker A was open for the planned outage.
- Upon further investigation, this contingency would not happen under normal conditions if breaker C1 at Member opened and breaker D1 at Clay failed to close.
- (*Utility name and station names are changed and redacted.*)

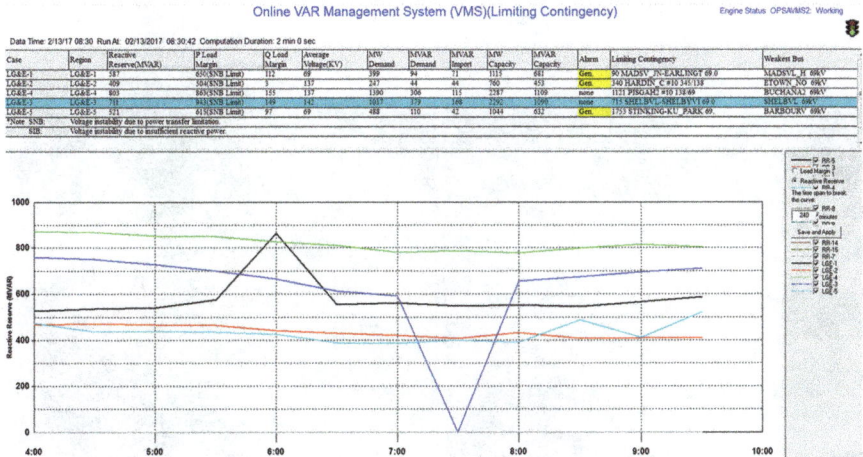

These types of results made it possible to fast-track funding for implementation of the features needed for VMS as a production tool in 2013. Since then, the application of VMS has been incorporated in TVA transmission system operations Standard Operating Procedures (SOP).

15.3.2.4 The Online VMS Application in the TVA Control Center

The VMS uses a different load flow solution (developed in the late 1990s) rather than that used by GE-Areva EMS applications: State Estimation (SE) and Real-Time Contingency Analysis (RTCA) tools based on 1970s methods. TVA RTCA performs contingency analysis every few minutes using a system snapshot and reports post-contingency values. RTCA may solve the load flow and obtain what appears to be a valid solution with acceptable voltage, when in actuality the solution may be very near the point of collapse, such that a marginal load increase would result in voltage instability.

The VMS takes the same real-time snapshot (every 15 mins) and scales the area load to determine the margin to instability. In fact, VMS normally solves the state beyond instability to determine a more exact collapse point (the "nose" of the P-V curve). The normal cycle takes less than 5 mins to determine a solution. VMS divides the TVA system into 15 regions based on load, reactive cohesion, and system topology. (Presently, only the TVA control area is managed; an extension to provide VMS results for the TVA Reliability Coordinator (RC) partners is in development.)

In the following figure, the main VMS display shows the 15 regions. VMS results, including the VMS region color-coding, are exported for display on the main Control Center Reliability Monitor. An area showing green has an adequate reactive reserve. For a yellow or red alarm, the VMS application is consulted for details.

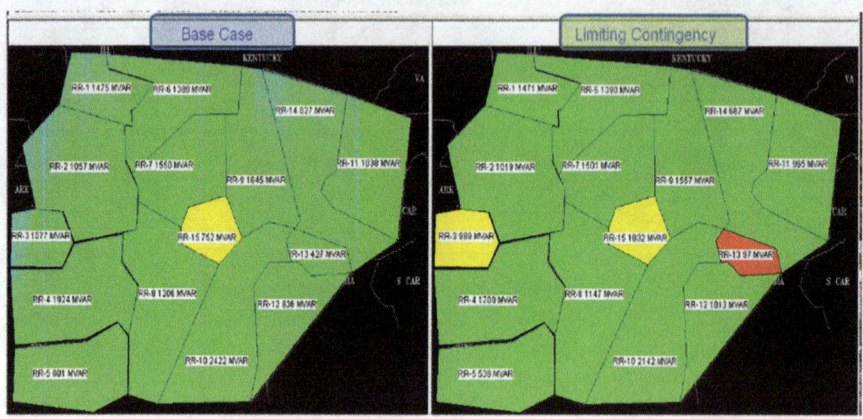

Top-level VMS Display

The VMS results present each region in a top-level summary display to facilitate monitoring in the control center and access to detailed reports for the regions. The panel on the left displays analysis results for the base case of the real-time snapshot with load increase. The panel on the right displays analysis results of the limiting contingency, that is, the next contingency with load Increase.

The following VMS Operating Guidelines were developed and instituted as SOP at TVA:

- Green-Shaded Region = Secure Region—This indicates normal operation.
- Yellow-Shaded Region = Reactive Deficient Region (MONITOR)

  – This indicates:
    - Low average voltage (<95%) on selected 161-kV buses or over 90% of online reactive resources being used
    - The operator should take action to return the region to normal operation.

- Red-Shaded Region = Unstable Region (TAKE ACTION)

  – This indicates:
    - Low average voltage (<95%) on selected 161-kV buses AND over 90% of online reactive resources being used or voltage collapse predicted by VMS

– Appropriate operator actions are dependent upon whether the red condition is due to a next contingency issue or a real-time issue.

VMS analyzes each region's voltage stability by assessing the ability of local reactive resources to serve the load with a stable voltage. VMS studies both the real-time (base case) and next contingency (all TVA contingencies) cases to determine the "limiting contingency." It scales the load in each region and plots the real power load against the voltage. The resulting shape of the P-V curve, which can be displayed by the operator, is an indication of the amount of VAR reserves in the region.

Operators can display detailed results for each region to show the buses selected for voltage monitoring, the 161-kV transmission connections (including transmission lines and 500/161-kV intertie transformers) at the region boundary, and regional MVAR sources (including generators and capacitors) within the region.

15.4 Conclusion

Voltage instability can be caused by several different factors, and making sure to evaluate these factors, such as reactive power deficiency, is essential. By partitioning the TVA service territory into reactive reserve zones and then performing real-time monitoring and analysis on each zone, operations can ensure maintenance and adequate reactive power to move real power across their bulk transmission system. The VMS tool was developed to meet a need and then continued to develop to handle new concerns with both real and reactive power.

Chapter 16
Use of Voltage and Transient Security Assessment Tools for Grid Operations at BC Hydro

Ziwen Yao and Djordje Atanackovic

16.1 Introduction of British Columbia Power Grid [1, 2]

British Columbia Hydro and Power Authority (BC Hydro) is a provincial crown corporation responsible for the generation, transmission, and distribution of electric energy in the province of British Columbia, Canada. BC Hydro generates over 54,600 gigawatt-hours of electrical energy annually to supply more than 2.0 million residential, commercial, and industrial customers within its service area that consists of most of the geographical area of the province. Over 90% of this energy is generated in hydroelectric power stations and is transmitted over large distances to the load centers. In addition, there are a number of independent power producers (IPPs), large industrials, and other utilities (collectively referred to as market participants) connected to the transmission system. Many smaller IPPs are expected to come on line in the future. This energy is delivered using an interconnected system of over 79,000 kilometers of transmission and distribution lines. Generation schedules are passed to Grid Operations by the generating companies, of which BC Hydro is by far the largest, and Grid Operations controls the generation in real-time.

The BC high voltage transmission system consists of 20,385 km of transmission lines, operating at voltages from 60 to 500 kV. The 500 kV bulk transmission network connects the major generation centers in the Northern and South Interior regions of the province with the major load centers in heavily populated southwestern BC. There are major interconnections with the United States of America (500 kV and 230 kV) and Alberta (500 kV and 138 kV as shown in Fig. 16.1).

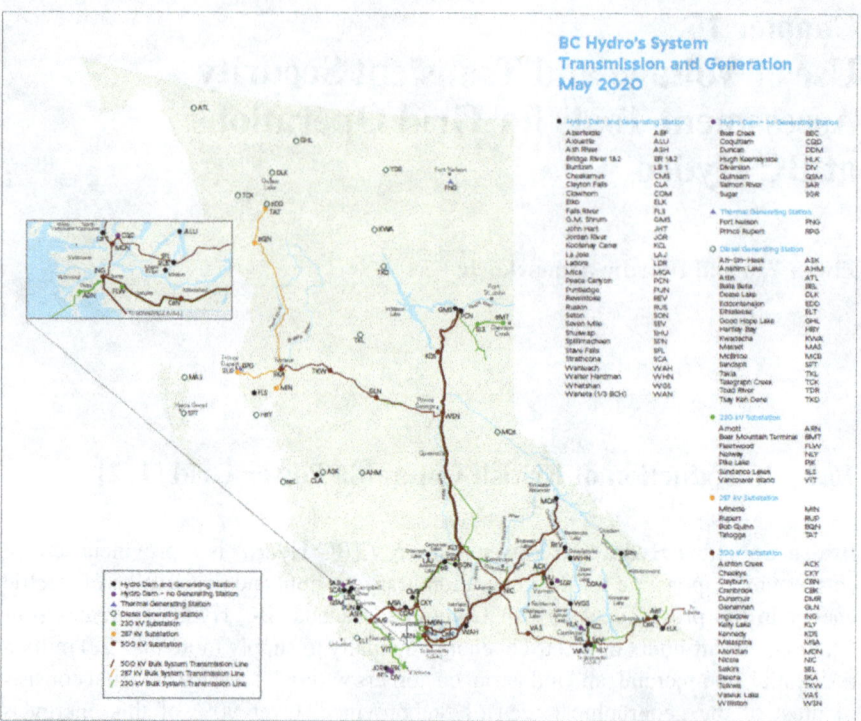

Fig. 16.1 British Columbia bulk transmission system [1]

Electricity is supplied to the two largest load centers, Lower Mainland and Vancouver Island, from the Peace River hydroelectric system through Kelly Lake Substation and from the Columbia River system through Nicola Substation. The relationship between installed generation capacity and electrical demand around the province drives the development and operation of BC Hydro's bulk transmission system. The distances between major generation and load centers are significant giving rise to transient and voltage stability problems that must be monitored and managed by the transmission operator.

BC Hydro's bulk transmission system is planned and operated so that at any time the system can withstand an outage of any single transmission line without compromising the supply to loads. There are exceptions to this rule, such as areas served by single radial lines (e.g., the north coast). The bulk transmission system complies with industry planning and operating standards stipulated by NERC (North American Reliability Corporation) to ensure a high level of reliability.

In BC, over 90% of electricity is supplied by hydropower that is transmitted from remote areas to the main load centers of the Lower Mainland (LM) and Vancouver Island (VI) via long distance transmission lines (Fig. 16.1). This attribute of the grid configuration with long distance transmission lines raises the typical security concerns as follows:

1. Thermal ratings violation;
2. Voltage instability;
3. Voltage limits violation;
4. Voltage deviation limits violation;
5. Frequency instability;
6. Transient instability.

On the other hand, hydro generating units provide flexible and effective ways for security control, such as generation shedding for improving transient stability, frequency stability, and other performance issues in the system. Hence, BC Hydro has installed extensive Remedial Action Scheme (RAS) within the provincial power grid to assure system security against various system disturbances. These individual RAS installed in arming substations are further integrated with EMS/SCADA applications to form and an integrated RAS system [3] that establishes RAS arming patterns periodically in real-time based on actual power system conditions.

The real-time assessment tools (RTATs) presented in this chapter are used in BC Hydro control centers to evaluate the security performance of the system with the integrated RAS in real-time operations.

16.2 BC Hydro's EMS and Integrated RAS Systems

16.2.1 Introduction of BC Hydro's EMS

Two redundant control centers are in operation within BC Hydro. In each control center, two redundant EMS are installed and are setup in a multi-host redundant environment providing the quad redundant operational architecture whereby one EMS instance runs as a primary with three other instances providing standby capability. All real-time data are fully replicated continuously among all four systems. Each EMS instance comprises core subsystem and platform applications. Core consists of core control system and Supervisory Control and Data Acquisition subsystem (SCADA) while platform comprises Automatic Generation Control (AGC) and suite of advanced applications for transmission network analysis. Network analysis applications include state estimator (RTNET), Transient Stability Analysis by Pattern Matching (TSAPM), Real-time Contingency Analysis (RTCA), Real-time Voltage Security Assessment (RTVSA), and Online Transient Security Assessment (OLTSA) which all execute in real-time applications sequence triggered periodically every 4 min or manually on demand.

State Estimator (RTNET) retrieves live data snapshot from SCADA and produces an estimation of the actual power system condition, including statuses of power equipment and control devices, power flow on transmission lines, and bus voltages. Other network applications receive the latest RTNET solution by copying the network database, NETMOM, from RTNET (or another up-stream application) to use the estimate as the starting point for their calculations. Therefore, NETMOM is the main database in EMS that contains power system network model including equipment ratings and the most recent solution of the network steady-state real-time condition.

The dynamic model for time domain simulation resides on a separate server where DSA application suite is located (DSA Master) and this model is combined with the basecases and scenario files assembled by EMS application in order to constitute full input case definition for TSAT application that performs transient security assessment using time domain simulation.

16.2.2 BC Hydro's Integrated RAS System

RAS schemes have been installed extensively in BC power grid to support real-time operations [3]. These distributed individual RAS schemes are fully integrated with EMS to constitute an integrated RAS system as shown in Fig. 16.2. Furthermore, the RAS arming patterns of all these individual schemes are determined by TSAPM that encompasses complete logic necessary to process a large number of patterns and match them against the actual power system condition to arrive at the optimal RAS arming pattern. Once determined, the RAS arming pattern is downloaded automatically from EMS to the relevant Programmable Logic Controllers (PLC) in the field at the arming stations to complete the arming process.

Such an integrated system is being used in the BC power grid to provide timely RAS arming patterns to arm RAS schemes in the field to match closely the ever-changing power system conditions. In addition, the RAS arming patterns are modeled in EMS-RTAT for evaluating the security performance of the system in real-time operations.

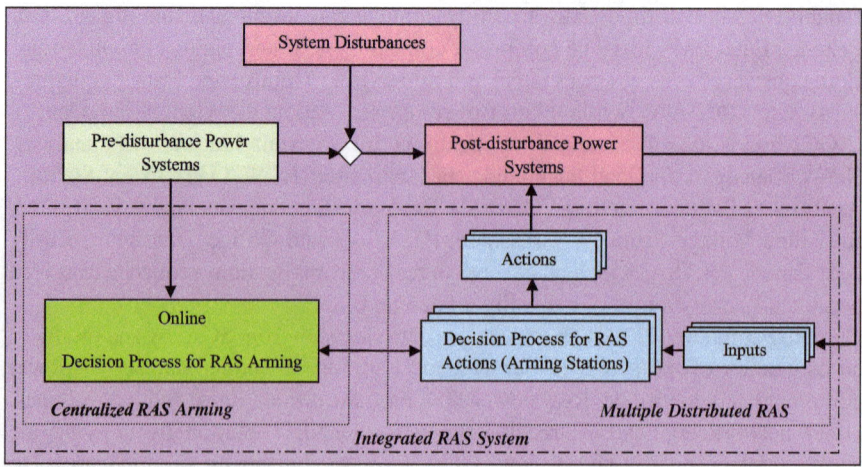

Fig. 16.2 Functional diagram of an integrated RAS system via EMS [4]

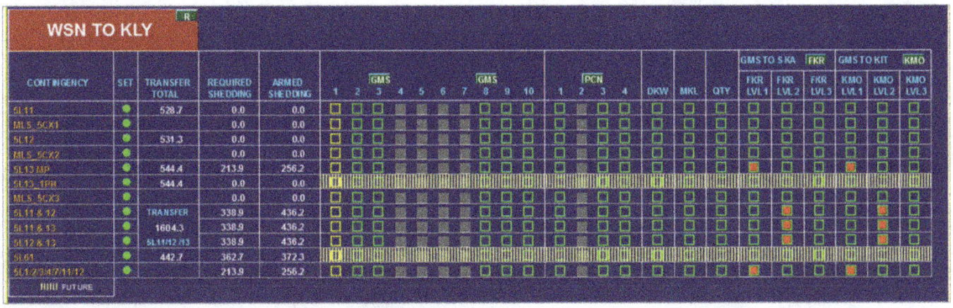

Fig. 16.3 Example of RAS Arming Pattern (or Matrix)

The composite boxes of "Multiple Distributed RAS" (Fig. 16.2) that represent the distributed individual RAS schemes are installed in "Arming Stations" in the provincial power grid. Each scheme or each "Contingency" in Fig. 16.3 was designed to address certain predefined security issues (a combination of issues 1–6 presented in the previous section). Moreover, each contingency contains a number of redefined RAS actions to be taken upon the occurrence of the corresponding contingency, which forms a row of a RAS arming matrix (Fig. 16.3). The RAS actions will be armed for certain contingencies in real-time based on current operating conditions. Those actions include: include generation shedding (GS), line tripping, load shedding, and shunt device switching.

Furthermore, all these individual schemes at "Arming Stations" are integrated via SCADA/EMS as an integrated RAS *system* (in other words, a wide area protection system), which determines the arming patterns and downloads them to the arming stations.

16.3 BC Hydro's RTAT Integrated with EMS

The suite of real-time assessment tools (RTAT) comprises contingency analysis (CA), transient stability analysis by pattern matching (TSAPM), voltage security assessment (VSAT), [5] and transient security assessment (TSAT) [6]. All of them have real-time (RT) versions and study (ST) versions in EMS environment; RT versions are running in online network sequence shown in Fig. 16.4 periodically every 4 min and/or on demand while ST versions are used by system operators and operations planning engineers.

The study version of the RTATs can be run in manual mode in the sequence depicted in Fig. 16.5.

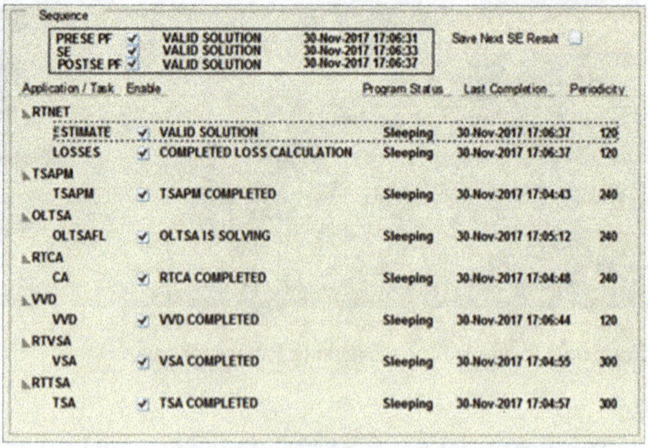

Fig. 16.4 Online network sequence

RTNET	State estimator
TSAPM	Transient stability analysis by pattern matching calculates RAS arming patterns based on operating orders
OLTSA	Online transient security assessment, contains 4 main processes (P0, P1, P2, and P3): 1. P0—Generates event sequences of disturbances in node-breaker models for P1 and P3 2. P1—Calculates generation shedding arming patterns for Peace area system 3. P2—Evaluates generation shedding arming patterns obtained from P1 by comparing with TSAPM results 4. P3—Dynamic security assessment, including VSA and TSA for the BC transmission system
RTCA	Real-time contingency analysis performs power flow based security assessment for the current operating conditions and generates event sequences of disturbances in node-breaker models for RTVSA
RTVSA	Real-time voltage security assessment performs power flow based security assessment for the current operating conditions and security regions

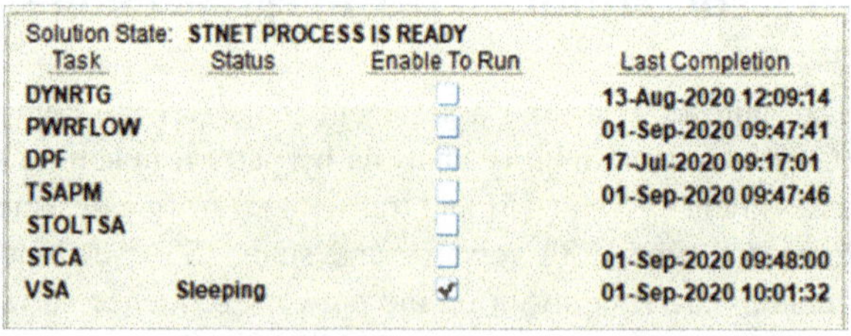

Fig. 16.5 Study network sequence

16.4 Use of VSAT and TSAT for Grid Operations at BC Hydro

As mentioned in the previous sections, the in-house application, TSAPM plays a role as an agent of decision-making for establishing RAS arming patterns in real-time according to the rules specified in system operating orders, which are mainly

Fig. 16.6 Illustration of bus configuration changes [7]

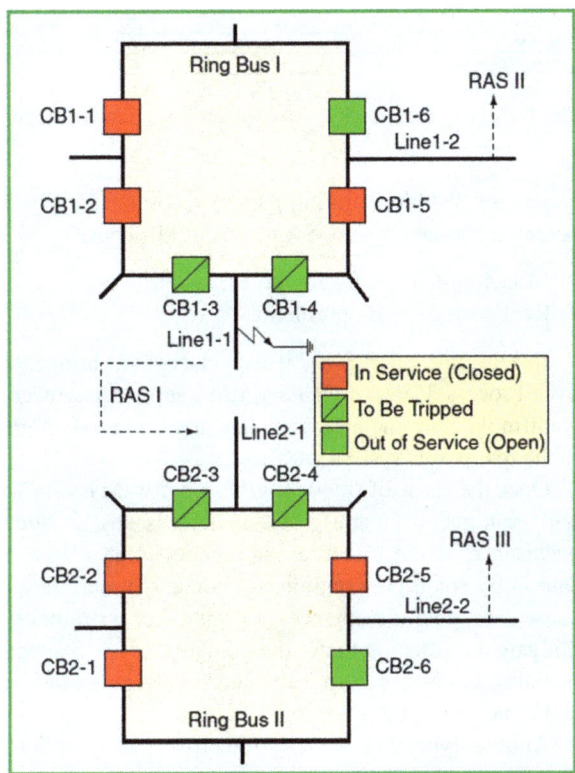

determined by extensive off-line planning studies that are usually performed using bus-branch models and a number of representative basecases of important system conditions.

Therefore, the main objectives of RTATs are to assess security performance of the real-time systems based on current system conditions using node-breaker model from state estimator. In addition, event sequences of disturbances due to predefined contingencies and subsequent RAS actions with detailed time delays are created by RTCA in a node-breaker format.

The node-breaker models for the representation of the power system and disturbances are essential for detecting potential cascading events in real-time assessment of security performance. Figure 16.6 illustrates how the same predefined contingency can trigger different event sequences from the same basecase with bus-branch model but with different underlying bus configurations. In the pre-contingency system, the OPEN/CLOSED statuses of CB1-6 and CB2-6 would not affect the power flow of the bus-branch basecase. However, they may cause different potential cascading events.

Figure 16.7 depicts a small portion of the display for dynamic security assessment results from Process 3 (P3) of OLTSA, where it indicated that the system would be "Insecure" due to the contingency of "C5L11_13" using the current

C5L11_12		0.0	0.0	0.0	0.0	Secure	225.6	338.6
C5L11_13		275.0	275.3	275.0	275.3	Insecure	1322.4	773.8
C5L12		0.0	0.0	0.0	0.0	Secure	0.0	0.0

Fig. 16.7 Dynamic Security Assessment Results from Process 3 (P3) of OLTSA

generation shedding arming pattern recommended by TSAPM. There may be two potential causes for the DSA results of "Insecure":

1. Modeling/DSA software issues;
2. Real power system problems.

In real-time operations, if any insecure contingencies (Fig. 16.7) are reported from Process 3 (P3) of OLTSA, EMS support engineers will investigate the case to confirm whether the insecurity is caused by modeling/DSA software issues or due to the real power system problem.

Once the cause of "modeling/DSA software issues" is excluded, the investigation will continue to identify the system issues as soon as possible prior to the occurrence of the insecure contingency. One of the common system issues was caused by some pre-contingency outages of certain circuit breakers, which might cause cascading disturbances. In this case, operators would need to take action to mitigate the situation under the guidance of EMS support engineers and operations planning engineers. Typically, these actions include reconfiguring bus topology and/or adjusting RAS arming patterns.

Another typical use-case of the online DSA applications in real-time operations is to adjust RAS arming pattern to increase security regions and transfer limits [8].

As shown in Fig. 16.1, BC power grid consists of two main generation sources (northern and eastern areas) that supply power to the BC load. BC power grid is connected to Alberta power grid and western power grid of the United States.

It is required for DSA applications to provide two-dimensional (2D) security regions of various combinations of these three main parameters (North Gen, East Gen, and BC Load) generated by varying three parameters while stressing the interface between BC and US grids as shown in Fig. 16.8 and transfer limits that take into account the 2D security region for the two generation sources.

Without losing generality, the results generated from VSAT will be used to illustrate various issues related to the limited search for 2D security regions.

The search technique implemented in online VSAT is based on the power flow tracking from the basecase operating point calculated by EMS state estimator. The fast decoupled power flow is used in VSAT to perform radial search from the operating point in a predefined number of directions in the 2D search spaces. At each of the search points, a set of credible contingencies is simulated to determine the limiting constraint that can result from thermal, voltage stability/voltage collapse, and voltage decline conditions. The search continues until the voltage stability limit is found for a particular direction. Once the search is completed for all 24 directions, the contours are constructed connecting limiting points for thermal, voltage decline,

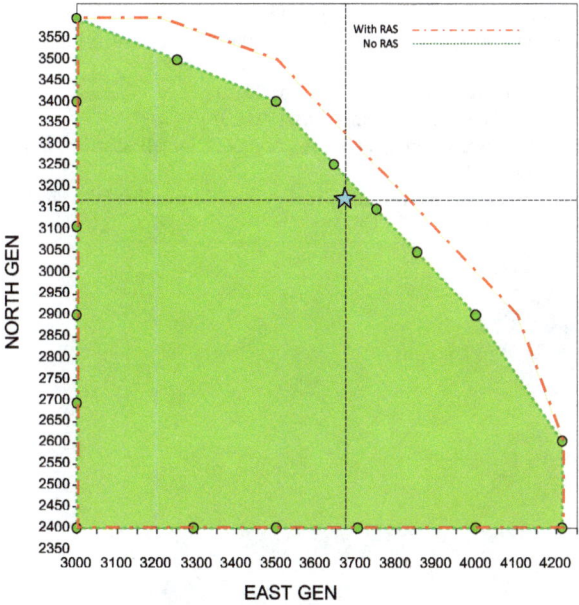

Fig. 16.8 Security Regions with and with no RAS [8]

and voltage stability limits. Those contours represent security regions with respect to the aforementioned conditions.

When the operating point is moving close to the security boundary of "No RAS" (or with the initial RAS arming pattern), operators can manually arm certain generating units for shedding in the RAS arming pattern (Fig. 16.3) to increase the security region to the boundary of "With RAS" (or with the adjusted RAS arming pattern).

Currently, Process 1 (P1) of OLTSA calculates generation shedding arming patterns only for the Peace area system. In the near future, P1 would be expanded to cover more areas of the BC grid.

References

1. https://www.bchydro.com/content/dam/BCHydro/customer-portal/documents/corporate/accountability-reports/financial-reports/annual-reports/BCHydro-System-Map-202005.pdf (2020)
2. https://www.bchydro.com/content/dam/BCHydro/customer-portal/documents/corporate/accountability-reports/financial-reports/annual-reports/BCHydro-Quick-Facts-20190331.pdf (2020)
3. S.C. Pai, J. Sun, BCTC's experience towards a smarter grid—increasing limits and reliability with centralized intelligence remedial action schemes, in *IEEE Canada Electric Power Conference* (2008), pp. 1–7

4. Z. Yao, V.R. Vinnakota, Q. Zhu, C. Nichols, G. Dwernychuk, T. Inga-Rojas, Forewarned is forearmed: An automated system for remedial action schemes. IEEE Power Energy Magazine **12**(3), 77–86 (2014)
5. https://www.powertechlabs.com/services-all/vsat (2020)
6. https://www.powertechlabs.com/services-all/tsat (2020)
7. Z. Yao, A. Steed, G. Dwernychuk, D. Cave, A model for all seasons. IEEE Power Energy Magazine **8**(1), 54–60 (2010)
8. Z. Yao, D. Atanackovic, Issues on security region search by online DSA, in *IEEE PES General Meeting*, July 2010

Chapter 17
Use of Voltage Stability Assessment Tools at San Diego Gas & Electric

Anita Hoyos, Kenneth Poulter, Robin Manuguid, Michael Vaiman, and Marianna Vaiman

17.1 San Diego Gas and Electric System Overview

SDG&E, a subsidiary of Sempra Energy, is a regulated investor-owned utility that provides electricity to 3.6 million customers in San Diego and Southern Orange Counties through 1.4 million electric meters. It is a summer peaking utility with an all-time peak of 4890 MW.

SDG&E's service territory covers 4100 square miles and includes a total 1103 miles of Bulk Electric System (BES) transmission comprised of 500 kV (251 miles), 230 kV (585 miles), and 138 kV (267 miles). Its transmission ties in with four other entities in the Western Interconnection: Southern California Edison (SCE), Imperial Irrigation District (IID), Centro Nacional de Control de Energia (CENACE), and Arizona Public Service (APS).

The company owns, operates, and is responsible for designing Remedial Action Schemes (RAS) needed to meet the requirements of its Planning Coordinator and Transmission Planner.

SDG&E is within the metered boundaries of the California Independent System Operator (CAISO) Balancing Authority area. There are three large Independent Power Producer generators (>500 MW) operating in the SDG&E service territory along with several quick start gas turbines and renewable energy generating facilities. Load is served by generation directly connected to its transmission system

---

A. Hoyos · K. Poulter · R. Manuguid
Electric Grid Operations Department, San Diego Gas and Electric, San Diego, CA, USA
e-mail: ahoyoss@gmail.com; kpoulter@sdge.com; rmanuguid@sdge.com

M. Vaiman · M. Vaiman (✉)
V&R Energy Systems Research, Inc., Los Angeles, CA, USA
e-mail: mvaiman@vrenergy.com; marvaiman@vrenergy.com

and imports from neighboring utilities located to the north, east, and south of San Diego. SDG&E owns and operates 1186 MW of generation that includes Blackstar Resources.

17.1.1 Voltage Stability Constrained Utility: Past and Present

SDG&E had traditionally been a voltage stability constrained utility. After close to 30 years of studies resulting in important additions to the system, the utility reached a point where it is no longer voltage stability constrained. To get to this point, many years of voltage stability studies and operating practices were performed using different tools.

17.1.2 Transmission System and System Changes

SDG&E's transmission system consists of four voltage levels: 69 kV, 138 kV, 230 kV, and 500 kV. For close to three decades, the interconnections to the neighboring utilities consisted of a 500 kV line to the east (to Arizona), five 230 kV lines to the north (to Southern California Edison), two 230 kV lines to the south (to CENACE), and a 230kV line to IID, see Fig. 17.1. The import capability into the San Diego area is constrained whenever the 500 kV line is forced out and could create a condition where the import is voltage stability limited. This condition also required increased unit commitment and a process for derating the import capability whenever select transmission equipment was out of service. SDG&E then deployed an aggressive capacitor installation project that covered capacitors from the 230 kV level to the distribution level. All the 230 kV to 69 kV capacitors were voltage-controlled and would come up staggered within seconds in response to the 500 kV line contingency. SDG&E then developed various QV routines to perform its voltage stability studies and became a leader in the Western Electricity

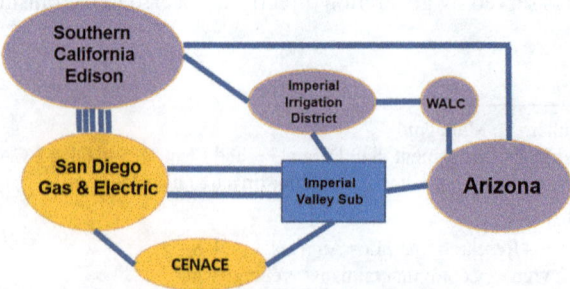

Fig. 17.1 Bubble diagram for the surrounding transmission system

Coordinating Council (WECC) in voltage stability studies and criteria development. Some of these tools are still used today, and many refinements have been added. Years later SDG&E added a 500 kV line, which increased the import level by relieving voltage stability constraints and some thermal concerns. This addition supported the decommissioning of the San Onofre Nuclear Generating Station. As a result of studies to increase its import capability in the last few years, SDG&E has added two parallel phase-shifting transformers and seven synchronous condenser units with almost 1600 MVar total reactive power capability.

Significant System Expansion:
- Addition of Sunrise Power Link 500 kV Line from Imperial Valley Substation.
- Installation of Synchronous Condensers:
  - 1125/−560 MVar, 5 units (northern area of SDG&E territory)
  - 450/−220 MVar, 2 units (southern area)
- Significant addition of renewable generation: from about 80 MW in 2013 to almost 1900 MW by 2020 and more being added in the Imperial Valley Area.
- Addition of Pio Pico Energy Center (3–100 MW peaker units) in the southern area of SDG&E's system.
- Addition of Carlsbad Energy Center (5–100 MW peaker units) in the northern part of the system with retirement of about 970 MW of generation at the same site.
- Addition of phase shifters (2–400 MVA transformers) in Imperial Valley on the 230 kV Imperial Valley—La Rosita (CENACE) Line.

17.2 Registered Functions and Relationship with CAISO BA/TOP and Peak Reliability

17.2.1 The Transmission Control Agreement

In 1998, the California Legislature enacted Assembly Bill 1890 ("AB 1890") that restructured the California electric industry and established the California Independent System Operator (CAISO) with centralized control of a state-wide transmission grid. The original Participating Transmission Owners (Pacific Gas & Electric, Southern California Edison, and SDG&E) under the Transmission Control Agreement transferred operational control of transmission assets to the CAISO for the purpose of allowing them to be controlled as part of an integrated Balancing Authority Area.

17.2.2 NERC-Registered Functions

Pursuant to the provisions of the Federal Power Act, applicable Federal Energy Regulatory Commission (FERC) Orders, and the North American Electric Reliability Corporation (NERC) Rules of Procedure, bulk power system users, owners, and operators who are responsible for performing reliability functions must be registered with NERC. SDG&E is registered with NERC as a Distribution Provider (DP), Generator Owner (GO), Generator Operator (GOP), Resource Planner (RP), Transmission Owner (TO), Transmission Operator (TOP), and Transmission Planner (TP). SDG&E has an agreement in place with the CAISO that results in the transfer or sharing of compliance responsibility for reliability standards related to the TOP function.

17.2.3 SDG&E's Relationship with the CAISO

The CAISO must comply with all requirements of NERC Reliability Standards applicable to its current registered reliability functions: Balancing Authority, Planning Coordinator, Transmission Operator, and Transmission Service Provider. The CAISO has five CFR[1] agreements that identify the reliability standards compliance responsibilities for each party registered under the TOP function for the transmission facilities within the CAISO balancing authority.

For example, the CAISO and SDG&E shall each separately operate or direct the real-time operation of devices to regulate transmission voltage and reactive flow to comply with VAR-001-5, R.3. However, SDG&E has sole responsibility for specifying a system voltage schedule as part of its plan to operate within System Operating Limits[2] (SOLs) and Interconnection Reliability Operating Limits[3] (IROLs) to comply with VAR-001-5, R.1. In the case of scheduling sufficient reactive resources to regulate voltage levels under normal and contingency conditions for compliance

---

[1] A Coordinated Functional Registration (CFR) is where two or more NERC-registered entities agree in writing upon a division of compliance responsibility for one or more reliability standards, requirements, or sub-requirements applicable to a particular function (see the NERC Rules of Procedure, Section 508, Appendix 5A).

[2] System Operating Limit is the value (such as MW, MVAR, Amperes, Frequency, or Volts) that satisfies the most limiting of the prescribed operating criteria for a specified system configuration to ensure operation within acceptable reliability criteria. System Operating Limits are based upon certain operating criteria. These include, but are not limited to: Facility Ratings (Applicable pre- and post-Contingency equipment or facility ratings), Transient Stability Ratings (Applicable pre- and post-Contingency Stability Limits), Voltage Stability Ratings (Applicable pre- and post-Contingency Voltage Stability), System Voltage Limits (Applicable pre- and post-Contingency Voltage Limits) (Source: NERC Glossary of Terms).

[3] A System Operating Limit that, if violated, could lead to instability, uncontrolled separation, or Cascading Outages that adversely impact the reliability of the Bulk Electric System.

with VAR-001-5, R.2, SDG&E and CAISO have split responsibility where CAISO schedules reactive resource facilities it has operational control over, including but not limited to generation and transmission lines. SDG&E is responsible for the scheduling of reactive resource facilities it has operational control over, including but not limited to transmission lines, reactive resource switching, and the use of controllable load.

It is important to note that while the CAISO has sole responsibility for compliance with the requirements of several TOP standards, SDG&E conducts and performs its own study studies and analyzes and coordinates with the CAISO. For example, while the CAISO has the sole responsibility to comply with TOP-002-4, R.1 by providing an Operational Planning Analysis, SDG&E shares its next day studies in advance with the CAISO for its review.

17.2.4 NERC, WECC, Peak Reliability, and the RC West

The Energy Policy Act of 2005 requires that FERC approves and enforces standards to protect and improve the reliability of the nation's bulk power system. NERC develops, revises, and implements standards under this statutory framework and delegates compliance monitoring and enforcement authority to various regional entities. In the Western Interconnection, the compliance monitoring and enforcement is delegated to the WECC by NERC.

WECC's role and scope of activities have grown with the introduction of mandatory reliability standards in 2007 and in 2009 and with the assumption of the Reliability Coordinator function for the Western Interconnection.

Peak Reliability (Peak) was formed on February 12, 2014, as a result of the bifurcation of the WECC into a Regional Entity (WECC) and a Reliability Coordinator (Peak). SDG&E was a Class 1-member (Electric Line of Business Entities owning, controlling, or operating more than one thousand (1000) circuit miles of transmission lines of 115 kV or higher voltage within the Western Interconnection).

On July 1, 2019, RC West took over the Reliability Coordinator function for the CAISO BA footprint and, following additional certification by NERC and WECC in early November 2019, the RC West will become the Reliability Coordinator for another 23 entities in the Western Interconnection, overseeing 87% of the load in the western United States, [1]. For a list of RC West entities and the RC West footprint, please refer to [2].

17.3 Current and Past Voltage Stability Analysis and Results

17.3.1 Off-line Voltage Stability (VS) Studies

SDG&E continues to perform, on a seasonal basis and in the Planning Horizon, as well as in the Operations Horizon,[4] voltage stability studies. For these studies, a combination of techniques is used, namely QV and PV. These are done to determine if any voltage stability limitation exists with elements out of service.

17.3.2 Online VS Study

In real-time, the CAISO and Peak have been using two different programs to monitor the import limits into the San Diego and CENACE areas. CAISO also runs day-ahead studies and provides the import limit for every hour of the day.

Accurate modeling of the internal as well as the external network model is key to obtain reliable results and will also minimize convergence problems.

The hurdle in doing either PV or QV analyses is convergence. Both analyses use power flow as their tool, and in most cases the system is taken to higher load levels, low generation under a set of critical contingencies which likely results in nonconvergence before the system has reached a collapse point. SDG&E engineers have developed a series of tools to be able to overcome these convergence issues. These tools include a "slow opening" of lines by increasing the line impedance or using fictitious generators at the line terminals, routines to switch reactive devices, and other creative ideas such as opening the Var limits to help solve and then restrict them in steps. It is of utmost importance that the engineers involved in such studies be able to determine if there is a convergence issue, or if the runs have reached a real voltage stability limit.

When observing the limits calculated in real-time, before taking any actions, it is important to have well-trained personnel that can validate the results so that convergence issues or problems in the modeling do not result in unnecessary operator action.

17.4 VSA Tool for Practical Online Implementation

From about 2015 to the retirement of Peak Reliability (Peak), two online VSA tools were running in real-time (as a service) at SDG&E: one in sync with the "primary"

---

[4]The RC West, the Reliability Coordinator for SDG&E TOP, defines Operations Horizons as: *A rolling 12-month period starting as Real-time (now) through the last hour of the 12th month into the future.*

VSA tool running at Peak (RC) and the other in sync with the "secondary" tool running at CAISO (BA, TOP). Solved state estimator/power flow cases are provided by CAISO and Peak at regular intervals.

As a participant of the Peak Reliability Synchrophasor Program (PRSP), SDG&E had access to the Peak-ROSE (Region Of Stability Existence) program/package. As of November 2015, a secure connection has been established with the Peak West-wide System Model (WSM) server to receive data needed every 5 minutes to run the real-time version of the Peak-ROSE program.

By the time Peak shuts its operation down by the end of 2019, SDG&E expects to have its VSA tool, SDGE-ROSE, configured to run using the CAISO's network model and will be considered the secondary tool to the CAISO's VSA tool. RC West intends to use SDG&E's real-time VSA results as the primary backup in the event of RC West's VSA tool failure.

The following ROSE capabilities are used by SDG&E:

- Continuously monitoring the electric grid;
- Identifying system stability limits under normal and contingency conditions in terms of MW margin across the interface;
- Alarming the operator if the operating point is close to system stability limit in terms of interface flows;
- Incorporating real-time Remedial Actions Schemes (RAS).

In the discussion that follows in Sect. 17.5, the focus is on mostly assumptions in the VSA tool as originally implemented by Peak and highlighting those assumptions that SDG&E changed in its implementation of the VSA tool.

17.4.1 Methodology of Computations

The main purpose of ROSE (both Peak-ROSE and SDGE-ROSE) is to perform voltage stability analysis.

SDGE-ROSE performs scenario-based voltage stability analysis. It computes one-dimensional stressing scenarios defined in the scenario files. For each scenario, the following analyses are performed:

1. Computation of the interface limit;
2. PV-curve analysis;
3. QV-curve analysis.

SDGE-ROSE incorporates a nonlinear power system model. The full Newton method is used to solve nonlinear power flow equations. The contingency analysis technique uses the full AC analysis.

ROSE uses State Estimator (SE) data in a node-breaker format for online calculation and visualization of the current operating point and its proximity to the steady-state stability limit. Relationship between the current operating point and the limit defines "health" of power system network state, and is the power system stability margin.

17.4.1.1 Computing VSA Import Limit and Limiting Contingency

SDGE-ROSE simulates a power transfer by changing user-defined generation/load in source/sink locations. At each transfer level, the tool computes pre-contingency flow on a specified interface. It also applies all user-defined contingencies at each transfer step. RAS actions are automatically triggered for each contingency when certain conditions are met. Power transfer is increased until one of the monitored constraints is violated. The following three constraints are monitored during SDGE-ROSE analysis:

1. Steady-state stability
   Steady-state stability is defined by convergence of power flow equations.
2. Power transfer reached the maximum
   This is a user-defined value, beyond which the transfer is not increased, even if constraint 1 has not been violated.
3. Source reached maximum
   If the source reached maximum prior to constraints 1 or 2 being violated, the computation stops.

After at least one of the contingencies causes a violation of the above constraints, the last "healthy" step is identified. This contingency is the **Limiting Contingency**. The last "healthy" step is one step less than the transfer step, at which a constraint violation caused by the **Limiting Contingency** occurs. The pre-contingency value of the interface flow at the last "healthy" step is considered the **VSA Import Limit**.

In addition, two stopping criteria can be enforced:

1. WECC Path 45 Interface Overload
   If this additional stopping criterion is enabled, the stressing is stopped when Path 45 flow is exceeded in the southbound direction (from CAISO to CENACE).
2. Sink Load Increase
   The stressing is stopped when an increase in sink load exceeds a user-defined threshold.

The **Margin (MW)** is the difference between the **VSA Import Limit** and the **Current MW** (e.g., base case interface flow):

$$\text{Margin (MW)} = \text{VSA Import Limit} - \text{Current MW} \qquad (17.1)$$

The transfer is increased with the user-defined power transfer step (default is 100 MW). After voltage stability violation is detected using the transfer step listed in the scenario file, SDGE-ROSE goes one step back, reduces the transfer step to 10 MW, and increases the transfer with a 10 MW step until voltage stability violation (e.g., the first "unhealthy" step) is identified.

SDGE-ROSE allows the user to perform stressing until:

1. The value of the **VSA Import Limit** for pre-contingency conditions is reached.
2. The first "unhealthy" step is determined.

In this case, stressing stops after the first "unhealthy" step is identified. All contingencies are applied at the first "unhealthy" step, and all contingencies causing a violation are identified. These are the **Limiting Contingencies**.

For each scenario, option **StoppingAt1stUnhealthyStep** defines whether stressing is performed until the first "unhealthy" step or until pre-contingency stability violation. At the last "healthy" step, SDGE-ROSE determines buses with the lowest voltage magnitude. These buses are called the **Weakest Buses**. The **Weakest Buses** are identified both pre- and post-contingency for each **Limiting Contingency**, and values of voltage magnitude at the **Weakest Buses** are written to the *EmsAlarm.csv* file.

17.4.1.2 Reverse Power Transfer

Reverse power transfer analysis is performed when SDGE-ROSE identifies a contingency or multiple contingencies that fail to solve at zero transfer level (base case condition); or if a margin is less than a user-defined percentage.

Case 1: Some Contingencies Fail to Solve at the Zero Transfer Level (Base Case Condition)
When some contingencies fail to solve at the zero transfer level, SDGE-ROSE starts the reverse transfer analysis for all contingencies in order to compute how much load may need to be shed such that no contingencies cause violations. Reverse transfer analysis starts if option **EnableReverseTransfer** is enabled.

The analysis is performed as follows:

1. Reduce load/generation with a user-defined step.
2. Apply all contingencies at that step.
3. Repeat items (1) and (2) above until either a transfer level is reached at which there are no post-contingency violations, or a user-defined threshold for maximum reverse transfer increase is reached.
4. After SDGE-ROSE determines the transfer level at which there are no post-contingency violations, increase the transfer using a new step size 10 MW until the transfer level at which contingency or contingencies fail to solve is reached. Further reverse transfer analysis to reach a user-defined margin is not performed.
5. Report how much load is reduced compared to the base case. The amount of load shed is written to the *EmsAlarm.csv* file.
6. If reverse transfer analysis, when some contingencies fail to solve at the zero transfer level is initiated, SDGE-ROSE issues alarm **"Reverse Transfer Analysis performed"** and writes it to the *EmsAlarm.csv* file (see Fig. 17.2).

Case 2: No Sufficient Margin
When Peak-ROSE can calculate the stability limit, but the margin is less than the user-defined requirement, reverse transfer analysis is triggered to calculate the limit which meets this user-defined margin requirement.

Fig. 17.2 Alarm **Reverse Transfer Analysis Performed**

Code	Alarm	State
G001	Unsolved Basecase	0
G002	POM Sequence Failed to Write Output	0
G003	New SE Case Not Received	0
G004	Reverse Transfer Analysis performed	1

The analysis is performed as follows:

1. Reduce load/generation with a 10 MW step. The analysis starts from the **Current MW** (e.g., base case interface flow).
2. Contingencies are not applied.
3. Repeat items (1) and (2) above until either the interface flow is reduced to meet the user-defined margin requirement or a user-defined threshold for maximum reverse transfer increase is reached.
4. The amount of load shed to meet the user-defined margin requirements is written to the *EmsAlarm.csv* file.
5. Alarm **"Reverse Transfer Analysis performed"** is not issued during this computation.

For reverse transfer analysis, the **Import Limit** is the stability limit after the reverse transfer analysis. The **Import Limit** is the last "healthy step." Since the **Import Limit** is less than the **Current MW**, the **Margin** is negative. The value of **% of Limit** is computed using the following formula:

$$\% \text{of Limit} = \frac{\text{Current MW}}{\text{Import Limit}} * 100\% \qquad (17.2)$$

Since the **Current MW** is greater than the **Import Limit** for reverse transfer analysis, **% of Limit** exceeds 100%.

17.4.2 Performing PV-Curve Analysis

After the **VSA Import Limit** has been computed, SDGE-ROSE plots PV-curves at the user-defined buses. PV-curves are plotted for all contingencies defined in the scenario file at all user-specified nodes.

The following four constraints are monitored during PV-curve analysis:

1. Steady-state stability

    Steady-state stability is defined by convergence of power flow equations.
2. Power transfer reached the maximum

    This is a user-defined value, beyond which the transfer is not increased, even if steady-state stability constraint has not been violated.
3. Source reached maximum

    If the source reached maximum prior to constraints (1) or (2) being violated, the computation stops.

Fig. 17.3 PV-curves display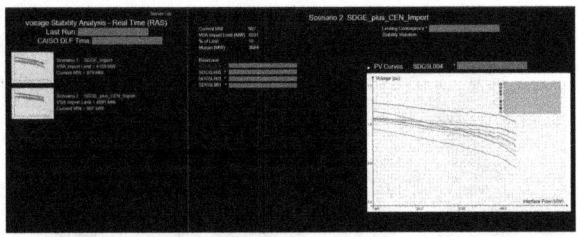

4. Additional stopping criteria:
   - WECC Path 45 Interface Overload;
   - Sink Load Increase.

SDGE-ROSE PV-Curve display is shown in Fig. 17.3.

Depending on the value of the option **StoppingAt1stUnhealthyStep**, the stressing is performed either until the value of pre-contingency stability violation is reached or the first "unhealthy" step is determined.

17.4.3 Performing QV-Curve Analysis

After the **VSA Import Limit** has been computed and PV-curves have been plotted, SDGE-ROSE computes QV-curves at the user-defined buses and at the **Weakest Buses**. While predefined buses do not change from one VSA run to the next one (unless modified by the user in an input file), the **Weakest Buses** are determined at each VSA run and may change from one VSA run to another.

QV-curves are plotted and the reactive margin is computed as follows:

- QV-curves are computed for the **Limiting Contingency** at the last "healthy" transfer step.
- If a **Limiting Contingency** is not identified (for example, source reached maximum or maximum user-defined transfer level is reached prior to post-contingency violations), QV-curves are computed for pre-contingency (N-0) conditions at the last "healthy" transfer step.

QV-curves are computed with the 0.005 p.u. (0.5%) step. When the solution diverges using the 0.005 p.u. (0.5%) step, SDGE-ROSE changes the step to 0.001 p.u. (0.1%), and performs the QV calculation.

SDGE-ROSE QV-Curve display is shown in Fig. 17.4. The reactive margin for the bus selected in Fig. 17.4 is equal to 55 MVAr. The margin is small because the computation is done at the last "healthy" transfer step for the **Limiting Contingency** (e.g., only 10 MW before the voltage stability limit is reached).

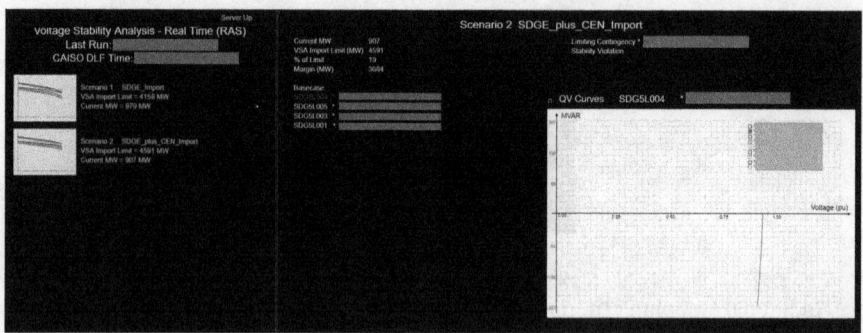

Fig. 17.4 QV-curves display

17.4.3.1 Performing dV/dQ Calculation

SDGE-ROSE has the capability to compute sensitivities dV/dQ:

1. Sensitivities dV/dQ are computed for those buses, for which PV- and QV-curves are plotted.
2. Sensitivities dV/dQ are computed at each transfer step.
3. Sensitivities dV/dQ are computed pre-contingency and after each contingency listed in the scenario file.

17.4.4 VSA Import Limit Alarms

SDGE-ROSE alarms are computed based on the MW margin along the monitored interface.

There are two levels of alarming thresholds:

- **Level 1**;
- **Level 2**.

Level 1 and **Level 2** thresholds are specified in percentage of the **VSA Import Limit**. Default values are:

- **Level 1** = 90% of the **VSA Import Limit**;
- **Level 2** = 95% of the **VSA Import Limit**.

For SDGE-ROSE alarms, the following approach is used:

- If the value of **Current MW** (i.e., base case interface flow) equals or exceeds the value of **Level 1** threshold, a **Level 1** alarm is identified and written to the *EmsAlarm.csv* file.
- If the value of **Current MW** (i.e., base case interface flow) equals or exceeds the value of **Level 2** threshold, a **Level 2** alarm is identified and written to the *EmsAlarm.csv* file.

17.4.5 Modes of SDGE-ROSE Operation

SDGE-ROSE works in two modes (see Fig. 17.5):

- Real-Time mode;
- Off-Line mode.

Real-Time mode works on a time schedule without user intervention, while Off-Line mode is initiated per user request.

SDGE-ROSE architecture in Real-Time mode is shown in Fig. 17.6.

When SDGE-ROSE is initiated in Real-Time mode, it performs flat start since the system state (i.e., voltage magnitude and phase angle) is not currently provided as a part of CAISO State Estimator (SE) solution. Every consequent VSA run uses the system state (i.e., power flow solution) from the previous VSA run as the starting point for solving the current State Estimator case.

In Real-Time mode, SDGE-ROSE is executed as a service running in MS Windows 2008 or later operating system. SDGE-ROSE performs voltage stability analysis and identifies limits under normal and contingency conditions using SE data as shown in Fig. 17.7.

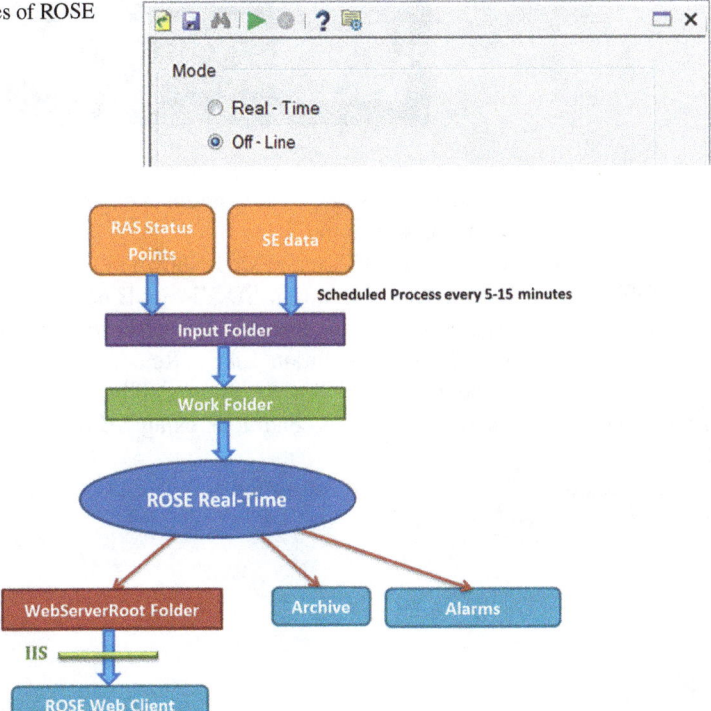

Fig. 17.5 Modes of ROSE operation

Fig. 17.6 ROSE Real-Time mode architecture

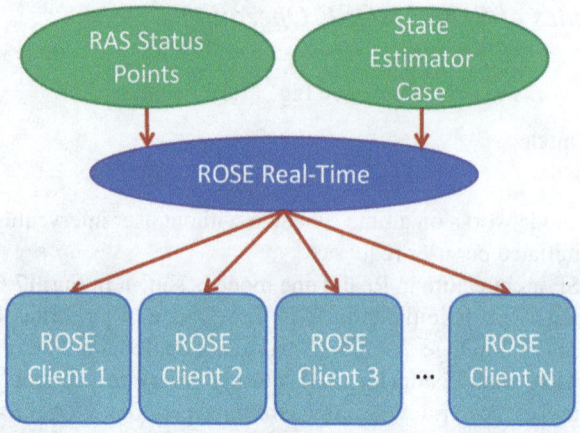

Fig. 17.7 SDGE-ROSE Real-Time Mode of Operation

Fig. 17.8 ROSE Web Client showing scenario summary in Real-Time mode

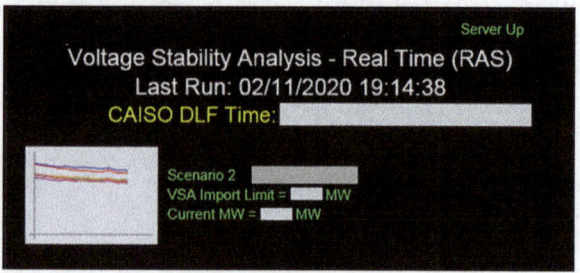

ROSE Web Client visualizes results of computational engine in Real-Time mode. Scenario summary display is shown in Fig. 17.8.

SDGE-ROSE architecture in Off-Line mode is shown in Fig. 17.9.

In Off-Line mode, initiated by the user, SDGE-ROSE performs flat start every time since the system state (i.e., voltage magnitude and phase angle) is not provided as a part of CAISO State Estimator solution. Like in Real-Time mode, when SDGE-ROSE works in Off-Line mode, it performs voltage stability analysis and identifies limits under normal and contingency conditions using SE data as shown in Fig. 17.10.

Computational results are visualized in the software interface (see Fig. 17.11) and Web Client (see Fig. 17.12).

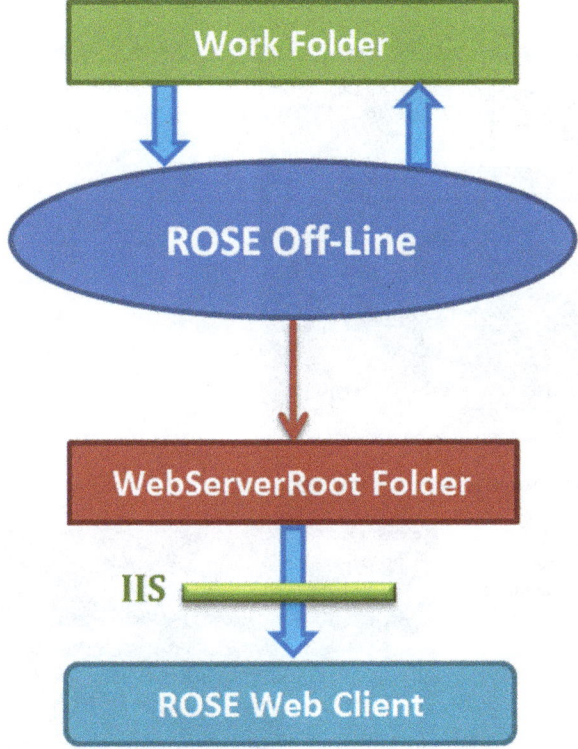

Fig. 17.9 ROSE Off-Line mode architecture

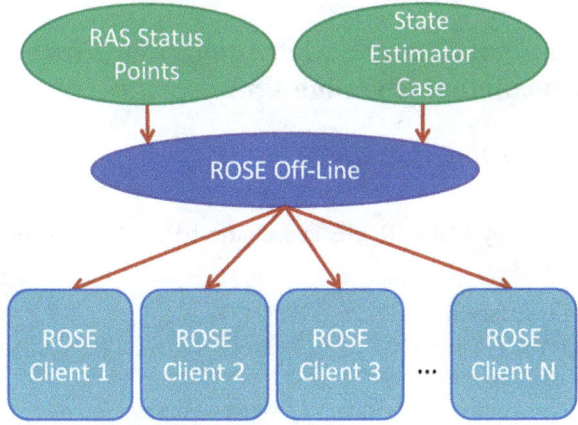

Fig. 17.10 SDGE-ROSE Off-Line mode of operation

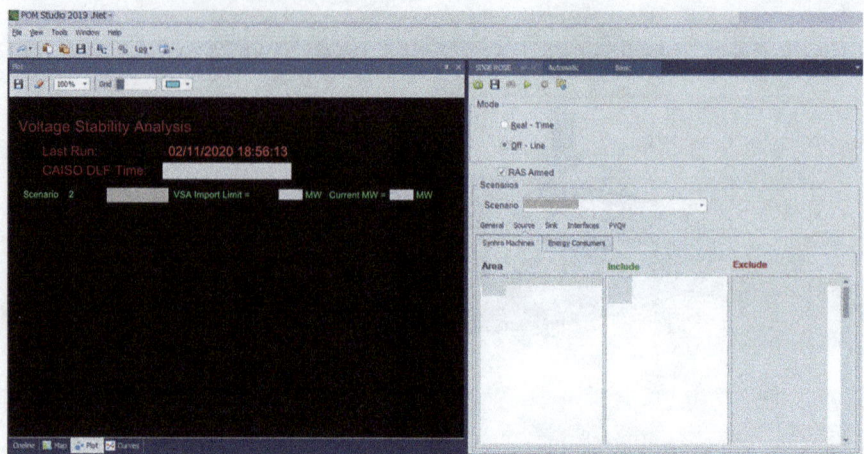

Fig. 17.11 Viewing scenario summary from ROSE interface

Fig. 17.12 ROSE Web Client showing scenario summary in Off-Line mode

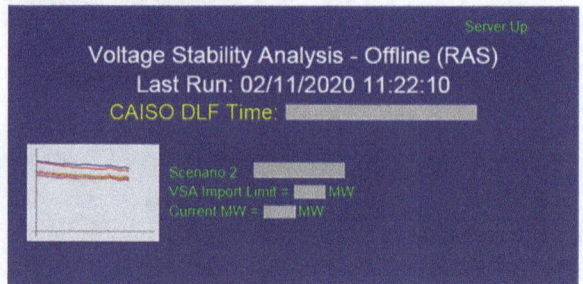

17.5 VSA Assumptions and Methodology for Practical Implementation of Online VSA

17.5.1 Scenario-Based Analysis

Analysis performed by SDGE-ROSE is scenario-based. Computations performed for each scenario are:

- Determining interface limits;
- Performing PV-curve analysis;
- Performing QV-curve analysis;
- Determining pre- and post-contingency weakest buses;
- Issuing an alarm if the current system state, in terms of interface flows, is close to the limit.

Each scenario input file contains the following information:

- Source and sink definitions;
- Monitored interfaces;
- Contingencies with RAS references;
- Buses for plotting PV/QV-curves;
- Alarming threshold levels;
- ROSE solution options and RAS options.

RAS actions are modeled in SDGE-ROSE as scripts, which are invoked through the scenario files. RAS actions are automatically triggered provided that certain triggering conditions are met.

There are two transfer scenarios running:

- Scenario 1: Import Into SDG&E;
- Scenario 2: Import Into SDG&E and CENACE combined.

The cut-planes or interface definitions for these two scenarios are shown graphically in Fig. 17.13.

The rest of the discussion will focus on Scenario 2—the SDG&E/CENACE Interface.

17.5.2 Stressing the Interface

The interfaces defined above are stressed by simulating a power transfer from the source area to the sink area, across the interface, by decreasing generation and/or increasing load in the sink area while simultaneously increasing generation and/or decreasing load in the source area.

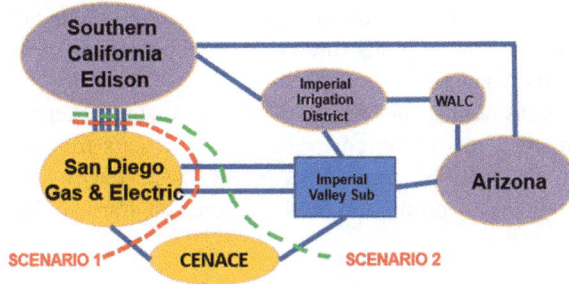

Fig. 17.13 Monitored Interface/Cut-Plane for Scenario definition

17.5.2.1 Source Area

Generating units that are online located in SCE (Southern California Edison) and APS (Arizona Public Service) are scaled up proportional to their Pmax. The real power limit of each generator is enforced. Provisions are made to exclude wind and solar plants and generators with negative output (i.e., pump load).

Scaling down the load or some combination of generation/load scaling in these two source areas has not been implemented but should be investigated to determine the impact on VSA results—particularly VSA limits which drop suddenly without any accompanying system change. Following one such event, an investigation by Peak found the solution diverged earlier than expected due to a 230 kV line outage, which radialized a generator in the source area. Some options to investigate include adding one or two more areas, using the EPF (Economic Power Factor adjusts the maximum real power output Pmax of the units in the source areas) for radialized generators, or excluding such generators in the source.

17.5.2.2 Sink Area

In Scenario 2, loads in SDG&E and CENACE are scaled up maintaining a constant power factor. For loads with a leading power factor (negative MVAR), the reactive power is frozen during stressing and only the real power is scaled up. All loads in these two areas are modeled as constant MVA.

Provisions are made to exclude nonconforming loads, such as plant auxiliary load, or small loads (less than 2 MW) but with a high reactive power component as estimated by the State Estimator. Nonconforming loads are provided by SDG&E and CENACE to Peak and CAISO for implementation in their VSA tools.

While both CAISO and Peak scale up the sink area loads uniformly in their VSA tools, SDG&E implemented a scenario which decreased the sink area generation for stressing. As explained below, monitoring for the constraint of voltage stability all the time results in significant differences in the limit results between the two VSA tools. This situation warrants further investigation to determine the optimal approach to rule out voltage stability concerns.

Furthermore, constraining the amount of transfer that can occur to a reasonably expected maximum seems preferred over reporting results from analyses having unrealistically high transfers. Any errors or inconsistencies introduced into the analysis are magnified in proportion to the amount of transfer driven across the interface.

17.5.2.3 Contingencies and Associated RAS

There are two major 500 kV paths from Imperial Valley to the SDG&E bubble where single line contingencies, two on each path, are currently modeled in the VSA tool (see Fig. 17.13). Initially, other 500 kV single line contingencies (east of

Imperial Valley) were included but later removed after concluding that their impact was considered outside of SDG&E and CENACE. These contingencies that are east of Imperial Valley are more appropriate when monitoring a much bigger interface (i.e., SCIT or Southern California Import Transmission). More discussion on this topic is provided in Sect. 17.6.

A RAS designed to trip a calculated amount of generation is enabled when all 500 kV lines west of Imperial Valley are in service. The total amount of generation that is tripped by the RAS is controlled by the CAISO and provisions are made to exclude generation from tripping, depending on the contingency.

When one of the four 500 kV lines is out of service, another RAS will be enabled and will trigger the tripping of all the generation connected in the Imperial Valley Area that is in the CAISO BA area.

17.5.2.4 Maximum Transfer Level and Transfer Increment

At each transfer level, the VSA tool computes pre-contingency flow for the interface defined in the scenario file and discussed above. Power transfer is increased up to a maximum transfer level, 5000 MW, in 100 MW increments. The maximum transfer level and increment values are user-defined.

At each transfer level, the VSA tool applies all contingencies and RAS, if enabled. Power transfer is increased until a violation of monitored constraints occurs. The following three constraints are monitored in the VSA tool:

1. Steady-state stability, which is defined by convergence of power flow equations.
2. Maximum power transfer level reached.
3. Source reached maximum.

If any one of the three constraints above occurs, then the computation stops. Because Peak and the CAISO decided to use load in the sink areas to increase the transfer, SDG&E initially used the same setup in its VSA tool implementation but later changed it from 100% sink area load participation increase to 100% sink area generation participation decrease.

After at least one of the contingencies causes a violation of constraint 1 above, the last "healthy" step is detected and the VSA tool returns to the previous step. The step increment is reduced from 100 MW down to 10 MW, and the analysis continues with 10 MW step increments until a voltage stability violation (i.e., the first "unhealthy" step) is identified. The last "healthy" step is one step less than the transfer step, at which a constraint violation caused by the **Limiting Contingency** occurs. The pre-contingency value of the interface flow at the last "healthy" step is considered the **VSA Import Limit**, [3].

17.6 Events that Led to Changes in Assumptions or Methodology

Processes were in place to monitor the VSA limits as telemetered by the CAISO and Peak to SDG&E. This included notifications generated when one or more of the following events occurs: a significant change in the limit (i.e., 1000 MW drop), a significant difference between the CAISO and Peak limits (i.e., more than 600 MW), or a limit which drops below a certain value (i.e., less than 3400 MW). This prompted a coordinated effort among Peak, CAISO, SDG&E, and others that may have been impacted to support validating the limit. Several events occurred that are worth highlighting as these events triggered changes to the real-time VSA assumptions and methodology.

1. August 16, 2016 (RT VSA limit dropped by 1600 MW)

    (a) Multiple 500 kV lines relayed south of Lugo 500 kV (due to the Blue Cut Fire).
    (b) Peak declared an IROL condition for the SDG&E Import.
    (c) VSA limit as calculated by Peak (see green line on Fig. 17.14) dropped to 3600 MW.
    (d) The 500 kV line contingency east of Imperial Valley was the limiting contingency and weak bus was located not in SDG&E but in SCE area.
    (e) Coordination among Peak, CAISO, SDG&E, SCE, IID, and CENACE resulted in agreement that load shedding, if needed, would be more effective if done in the SCE area.

    Figure 17.15 shows another PI display highlighting the limiting contingency and weak bus (circled). These two pieces of information were later

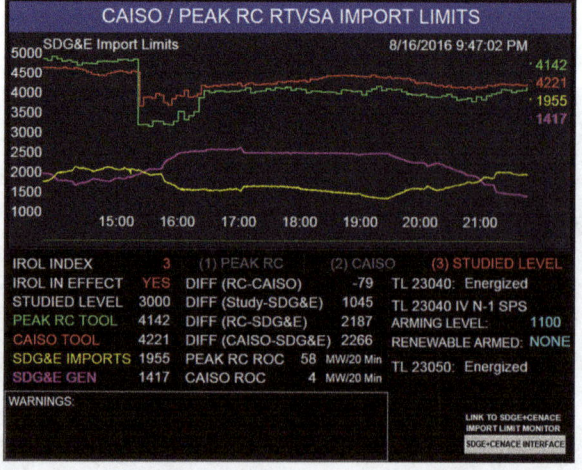

Fig. 17.14 Telemetered Limits provided by CAISO (Red Line) and Peak (Green Line) are displayed on OSIsoft PI display for monitoring in the control room

Fig. 17.15 VSA Limit calculated by Peak and SDG&E where SDG&E shows the Limiting Contingency and Weak Bus (Circled)

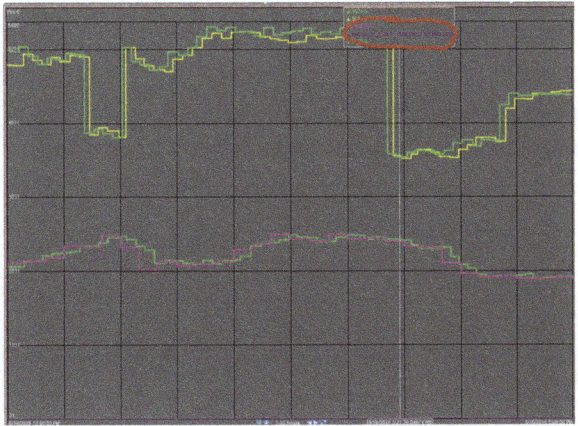

added to the PI display shown in Fig. 17.14 to provide awareness of critical contingency and weak bus in real-time.

Off-line studies, using Peak's WSM model, revealed that the multiple line outages to the south of Lugo lines stressed the SCE area. PV studies designed to stress the SDG&E area produced results limited by the stressed SCE network, which was the most impacted entity for the given limiting contingency. Thus, any operating action to mitigate IROL exceedance needed to be implemented in the SCE area, not in SDG&E or CENACE, to be effective. It was also ascertained that contingencies east of Imperial Valley had a wider impact beyond the monitored interface for SDG&E/CENACE combined. These contingencies were later removed. It was also determined that another interface with wider boundaries (i.e., Southern California Import Transmission System, Path 46 West of Colorado River) can help identify IROL conditions and can provide effective targeted mitigating actions (i.e., switching static reactive devices, increasing and dispatching generation, curtailing imports, and load shedding as a last resort in the most effective location).

2. March 17, 2017 (RT VSA Limit dropped by 1000 MW)

   (a) SDG&E's 500 kV line was scheduled to be out of service at about noon with an expected drop in the VSA limit.
   (b) About 1000 MW drop occurred *but* sooner than expected (see Fig. 17.16).
   (c) A concern was brought up that the planned outage on the 500 kV line will further drop the VSA limit.

   After investigation by Peak staff, the cause of the earlier drop in the VSA limit was due to switching of a 115 kV line in a remote area in CENACE. This investigation showed the need for a realistic substation MW load profile for a local radial subarea weakened by a planned transmission outage and further exacerbated by the increasing load driving the PV transaction. Many

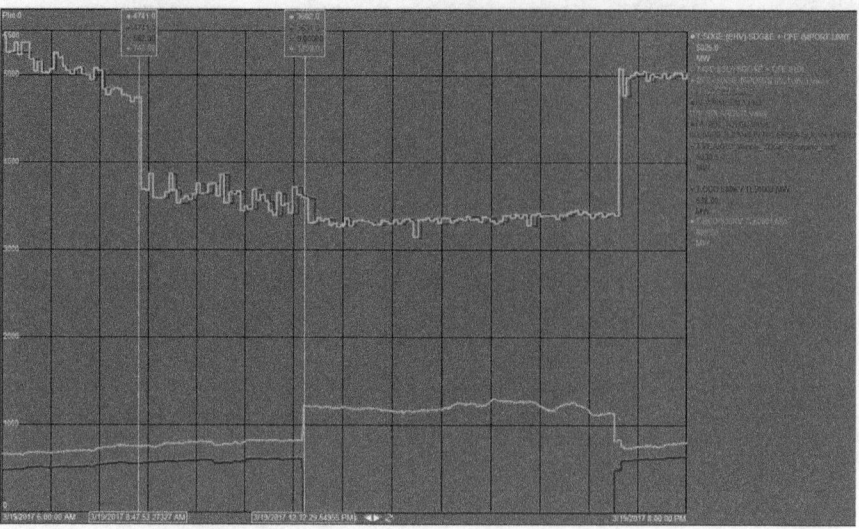

Fig. 17.16 VSA Limit dropped prior to expected drop associated with a planned 500kv line outage

instances of similar VSA limit drops have occurred which prompted SDG&E to revise its scenarios to use generation in the sink areas for stressing the interface. This is another way to validate the VSA limit calculated by the CAISO and Peak. Most of the time, the source reaches the maximum (i.e., no more generation to ramp down in the sink areas) before divergence of the power flow equations. Thus, most of the time, voltage stability can be ruled out.

3. December 3, 2018 (RT VSA limit dropped by about 1500 MW in Peak's VSA tool but not in CAISO's tool)

   (a) A 230 kV line was de-energized in the source area resulting in a radialized generator; increasing the generation resulted in solution divergence sooner than normal.
   (b) CAISO has a way to limit generation output when the particular generator is radialized.

   This highlights the importance of sharing special operating conditions among entities running RT VSA to minimize significant differences between the results.

4. September 20, 2015 (SDG&E shed load for management of an IROL following loss of a combine cycle plant)

   A Confidential Brief Report was submitted for this event and thus many specifics about the event are confidential. The event was also processed by NERC through its Cause Code Analysis Process (CCAP) for effective labeling, collection, and trending of causes. However, as highlighted in the events described above, the limits calculated by the two VSA tools on September 20 also differed. Both VSA results can become unreliable for many reasons. Often the reasons are determined only after an investigation of the modeling assumption (i.e., switching of automatically controlled shunt devices in one tool but no switching in the other), exclusion of a RAS model which can significantly impact (in this case positively impact) the limit calculated by the VSA tool.

The modeling and assumption issues were addressed in the weeks following the event. The assumptions going into the tools were discussed in detail and put into a document to ensure consistency between the tools. Finally, periodic conference call among the relevant parties is put together to discuss observed significant difference in limits, ongoing activities, and improvements to the tools.

This event highlighted the importance of having two independent sources, a primary and backup, for calculating real-time VSA limits. It is also important to have a solid process in place for quickly evaluating the integrity of each tool and deciding which tool to use as the primary for producing the operating limits. The models and VSA assumptions must be consistent between the two tools to have confidence in the results.

17.7 Lessons Learned

SDG&E has learned a lot in 4 years running real-time voltage stability analysis (RT VSA). Here are some of the lessons learned when running RT VSA:

1. *Network model should be accurate.* Ensure periodic reviews due to system additions and retirements.
2. (TOP) can update its EMS model more frequently than the CAISO and RC can update theirs, so it is important to share model data in advance, and notify the CAISO and RC of the status of the work.
3. Be proactive about reviewing vendor patches or program updates to determine which VSA assumptions may be impacted. This includes subscribing to vendor notifications and attending vendor user group meetings.
4. Engaging vendor and *IT support* early will help identify and resolve problems faster.
5. Users (i.e., operating engineers), particularly new users, should review and inspect modeling/method assumptions periodically, particularly after a system event occurs.
6. Monitoring of the program output and log file and processes will help diagnose issues sooner.
7. Consider running additional scenarios and flexible contingencies (beyond N-1 criteria) during a hot weather alert.
8. Having two real-time VSA tools (and models), monitoring and investigating significant limit differences and huge limit drops help discover issues in the VSA stressing methodology, VSA assumptions, and the network model.

9. System operators and operating engineers should be trained on the importance of understanding early the resulting limiting bus(ses) and limiting contingencies for situational awareness particularly in infrequent operating conditions.
10. Ensure procedures provide clear guidance for analyzing, monitoring, validating, and dealing with real-time VSA IROLs.

    (a) Benchmarking, communicating, and coordinating among impacted entities are required.
    (b) With regular training be able to run off-line studies to evaluate the validity of the near real-time result.

11. Real-time VSA tools currently lack the ability to dynamically conduct predictive evaluation—something that is arguably more important than the fixed real-time analysis. This is especially true for San Diego, where no voltage stability concerns exist without multiple elements out of service.
12. Real-time in this context is not quite real-time. There can be up to a 15-min delay between the real-time system changes and producing valid stability limits. It remains critically important to continue conducting seasonal off-line studies to have preestablished limits, so operating engineers know what system adjustments to conduct while waiting for results to refresh. Furthermore, off-line studies provide insight into what valid results should look like, and what changes need to be made to the VSA assumptions as the system evolves.

17.8 Future Work and Conclusion

Over the past 4 years, SDG&E has learned a lot about the implementation of the online VSA tool. As long as the CAISO and RC are running a VSA tool for monitoring IROL conditions in the SDG&E and CENACE combined areas, SDG&E will continue its own implementation of the VSA tool and will apply lessons learned. Operating Engineers among the CAISO, RC, SDG&E, and other potentially affected entities will continue to coordinate and discuss any anomalies in the results of the VSA tool as the network model including RAS model continues to change. This is important to ensure the correct and effective operating decision is executed.

Future work involves using CAISO's export of its network model in CIM15, which is currently being tested. The results will be compared with the results from the CAISO's VSA tool (considered the primary) before it can be released for production as the secondary VSA tool. Furthermore, other constraints will be implemented (in addition to the three monitored constraints mentioned in Sect. 17.4).

Acknowledgments The continued support from CAISO staff to provide their real-time model in CIM format for input into SDG&E–ROSE is greatly appreciated. In addition, the authors wish to acknowledge Alex Ning (formerly with Peak Reliability) for his continued support while Peak

Reliability transitions to cease operations. And, lastly, the authors would like to acknowledge leadership of Tariq Rahman and support of Subburaman Sankaran , both from SDG&E.

References

1. http://www.caiso.com/Documents/ReliabilityCoordinatorFAQ.pdf
2. http://www.caiso.com/Documents/RCWestEntities.pdf
3. Peak-ROSE (2017) Manual, V&R Energy

Chapter 18
Stability Applications in the Dominion Energy System Operations Center

Katelynn Vance and Gilburt Chiang

18.1 Introduction

The National Academy of Engineering named electrification as the greatest engineering feat of the twentieth century [1]. The bulk electric system (BES) is an immensely complex system that requires the balancing of many different components to deliver power continuously and in compliance of the North American Electric Reliability Corporation (NERC) requirements. As the landscape of the BES evolves more rapidly, the ability to monitor and understand how the power system will react in stressed situations continues to be paramount.

This discussion will focus on the stability assessment of the electric transmission system at Dominion Energy Virginia and how its inclusion, alongside those traditionally performed, will help meet the needs of the modern power system. However, there will first be an introduction of Dominion Energy as a company and basic information about the System Operations Center (SOC).

18.1.1 Dominion Energy Overview

More than seven million customers in 20 states energize their homes and businesses with electricity or natural gas from Dominion Energy (NYSE: D)*. The com-

---

K. Vance (✉)
Dominion Energy, Inc., Richmond, VA, USA
e-mail: katelynn.a.vance@dominionenergy.com

G. Chiang
Bigwood Systems Inc., Ithaca, NY, USA
e-mail: gilburt@bigwood-systems.com

Table 18.1 Miles of transmission line by voltage

Voltage level	Miles
500 kV	1312.42
230 kV	2928.54
138 kV	63.71
115 kV	2306.80
69 kV	78.55
Total	6690.02

pany, headquartered in Richmond, Virginia, is committed to sustainable, reliable, affordable, and safe energy and to achieving net zero carbon dioxide and methane emissions from its power generation and gas infrastructure operations by 2050. Dominion Energy (DE) is the holding company of Dominion Energy Virginia (DEV), a public utility with regulated generation and electric service in Virginia and North Carolina.

As of December 2019 [2], Dominion Energy Virginia is comprised of 21,100 megawatts of generating capacity and 65,100 miles of distribution and transmission lines in Virginia and North Carolina. This portion of the company provides dependable service to some of the largest military installations and data centers in the world. The breakdown of transmission line miles by voltage is provided in Table 18.1. Dominion Energy Virginia has transmission interties with Appalachian Power, First Energy, Duke Energy, and Potomac and Electric Power Company. A transmission and distribution system map of the Dominion Energy Virginia footprint can be seen in Fig. 18.1.

Dominion Energy Virginia is part of PJM Interconnection, LLC ("PJM") which functions as the regional transmission organization that provides service to a large portion of the eastern United States. PJM does not own any transmission, distribution, or generation equipment. Their primary objectives are to maintain overall system security while functioning as a market for generation dispatch throughout the day. All or parts of Delaware, Illinois, Indiana, Kentucky, Maryland, Michigan, New Jersey, North Carolina, Ohio, Pennsylvania, Tennessee, Virginia, West Virginia, and the District of Columbia are included in their footprint. This service area has a population of approximately 65 million and on August 6, 2006, set a record high of 166,929 megawatts (MW) for summer peak demand, of which Dominion Energy Virginia's load portion was approximately 19,256 MW serving 2.4 million customers. On July 22, 2011, the Company set a record high of 20,061 MW for summer peak demand. On February 20, 2015, the Company set a winter peak and all-time record demand of 21,651 MW. Based on the 2020 PJM load forecast, the Dominion Energy Zone is expected to be the fastest growing zone in PJM with average growth rates of 1.2% summer and 1.4% winter over the next 10 years compared to the PJM average of 0.6% and 0.6% over the same period for the summer and winter, respectively. A map of the PJM service territory is presented in Fig. 18.2 [3].

As a part of their reliability coordinator function, PJM and Dominion Energy Virginia work together to plan, study, and monitor the electric grid. PJM's transmis-

18 Stability Applications in the Dominion Energy System Operations Center

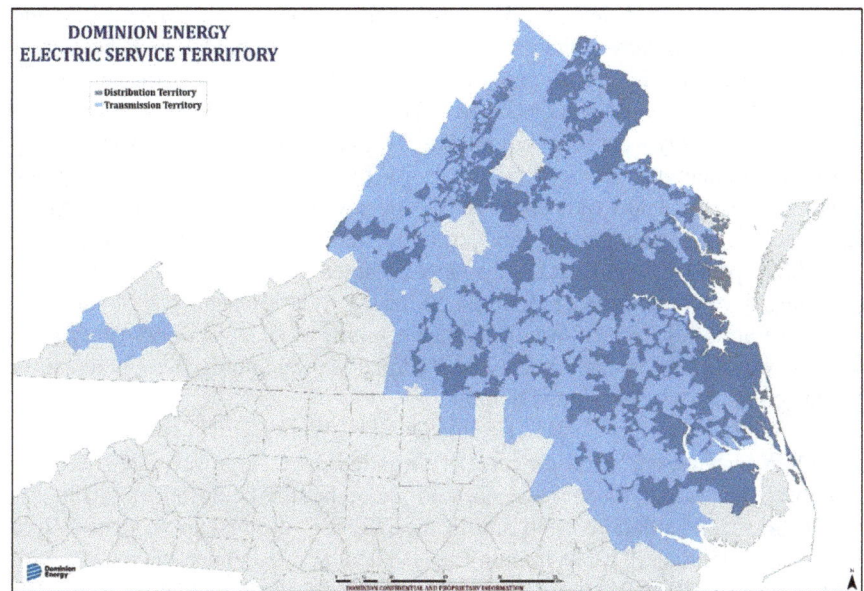

Fig. 18.1 Dominion energy distribution and transmission territory

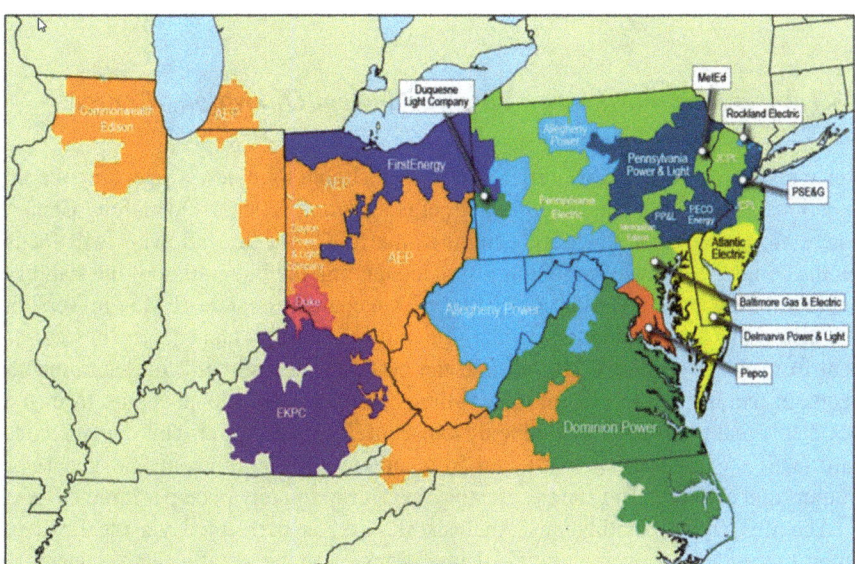

Fig. 18.2 PJM service territory

sion and generation roles lead to close contact and communication between PJM and each individual company in their footprint for planning and operating discussions.

Solar generation is a vital part of Dominion Energy's comprehensive clean energy strategy to meet standards outlined in the Virginia Clean Economy Act and to achieve the company's net zero commitment. As part of the Virginia Clean Economy Act's requirement for zero-carbon electricity by 2045, over the next 15 years, Dominion Energy plans to add about 16,000 megawatts of solar generating capacity through company-owned projects and power purchase agreements signed with third-party developers in Virginia. It has also met its stated 2018 goal of bringing online, beginning development on, or signing contracts for, 3000 megawatts of solar and wind generating capacity in Virginia by the beginning of 2022. The company's solar portfolio was recently ranked third by S&P Global Market Intelligence among utility holding companies in the United States.

Dominion Energy has Coastal Virginia Offshore Wind (CVOW) project to bring two 6 MW turbines online off the coast of Virginia Beach. As of July 2020, turbines have been installed, and the line back to shore is still being added. Although this is the second offshore wind site in the US, it is the first to be owned by a utility. In September 2019, Dominion Energy filed an interconnection request with PJM to bring 2640 MW of offshore wind online. Based on regulatory approvals, construction is planned to start in 2024 [4].

Information above provided is as of Dec 2019 and prior to the sale of Dominion Energy's Gas Transmission and Storage business scheduled to be effective Q4 2020.

18.1.2 Dominion Energy Virginia System Operations Center

Since many companies work differently, it is desirable to review a selection of core functionalities within the Dominion Energy Virginia System Operations Center. There are many other functions central to the SOC, but the following will focus on the components most important for understanding the integration of the stability analysis. DEV uses a GE/Alstom Energy Management System (EMS) in Version 2.6. Known system parameters, in addition to measurements and equipment statuses sent from the substations, are fed into the system state estimator to determine an accurate picture of the system in real-time. There are additional inputs from the ICCP link that runs between PJM and Dominion Energy Virginia. This data includes information on measurements taken from other utility equipment and generator outputs and statuses. The state estimator runs approximately once per minute.

The Real-Time Contingency Analysis (RTCA) is performed via the EMS to study the outage of every individual line, transformer, generating unit, capacitor, and reactor. RTCA, which is also performed once a minute, analyzes approximately 1100 contingencies. At the end of each cycle, there is a report for all thermal, low voltage, and high voltage violations, as well as islanded load or generation. System operators monitor these results and act on any violations that occur by taking preventative measures including but not limited to inserting or removing reactive devices. Depending on the results, operators may contact PJM to verify and discuss possible mitigation strategies. Additionally, operators will confer and strategize with the Reliability Engineers (REs) who are a part of the Operations Planning team if they are on shift at the time. Currently, there is at least one RE on shift in the control room from the hours of 06:00–23:00 every day. NERC sets the rules surrounding N-1 security and failure to comply can result in fines [5].

In addition to the real-time applications, there are two different types of study platforms which take real-time snapshots from the EMS to be evaluated for N-1 security under conditions that differ from real-time. These tools are predominantly used by the Operations Planning team at the SOC. The engineering component of the Operations Planning team consists of REs and long-term planners. As mentioned above, the REs work a shift schedule in the control room and help support daily operations. They also perform outage studies from 10 days prior to and up until 1 day before the outage takes place. The long-term planning engineers study outages and work with project managers in the field to schedule them in the 1 month to 3 years out time period. Additionally, system operators are required to study an outage before removing the equipment by completing a switching order.

The first platform functions within the GE/Alstom EMS and is referred to as STNET. STNET allows the user to take a real-time snapshot or retrieve a case previously autosaved from the last several months. Autosaved cases are generated at 04:00, 08:00, 12:00, and 17:00 hours so that they roughly obtain the peak and valley loads each day for both winter and summer load profiles. STNET will also retrieve load schedule data for a time of interest, and then perform an N-1 computation. STNET works with another EMS program called Outage Scheduler which imports the planned outage data via Sun-Net's Transmission Outage Application program. The temperature is set by the user, and the EMS will use that data to pull in the correct ratings for the given temperature. The STNET powerflow case can be manipulated depending on how the user wishes to solve any N-1 voltage, thermal, load loss, or generation loss violations that may appear. The user can choose to increase or decrease generation, insert or remove capacitor banks, and close normally open switches to energize dropped load. The other platform for outage review was built internally to DEV and is called ANalysis On DEmand (ANODE). It takes system snapshots of the EMS every 10 min but performs the studies outside of the EMS. The platform will be discussed in much greater detail in Sect. 18.5 because it provides the foundation for data acquisition used in the stability analysis.

18.2 Decision Factors for Including Stability Analysis in the Control Room

Power system operators and supporting staff want to ensure a high level of reliability in their day-to-day operations. As stated by the IEEE/CIGRE joint task force, the "*reliability of a power system refers to the probability of its satisfactory operation over the long run. It denotes the ability to supply adequate electric service on a nearly continuous basis, with few interruptions over an extended time period"* [6]. Traditionally, at DEV, reliable system operation has been achieved through studying for thermal, voltage, and load and generation loss violations. The correct management of capacitors, reactors, flexible AC transmission (FACTs) devices, load tap changing transformers (LTCs), and generator voltage setpoints allow for the power system voltage to operate within a bandwidth of approximately 0.95–1.05 pu. PJM uses economic dispatch and system conditions to run generation in such a way that thermal violations do not occur. PJM takes the reactive reserve into consideration when requiring generation to come online or when approving outages on reactive devices. However, the assurance of system reliability requires knowing when the power system is operating in a secure region which is sufficiently distanced from boundaries which would cause system violations or instability conditions.

Even though steady-state analysis provides important information on the power system operating conditions, the BES is a highly nonlinear system that changes constantly as the load, generation, and other operating parameters vary. Since not all of the phenomena associated with the power system can be captured with a linearized steady-state analysis, a stability analysis provides further insight. The stability definition being used for this chapter is the following: *"Power system stability is the ability of an electric power system, for a given initial operating condition, to regain a state of operating equilibrium after being subjected to a physical disturbance, with most system variables bounded so that practically the entire system remains intact"* [6].

18.2.1 Generation Retirement

With large amounts of generation retiring in a short time frame, system load continuing to grow in the Dominion Energy Virginia territory, and the continual upgrades and maintenance which must be completed, the need for a stability analysis has become more apparent. Although Long-term System Planning studies the system for locations where improvements need to be made based on an N-1 and N-1-1 analysis, the model used for these studies is different from the operational model. In long-term studies, a bus-branch model is used rather than a node-breaker model and all equipment assumed to be in service. Additionally, nearly all generation is available for dispatch, and the cases are only evaluating peak or valley loading conditions. These studies are meant to represent the most extreme conditions for the system during the most stressed times so that equipment can be upgraded or added. The operational model represents the actual system in real-time. Maintenance outages are always taking place throughout the system to maintain safe and reliable equipment, so these open points are shown in the system model. Additionally, there are variations in generation outputs and interface transfers with other utilities. To comply with NERC standards and PJM requirements, when the traditional study tools show outages causing thermal or voltage violations under the predicted loading conditions, the outage is not taken. However, that does not mean that the system could not be extremely stressed and may be susceptible to stability issues.

The changes in the generation profile are also significant for understanding how the differences in system inertia will change dynamic responses. The influx of large amounts of solar penetration has fundamentally changed how the physical machines in the power system will react during an event. This is especially true during operational situations where generation can be electrically far away or disconnected from normal connections which could provide damping. There have already been previously identified scenarios by PJM for transient stability concerns [7]. Monitoring them in real-time can provide a less conservative and more precise approach.

18.2.2 Ability to Compare Results with PJM

As Dominion Energy Virginia's RTO, PJM runs both a voltage and transient stability analysis that is discussed in a separate chapter of this book. Both PJM and DEV perform very similar state estimation and RTCA analysis. If there is a discrepancy between the two results, the system will be operated to the more conservative of the two until further investigation can be completed. In the case of stability analysis, DEV has been unable to have a discussion with PJM about the dynamic results. Similarly, for operations planning studies, there was no ability to compare information or outcomes. The ability to have a redundant analysis on another facet of power system reliability is part of why DEV implemented a real-time and study stability application.

18.2.3 Regulatory Considerations

Increasing regulatory involvement also drives the industry to include power systems stability as a larger component of their process. In October 2013, NERC approved Reliability Standard TPL-001-04 as a transmission planning standard [8]. Even though this standard is technically for Long-term Planning studies, it indicates that NERC is beginning to notice and become involved in setting uniform stability criteria across the industry. Among other specifications, the standard requires that all planning beginning after January 1, 2016, include further stability analysis as well as a complicated dynamic load model due to concern for the future impact of stability in the power infrastructure. Furthermore, ISOs and RTOs are required to set System Operating Limits (SOLs) for voltage and transient stability [5]. These limits can be determined offline, but many ISOs and RTOs have started to utilize a real-time stability analysis to avoid overly restrictive limits when they may not be necessary. Although it may never be mandated, the trends indicate that that stability analysis should find its way into the control room.

18.2.4 Increased System Stress and Upgrade Delays

In addition to the change in generation, Dominion Virginia Energy has several ties to other utilities through which a large amount of power is transferred daily. The very basics of voltage stability start with an analysis using the power-voltage (PV) curve. There is a concern that the large transfer of power could create a scenario where the system could move beyond the nose of the PV curve making a voltage collapse possible. These issues are compounded with generation retirement creating large through fares of power flow.

In a traditional steady-state power flow analysis, voltage instability can often materialize as the inability of the power flow equations to be solved numerically. There had been several instances of an unsolved N-1 violation when equipment was switched out of service despite the studies indicating that there would be no violation. In these cases, the equipment was immediately switched back into service until further studies could be completed and mitigation strategies determined. These events further solidified the need for a voltage stability analysis.

Finally, more attention was brought to voltage stability analysis at DEV through a very tenuous set of outages to build a transmission line across the James River. The need for a 500 kV transmission line down the Virginia peninsula was identified in 2011 [9]. However, due to many legal and permitting hurdles, the construction of this line from Surry to Skiffes Creek was delayed until 2017. The load growth in the area along with the retirement of two generating units at Yorktown Power Station could possibly create a voltage collapse in the region for certain operating conditions. As a result, a remedial action scheme (RAS) was developed for the North Hampton Roads area to be armed under specific conditions. If the RAS scheme were to have operated, it would have dropped 150,000 customers but would have prevented a widespread voltage collapse [9]. An emergency request was filed through PJM in accordance with Section 202(c) of the Federal Power Act with the US Department of Energy that would allow for the units at Yorktown to stay online during construction [10]. This request had to be renewed every 90 days until the completion of the transmission line but prevented the RAS from needing to be armed nearly every day during hot and cool months until the construction was completed. This was an extremely difficult situation for both operations staff at DEV and PJM and brought to everyone's attention that the quickly changing landscape was creating voltage stability problems now and certainly could in the future.

18.3 Stability Tool Requirements

One of the first steps to this process was to determine what Dominion Energy Virginia requires from a voltage stability application. In the SOC, there are two groups with which this application will interact. These groups are the real-time system operators and the SOC Operations Planning group. The operators respond to real-time issues and execute switching orders to reenergize or deenergize equipment in the field. The SOC Planning group studies scheduled transmission and generation outages from 3 years in advance to the day before an outage were to take place. Therefore, the stability analysis requires the ability to perform real-time analysis as well as a study functionality.

For the purpose of real-time analysis, it has been determined that every 10 min is a sufficient interval for the stability application. For system outage studies, the software must be able to import outages to the study case to evaluate new violations. Additionally, the cases must be structured to accept load forecast data. All of the programs need to be able to have both a base-case and N-1 contingency analysis. For a transient analysis, faults are applied rather than just the removal of a line. For the voltage stability program where transfer patterns across the system are monitored, the application needs to calculate the steady-state system limits as well. This helps keep the voltage stability limit in the perspective of the thermal or low voltage operational limits so that unnecessary action is not taken.

In addition to ensuring that the software package could adequately solve stability problems, it is necessary that the results be displayed in a useful and intuitive manner. There was also a desire for a visual map of the system to show how different regions and tie lines are affected by the voltage stability analysis. This visualization uses green, yellow, and red semi-circles to act as gauges showing how at risk an interface is. The transient stability analysis shows a small map of the DEV footprint as red or green based on the output of the real-time results.

The EMS is characterized by NERC as a Critical Infrastructure Protection (CIP) asset. This means that there are many requirements placed on hardware and software associated with EMS. Due to this and the scale of the EMS already, the decision was made to keep any stability analysis outside of the EMS for development and implementation. There is currently no plan to integrate into the EMS, but it could be possible in the future. As a result of this decision, the responsibility for the implementation of this software has stayed within the Operations Planning team at the SOC. If this were to integrate with the EMS, it would transfer to the SOC Engineering group with support from the EMS IT team. If this asset becomes a part of the CIP infrastructure, but remains outside of the EMS, it is yet to be determined who would maintain the system.

From the start of the project, it was known that the software would run and store results on an HP Superdome server with 5 TB of storage. The HP Superdome has up to eight slots with 36 CPUs per slot. Currently, the DEV implementation has 144 cores dedicated for production. This does allow for fast computing times despite the volume of the processes already occurring on the server for other applications. In the future, other hardware configurations may be utilized for redundancy.

After a long and comprehensive Request for Proposal (RFP) process, DEV selected Bigwood Systems Inc. (BSI) as the vendor for the stability analysis platform. They provide real-time and study analysis for voltage and transient stability. There is also a small signal stability application that is utilized only for operational planning system studies.

18.4 Stability Analysis Tools at DEV

To perform online stability analysis, the Dominion Energy Virginia tools use a fast, and more importantly, accurate, contingency screening and ranking method before performing detailed analysis. This approach significantly speeds up the computation time thus allowing for closer to real-time analysis.

18.4.1 Transient Stability Analysis

For transient stability, the conventional approach has been to use time-domain simulation, which involves step-by-step simulations of each contingency to filter out very stable or very unstable contingencies. However, this strategy is time-consuming and may not be able to fully identify multi-swing stable or unstable contingencies.

The BSI transient stability tool employs the "direct method" called BCU for fast screening and ranking is used in the Transient Stability Assessment and Enhancement (TSA&E) tool. The BCU approach combined with time-domain simulation can perform transient stability analysis in real-time and differentiate between critical and noncritical contingencies. Given a set of credible contingencies, this strategy would be used in two stages of assessments:

- *Stage 1 Screening*: Perform the task of dynamic contingency screening to quickly separate contingencies that are "definitely stable" from a set of credible contingencies, based on the energy margin.
- *Stage 2 Detailed Analysis*: Perform a detailed assessment of the dynamic performance for each contingency that was not screened out in Stage 1.

This approach effectively screens out a large number of "definitely" stable contingencies, captures the remaining critical contingencies, and then applies detailed simulation programs only to potentially unstable contingencies.

18.4.2 Voltage Stability Analysis

The Voltage Stability Analysis and Enhancement (VSA&E) tool is being utilized at DEV. The strategy used in voltage stability is similar to the approach in TSA&E with screening out stable contingencies based on sensitivity when applying the contingency to the base-case P-V curve and performing further assessment on potentially unstable contingencies. The contingency evaluation method is summarized below.

- *Stage 1 Screening*: Perform the task of contingency screening to separate contingencies that are "definitely voltage stable" from a set of credible contingencies.

- *Stage 2 Ranking*: Perform the task of contingency ranking in terms of load margin for each contingency remaining in Stage 1.
- *Stage 3 Detailed analysis*: Perform detailed analysis via computation of the P-V, Q-V, and P-Q-V curves for the top-ranked contingencies.

In a typical instance of a power system losing stability, a bifurcation occurs. The VSA&E tool computes both a saddle-node bifurcation (SNB) and a structure-induced bifurcation (SIB) as well as their corresponding load margins and sensitivities. The SNB is due to transmission line limitations while the SIB is due to generator reactive power limitations.

The VSA&E program analyzes typical thermal and voltage violations as well as voltage stability limits. By enacting these traditional checks, as well as calculating stability limits, the program provides a comprehensive system overview. This prevents the operator or study engineer from spending unnecessary time and effort pursuing stability specific solutions to problems which will be remedied when other violations are mitigated.

The assessment will identify and alert users when there is a critical contingency or potential instability. Once instabilities have been identified, both the voltage stability and transient stability tools will determine actionable control recommendations to mitigate these contingencies. The control engine will consider all active devices and controls (transformer load tap changers (LTC), shunt capacitors, and generator reactive outputs) based on user preferences. The control adjustments will seek to reestablish stability under severe contingencies or increase the maximum power transfer before voltage issues are observed. Dominion Energy will use preventive control actions to analyze how they will affect the load margin before implementing the control switches.

18.4.3 Small Signal Stability Analysis

In addition to the real-time voltage stability and transient stability assessment tools, a small signal stability assessment (SSA) was also implemented to round out the full suite of monitoring and analysis tools. SSA is designed to perform small signal stability analysis for large power systems through analyzing the eigenstructure of the system state matrix. Cases can be based on existing simulation files or simulation files created from scratch using the interactive parameter setting interface. The tool will serve to calculate small signal stability limits under various contingencies and operating conditions and alert users of potential instabilities. The program can scan for dominant eigenvalues or perform a full analysis for the entire state matrix. Although the participation factors and right eigenvectors can be determined from post processing of the state matrix, the program lists them for ease of reference in the results. Further developments include subsynchronous oscillation screening and mitigation.

18.5 Implementation

Dominion Energy Virginia has installed a comprehensive package supplied by Bigwood Systems for stability analysis including voltage stability assessment & enhancement (VSA&E), transient stability assessment & enhancement (TSA&E), and small signal stability assessment. The VSA&E and TSA&E tools run in both real-time and study mode. The VSA&E tool also includes a look-ahead mode that considers outage scheduling and load forecasting data to perform voltage stability analysis for the next 30 min up to the next 24 h ahead. In addition, a control engine will recommend device adjustments to mitigate any instability found in the assessment, which can include adjustments to transformer LTCs, capacitor bank switching, and generator reactive outputs.

Before diving into the stability specific implementation, it is necessary to explain the data acquisition and file development of all the inputs. An architecture diagram which represents the information below is shown in Fig. 18.3.

18.5.1 Data Acquisition: General

As part of a separate initiative that began in 2014, a platform was developed to automate outage planning analysis which, as part of its baseline functionality,

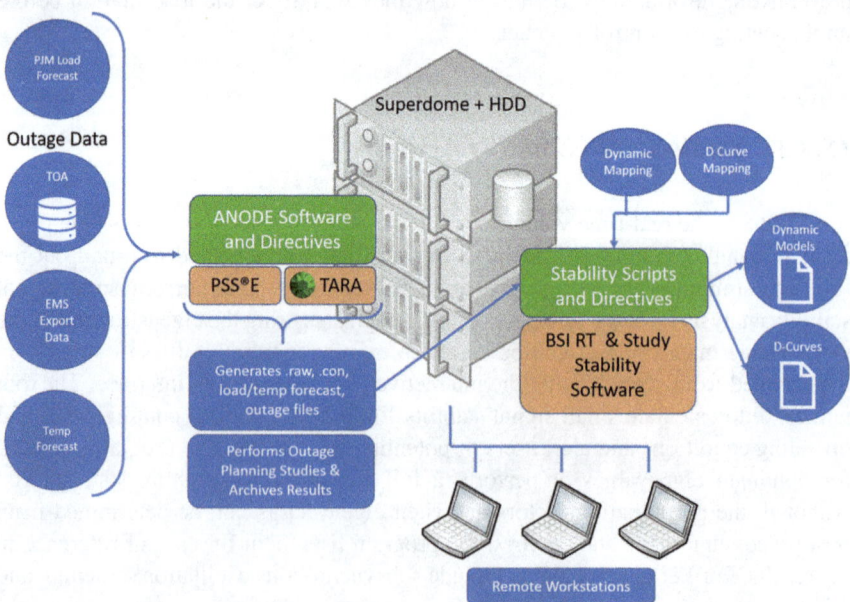

Fig. 18.3 Architecture diagram

aggregates data from disparate systems including EMS network models and snapshots, contingency definitions and results, temperature forecasts, and planned outage data [11]. As previously mentioned, the other platform, ANalysis On DEmand (ANODE), is used for running templated and custom outage planning analyses. Because much of the data required for both a stability analysis overlaps with what is required for an operational outage planning analysis, ANODE can be leveraged to provide the necessary data and models for stability analysis. ANODE was also extended to create models only needed for stability analysis. ANODE and its constituent components, including the extensions for stability analysis, were developed using Python 2.7 due to its ease of use and prevalence in the power systems simulation.

The exported EMS data includes the network model database, line limits database, and contingency database. In order to retrieve this required data, there are GE/Alstom EMS-specific utilities which are executed. The export service is executed every 10 min and sends the model information as comma separated value (CSV) files via a SFTP to a location on the Enterprise network. ANODE takes these files and manipulates the data into powerflow, contingency, monitor, and subsystem files in PSS/E Version 33. These newly created outputs allow Siemen's PSS/E or PowerGEM's Transmission Adequacy and Reliability Assessment (TARA) software to run power flow and contingency analysis outside of the EMS. This is a key component of the outage planning workflow because it allows for much greater control and flexibility with the cases than what is allowed within the EMS. The PSS/E Version 33 powerflow file and contingency file can be read into VSA&E and TSA&E for real-time and study applications.

In addition to the EMS files, the ANODE platform includes directives that collect temperature, outage, and load forecast data from various sources. The expected temperature is sourced from darksky.net at a rate of once per hour. The transmission outages are pulled from a database once per hour. This database is called the Integrated Transmission Outage Application (iTOA) which is developed by Sun-Net. The load forecast is retrieved from PJM Oasis every 15 min. Finally, generator outages are collected from NERC System Data Exchange (SDX) once per day. The ways that the stability programs utilize and interact with these files will be discussed in further detail below.

18.5.2 Data Acquisition: Stability

The BSI stability assessment tools are configured to receive data most of the inputs from the sources listed above and then perform the associated analysis. It is noted that most files can be input, but that temperature is not one of them. In the case of a study, the temperature settings will need to be modified using the ANODE platform before running in a stability program. For real-time, the temperature rating sets are continually updated by the operators and will reflect those values through the direct transfer of EMS data to the stability tools. The load forecast, outage, contingency data can all be input directly.

There are additional files that are necessary for the transient, voltage, and small signal analysis. For the transient and small signal analysis, a dynamic data file for the generators and FACTs devices is required. The dynamics file is generated using a static mapping file between the Eastern Interconnection Multiregional Modeling Working Group (MMWG) model names and bus numbers and the generator names in the EMS. The mapping file requires an update several times per year based on new generation. For the voltage stability assessment, the generator D-curve file is created as well as a preferred generator dispatch file. The generator D-Curve information is provided by PJM via their eDart interface. The D-Curve mapping file uses the same logic for comparing generator names in the MMWG and EMS cases and is updated once per year.

It is important to note here that every time the EMS solution runs, if there are any changes to the topology, all of the bus numbers will change. This means that there cannot be a one-to-one mapping based on static bus numbers, and a table with unit names must be used to properly assign the correct bus number to the relevant file data. The above input files are generated using scripts that extend the ANODE platform in Python 2.7. There is an architecture diagram in Fig. 18.3.

For the VSA&E program, the defined interfaces are maintained by mapping internal to the Bigwood Systems program instead of requiring a script to ensure bus number continuity. Additionally, the busses monitored for each interface are maintained internally to the program as well. Each interface has a predefined set of contingencies to run which were set during the initialization of the program. The interface definitions, monitored busses, and contingencies to be run can all be manipulated based on user needs. The interface definitions and monitored bus numbers change with each run of the state estimator, but the data bridge set up by BSI takes that into account and matches the components based on their bus names. In the event of large changes in system topology, some manual changes to these would need to be set again. In thinking through the implementation and maintenance process, much care was taken to reduce the amount of continually required user intervention. However, it is not possible to avoid all of it as new units are brought online and new substations are added.

18.5.3 Stability Program Setup and Use

The intended setup will have the real-time VSA&E and TSA&E run every 10 min. There is a screenshot of the VSA&E Real-Time results summary shown in Fig. 18.4. Please note that this figure does not include any of the actual scenarios for the DEV system. They are interfaces chosen at random for the purposes of illustration and do not reflect actual system stability boundaries. As seen below, there is a PV curve generated for each of the defined interfaces. In cases where it is necessary, there are suggested controls for increasing this margin. The busses which are most sensitive to

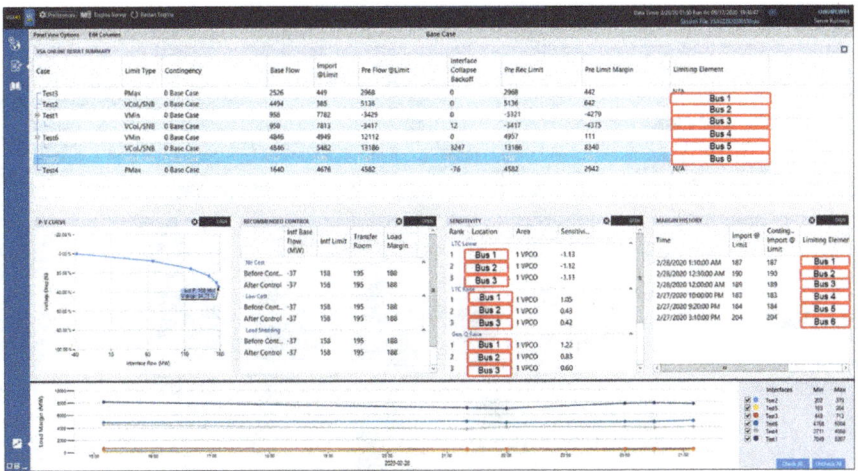

Fig. 18.4 VSA&E real-time example image

the changes in the interface are listed as well as how they can be altered to help in the event that a controlling action is needed. Additionally, there is historical information presented on which busses have been the most limiting factor over the previous runs and the trends in the margin for each defined interface. The trending graph would normally show a new point every 10 min, but for the book example, the cases were created at slightly different intervals. This analysis view is available for both Base-case and Contingency Scenarios. Additionally, there is a map view with gauges showing the distance to the margin in green, yellow, or red which is overlaid on the transmission system map. However, an image of this cannot be provided due to the sensitive nature of physical locations of equipment in the power system. Figure 18.5 shows a much less detailed map that is utilized in the TSA&E real-time display. At the bottom of the display, bar graphs are provided to indicate how many of the last runs have had any unstable contingencies. Since this is just an example of several cases, it does not show as many as would be present if it were running in real-time. The example provided is also not a representative case, but just used for illustrative purposes in this chapter.

The VSA&E look-ahead tool will forecast the voltage stability margin for the next 6–8 h to determine if the scheduled outages or changes in loading. Additionally, the look-ahead tool may extend its forecast to 24 h ahead of real-time, in which case the run frequency will be once an hour.

As the application integration at DEV is nearing its final stages, the intended use cases are both real-time and operations planning. Initially, real-time will be utilized more by the Reliability Engineers who have daily shifts in which they perform the studies in the next day to 10-day window and assist operators with questions or concerns. Eventually, operators would be trained on using the tool as they talk to PJM and take steps to mitigate issues. For the VSA&E tool look-ahead feature, the

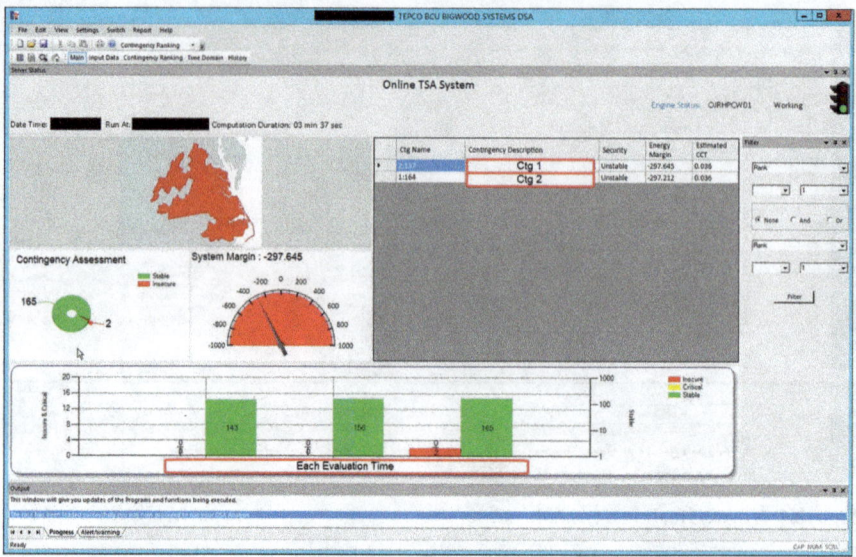

Fig. 18.5 TSA&E real-time example image

reliability engineers would monitor the results and work with operators to determine mitigation strategies for the next 8 h.

As previously mentioned, both the VSA&E and TSA&E applications provide feedback on mitigation strategies to increase the margins measured in each respective application. The programs can recommend changing the real power output of generators or their voltage setpoints, changing LTCs, and utilizing reactive and FACTs devices. These recommendations are suggested based on user input. When the initial files are generated, the user can set weights to which controls they would like to utilize first. For example, changing LTCs, utilizing capacitors and reactors, or changing generator voltage set points could be categorized as being of "No Cost." Optimization through generation real power redispatch could be categorized as "Low Cost." These suggestions are made for both real-time and study cases. An example of the recommendations for increasing the stability margin for a fictional study case is seen in Fig. 18.6. The expected gain in stability margin when these controls are applied is shown in Fig. 18.7. In this example, the required increase in margin is set to 100 MW, and could be set higher or lower, by the user.

The program can also evaluate the impact of load shedding or placing a reactive resource in a location where it does not currently exist as extremely "High Cost" solutions. Although the recommendation of placement of a new reactive resource is not helpful in real-time, it could be extremely helpful in long-term studies of outages that can last from 6 months to 2 years. DEV utilizes mobile reactive resources to support long-term outages across the system.

For longer term operations planning studies, both applications can import cases which have had outages applied to them in the steady-state analysis completed

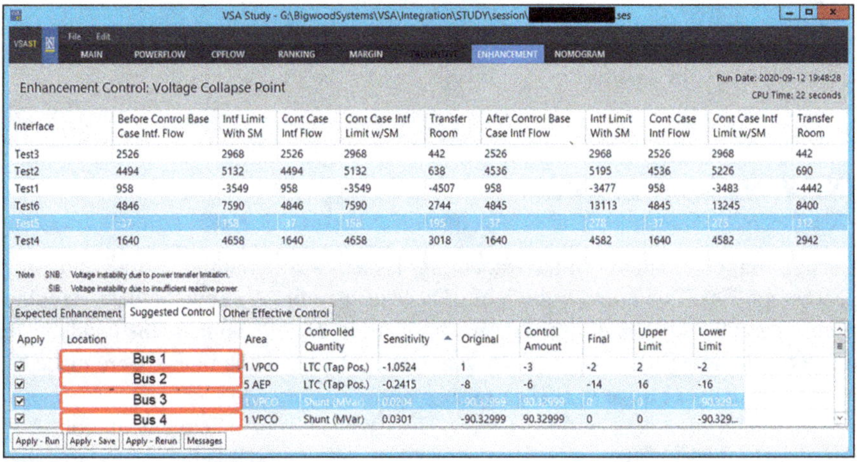

18 Stability Applications in the Dominion Energy System Operations Center

Fig. 18.6 Recommended controls

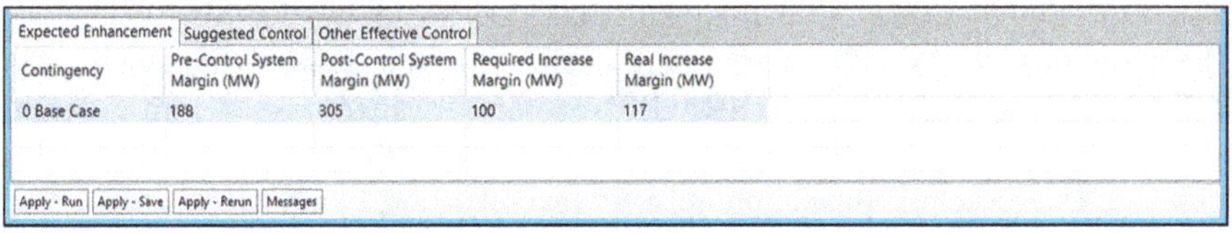

Fig. 18.7 Expected increase in stability margin for suggested controls

using the ANODE platform that utilizes PSS/E and TARA powerflow engines. As a function of the way ANODE works, these study cases will have the appropriately scaled load, generation, and temperature ratings set for the given study time. The VSA&E program runs a contingency analysis along the defined interfaces and power transfer direction to evaluate their security. In VSA&E, there is a ranking feature which will rank the severity of the contingencies to help understand the nature of the violation and how it is impacted by contingencies across the system. The TSA&E simulation performs a fault analysis to determine the transient stability of the generators. Based on the results and the ability of the corrective actions to fix issues for either type of analysis, an outage will be rescheduled to avoid instability.

For both real-time voltage stability and transient stability, every real-time case that is computed is archived on the server. All cases can be retrieved via laptop by users and brought into the study mode tool where model parameters can be manipulated for new simulations to be run.

A small signal stability analysis application was included in the software package that is used only for study purposes. The application provides useful information about the different dynamic components in the system and can be utilized as needed based on the results of the transient studies. This type of analysis provides valuable insight for understanding a particular generator mode or how modes interact with each other across the system. The program's ability to provide system matrices and eigenstructure information can also be used to study more intricate possible control applications on the system.

18.5.4 Lessons Learned

During implementation, initial modeling issues were mainly around mapping generators, FACTs busses, and circuit IDs. The specifics on how the data mapping occurred on the DEV side were discussed in Sect. 18.5.2. Some adjustments were made by changing the scripts which generated the files while others were made by adjusting the data bridge in BSI software.

Additional modeling issues surround the dynamic model implementation. It is notoriously difficult to ensure that a reduced dynamic model behaves in a way that is representative of the actual system response. However, the transient and small signal analysis require much more care in ensuring that the DEV model accurately represents the dynamics shown in the case developed by the MMWG cases. The cases that the MMWG develop are overseen by the Eastern Interconnection Reliability Assessment Group (ERAG). Not only are there issues with different generators which did not have a one-to-one match with the MMWG case, but also with FACTs device mapping. Up until very recently, FACTs devices were modeled as their own generating units in the EMS. They are now modeled as FACTs devices in the EMS and are easier to translate with the dynamic model. Finally, the ties to external utilities are modeled as large generators in the EMS. However, those are not actually generating units and do not have any dynamics as such. In an effort to simplify the reduction, they have been modeled as classical generators for the current implementation. That is subject to change based on continued testing. A lesson learned from this process would be to involve Transmission Planning which already does a stability analysis on the larger scale at the beginning of testing and data collection. This will help provide a general understanding for what would normally be found in a transient analysis.

18.6 Next Steps

In the coming months, the rollout of VSA&E in real-time operations will be completed. There will be training developed for reliability engineers performing the operations studies first. Then, training will move onto real-time operators. As this analysis is rolled out, there will be further refinement of the monitored interfaces and areas. When this tool reaches a point of acceptance and understanding with system operators such that they can make informed decisions based on the output, there will need to be discussion about moving the hardware and software into a CIP environment.

Further testing needs to be conducted on the TSA model and environment. Because of the modeling issues associated with the reduced model, there have not been enough tests run to ensure that the model has been reduced properly. This will be an effort that will involve close work with long-term transmission planning and PJM. The small signal analysis tool has and will continue to play an important role in this testing because of the information that it provides about generator modes and how they interact with each other. Additionally, Dominion Energy Virginia has been working with PingThings to implement a robust synchrophasor platform for data analytics [12]. As synchrophasor data is collected and analyzed, it will provide valuable insight into the validity of the models.

As both systems continue to integrate into the operations planning project cycle, DEV and BSI will need to work together on an API that will allow for the aggregation of the results so that they can be placed into already existing ANODE reports for the long-term study engineers. Currently, the developers of ANODE have been working on a user interface (UI) to report the operational planning study results in an easily digestible and incredibly useful manner. The new results of the stability analyses will need to be included as part of this. Upon reviewing the high-level results of these combined studies, the engineer can then go back to the tool to further analyze and assess the results.

References

1. N. A. Armstrong, The engineered century. The Bridge, pp. 14–18 (2000)
2. FERC, [Online] (2017), https://www.ferc.gov/industries-data/electric/general-information/electric-industry-forms/form-1-electric-utility-annual. Accessed 20 Aug 2020
3. PJM, [Online] (2020), https://www.pjm.com/-/media/library/reports-notices/weather-related/20180226-january-2018-cold-weather-event-report.ashx?la=en. Accessed 20 Aug 2020
4. Dominion Energy, [Online] (2020), https://www.dominionenergy.com/projects-and-facilities/wind-power-facilities-and-projects/coastal-virginia-offshore-wind. Accessed 20 Aug 2020
5. NERC, [Online], https://www.nerc.com/pa/Stand/Prjct201403RvsnstoTOPandIROStndrds/2014_03_third_posting_white_paper_sol_exceedance_20141001_clean.pdf. Accessed 23 Sept 2020
6. P. Kundur et al., Definition and classification of power system stability IEEE/CIGRE joint task force on stability terms and definitions. IEEE Trans Power Syst **19**(3), 1387–1401 (2004)
7. PJM, PJM [Online], (2020), https://www.pjm.com/~/media/documents/manuals/m03.ashx. Accessed 5 Oct 2020
8. NERC, [Online], https://www.nerc.com/files/TPL-001-4.pdf. Accessed 1 Sept 2020
9. Energy.gov, [Online] (2018), https://www.energy.gov/sites/prod/files/2017/09/f36/North%20Hampton%20RAS%20Presentation%20to%20PJM.pdf. Accessed 20 Aug 2020
10. Dominion Energy Virginia, [Online] (2019), https://cdn-dominionenergy-prd-001.azureedge.net/-/media/pdfs/global/electric-projects/power-line-projects/skiffes-creek/2018-irp.pdf?la=en&rev=6ccb3a801b7449fca7d050df3cd41422&hash=CC97DA13D0E24BE37CDB2A1D82612DAA. Accessed 21 Aug 2020
11. K. Jones, M. Parker, M. Jalapour, Design and implementation of an extensible platform for large scale, programmatic transmission outage planning analysis, in *CIGRE US National Committee 2018 Grid of the Future Symposium*, Reston, VA, 2018
12. K. Jones, Utilities tech outlook, [Online], https://transmission-and-distribution.utilitiestechoutlook.com/cxoinsight/the-right-tool-for-the-data-job-nwid-259.html. Accessed 2 Sept 2020

Chapter 19
TEPCO-BCU for Transient Stability Assessment in Power System Planning Under Uncertainty

Ryuya Tanabe, Hsiao-Dong Chiang, and Hua Li

19.1 Introduction

Japan has a maximum electric power usage of approximately 160 million kW, with the demand for electricity within the area covered by the TEPCO Power Grid being equivalent to an area greater than that of some countries such as the UK or Italy. TEPCO is the largest electric utility in Japan and the fourth largest electric utility in the world. The TEPCO power grid is one of the most reliable power transmission and distribution networks with a power outage of 0.06 times/year, duration of a power outage of 6 min/per year, and a power distribution loss of 4.2%.

For many utilities around the world, there has been considerable pressure to increase power flows over existing transmission corridors, partly due to economic incentives (a trend towards deregulation and competition) and partly due to the practical difficulties of obtaining authorization to build power plants and transmission lines (environmental concerns). This consistent pressure has prompted the requirement of extending EMS to take into account the dynamic security assessment (DSA) and control. Such an extension, however, is a rather difficult task and requires several breakthroughs in measurement systems, analysis tools, computation methods, and control schemes. TEPCO and BSI have jointly developed the TEPCO-

R. Tanabe
System Planning Department, Tokyo Electric Power Company, Tokyo, Japan

H.-D. Chiang (✉)
School of Electrical and Computer Engineering, Cornell University, Ithaca, NY, USA
e-mail: chiang@ece.cornell.edu

H. Li
Bigwood Systems, Inc., Ithaca, NY, USA

BCU system to address the online DSA concern and the off-line transient stability assessment of unreduced large-scale power grids with a large contingency list.

After decades of research and development in the direct methods, it has become clear that the time-domain method approach in stability analysis cannot be completely replaced. Instead, the capabilities of the direct methods and the time-domain method should be used to complement each other. The current direction of development is to combine a direct method and a fast time-domain method into an integrated power system stability program to take advantage of the merit of both methods. TEPCO-BCU was developed in this direction by integrating the BCU method, improved BCU classifiers, and the BCU-guided time-domain method. TEPCO-BCU has been evaluated on several practical power system models. The evaluation results indicate that TEPCO-BCU works well with reliable transient stability assessment results and accurate energy margin calculations on several study power systems, including a 50,000-bus planning case system. Detailed study of TEPCO-BCU and of a commercial time-domain package on the planning cases is conducted.

19.2 Dynamic Contingency Screening

The strategy of using an effective scheme to screen out a large number of stable contingencies and capture critical contingencies and to apply detailed simulation programs only to potentially unstable contingencies is well recognized. This strategy has been successfully implemented in online SSA. The ability to screen several hundred contingencies to capture tens of the critical contingencies has made the online SSA feasible. This strategy can be applied to online DSA. Given a set of credible contingencies, the strategy would break the task of online DSA into two stages of assessments [1, 2]:

Stage 1: perform the task of dynamic contingency screening to quickly screen out contingencies that are definitely stable from a set of credible contingencies.
Stage 2: perform a detailed assessment of dynamic performance for each contingency remaining in Stage 1.

Dynamic contingency screening is a fundamental function of an online DSA system. The overall computational speed of an online DSA system depends greatly on the effectiveness of the dynamic contingency screening, the objective of which is to identify contingencies that are definitely stable and thereby avoid further stability analysis for these contingencies. It is due to the definite classification of stable contingencies that considerable speed-up can be achieved for dynamic security assessment. Contingencies that are either undecided or identified as critical or unstable are then sent to the time-domain transient stability simulation program for further stability analysis.

The following requirements are essential for any candidate (classifier) intended to perform online dynamic contingency screening for current or near future power systems [2]:

1. (Reliability measure) absolute capture of unstable contingencies as fast as possible; i.e., no unstable (single-swing or multi-swing) contingencies are missed. In other words, the ratio of the number of captured unstable contingencies to the number of actual unstable contingencies is 1.
2. (Efficiency measure) high yield in screening out stable contingencies as fast as possible, i.e., the ratio of the number of stable contingencies detected to the number of actual stable contingencies is as close to 1 as possible.
3. (Online computation) little need of off-line computations and/or adjustments in order to meet with the constantly changing and uncertain operating conditions.
4. (Speed measure) high speed, i.e., fast classification for each contingency case.
5. (Performance measure) robust performance with respect to changes in power system operating conditions.

19.3 The Architecture of TEPCO-BCU

TEPCO-BCU is an integrated package developed under joint multiyear efforts between Tokyo Electric Power Company, Tokyo, Japan and Bigwood Systems, Inc., Ithaca, NY, USA, for fast and yet exact stability assessment and control (including accurate energy margin calculation and controlling UEP calculations) of large-scale power systems for online mode or online study mode, or off-line planning mode [3–5]. The algorithmic methods behind TEPCO-BCU include the BCU method [6, 7], BCU classifiers [8, 9], improved energy function construction [10], and the BCU-guided time-domain method [3]. Several advanced numerical implementations for the BCU method have been developed in TEPCO-BCU. The improved energy function construction has been developed to overcome the long-standing problem associated with the traditional numerical energy function that has been suffering from severe inaccuracy.

The main functions of TEPCO-BCU include the following:

- Fast screening of highly stable contingencies
- Fast identification of insecure contingencies
- Fast identification of critical contingencies
- Computation of the energy margin for transient stability assessment of each contingency
- BCU-based fast computation of the critical clearing time of each contingency
- Contingency screening and ranking for transient stability in terms of energy margin or critical clearing time
- Detailed time-domain simulation of selected contingencies

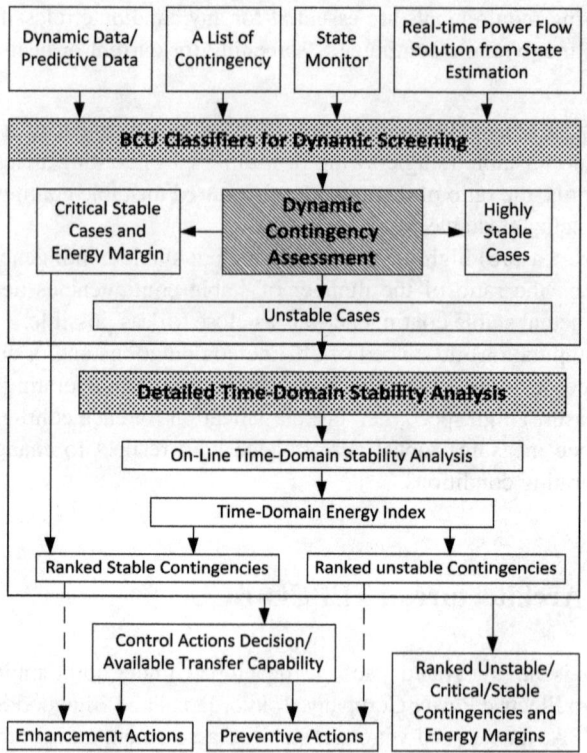

Fig. 19.1 TEPCO-BCU for online dynamic security assessment and control

BCU classifiers are designed to meet the above five performance measures. These five measures are essential for any dynamic contingency screeners intended for online dynamic security screening. The BCU-guided time-domain method is a time-domain-based, BCU-guided method for stability assessment and computing the critical energy value. The BCU-guided time-domain method is combined with the improved BCU classifiers for screening out contingencies to meet the above five measures. The integrated system, which is composed of the improved BCU classifiers and the BCU-guided time-domain method, is called TEPCO-BCU, and the architecture of this system is shown in Fig. 19.1. The system is reliable and yet fast for calculating the energy margin for every contingency. In addition, the critical energy value computed by the TEPCO-BCU method is compatible with that computed by the Controlling Unstable Equilibrium Point (CUEP) method [5, 6, 10, 12].

When a new cycle of DSA is warranted (say every 15 min), a list of credible contingencies, along with information from the state estimator and topological analysis, are applied to the dynamic contingency screening program whose basic function is to screen out contingencies that are definitely stable. Contingencies that are classified as definitely stable are eliminated from further analysis. The block function of control action decisions determines if timely post-fault contingency

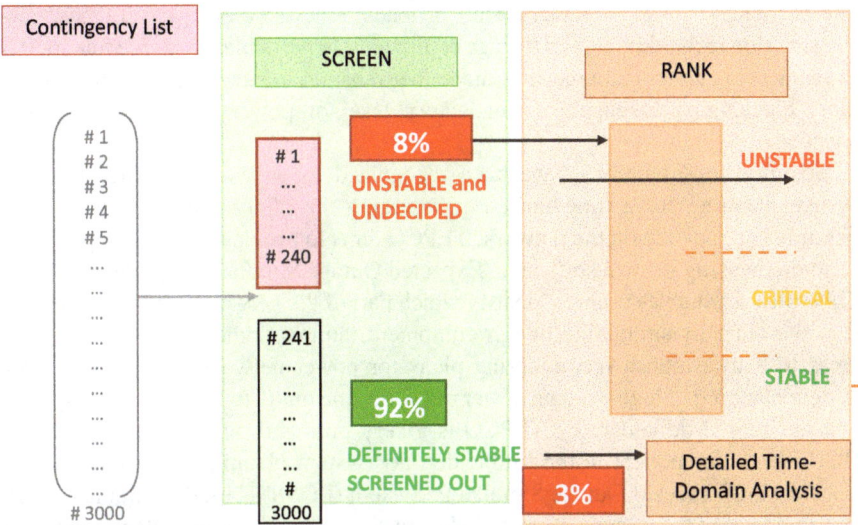

Fig. 19.2 A contingency list is screened and ranked by the TEPCO-BCU and the remaining unstable/undecided contingencies, say 3% of the total contingencies, are sent to the time-domain simulation program for detailed time-domain simulation for final classification into unstable cases and critical cases

corrective actions such as automated remedial actions are feasible to steer the system from unacceptable conditions to an acceptable state. If appropriate corrective actions are not available, the block function of preventive actions determines the required pre-contingency preventive controls, such as real power re-dispatches or line switching to maintain the system stability, should the contingency occur.

The TEPCO-BCU architecture includes the capabilities of the BCU method and the time-domain method in which two methods complement each other. For a given list of contingencies, say 3000 contingencies, the BCU screens out a vast majority of contingencies, say 92%, that are definitely stable and removes them from further analysis. The remaining 8% of the contingencies are then sent to the ranking stage for classification as unstable contingencies, critically stable contingencies, and stable contingencies. The set of unstable contingencies classified by the BCU method needs to be sent to the time-domain simulation program for final stability assessment; in other words, 90 contingencies out of the original 3000 contingencies require detailed time-domain simulation (see Fig. 19.2).

19.4 Planning Under Uncertainty Based on TEPCO-BCU

Transmission expansion planning seeks to make the optimal decision on when and where to make network expansion and the related assets, taking various factors into account, such as the uncertainties of loads, renewable energy resources, new transmission technologies, and storage equipment. In particular, the large integration of renewable energy resources has made the transmission expansion problem even more challenging to ensure a satisfactory level of power system security and adequacy.

Stability considerations have been recognized as an essential part of power system planning for a long time. For assessing the effect and seriousness of a possible contingency on the network, TEPCO aims to provide the most appropriate solution by using its own software, "Expected Outage Simulating System (EOSS)." This software enables users to identify which part of the network is more vulnerable than the other by simulating line or equipment failures, indicating heavily loaded areas to help establish reconfiguring plans for power network improvement. For system planning of high-voltage power networks, the multi-functional power system analysis tool (Midfielder) by TEPCO is a very powerful planning tool. TEPCO-BCU is one component in the Midfielder. For system planning of regional power systems, the Expected Outage Simulating System (EOSS) is used to simulate every possible fault of the system automatically and select the minimum-outage supplying power system. As the TEPCO system continually grows in size and in complexity due to renewable penetration, it is becoming increasingly more difficult to examine transient stability in power system planning.

In transmission planning studies, robust system performance such as transient stability assessment (TSA) must be demonstrated during credible normal system conditions and in response to contingencies of the following:

- Network contingency list (i.e., the traditional contingency list)
- Renewable contingency list (i.e., capturing the uncertainty of renewable energy)

To derive the renewable contingency list, a flexible scenario generation tool is needed to address two major issues: (a) characterizing the dependence among renewable sources and (b) effectively generating the renewable scenarios that capture the dependence. To this end, we have applied an effective tool, taking advantage of a copula and a Latin Hypercube with dependence method, that consists of the following stages:

- **Stage I:** (Distribution Modeling) Historical measurements and data of different renewable energy sources are used to compute the marginal distribution functions.
- **Stage II:** (Dependence Modeling) Assuming a structure of dependence, a multivariate joint distribution is constructed using a copula to represent dependence among the renewable sources. This contains the information regarding both the degree of dependence and the structure of dependence.
- **Stage III:** (Ranking and selection) Compute the fit indices (e.g., the Spearman's rank when studying spatial dependence) and select the best dependence model for the sampling stage.

- **Stage IV**: (Scenario Generation) Apply a Latin Hypercube with dependence method to generate scenarios, according to the multivariate joint distribution and dependence model obtained in the preceding stage.
- **Stage V**: (Output Analysis) Analyze the produced scenario.

The tool possesses several desired features. First, it can better represent the uncertainties because the model in Stage II additionally takes into consideration the dependence among renewable sources, which can accurately capture the uncertainties and provide comprehensive information about the dependence, including the degree of dependence, as well as the structure of the dependence. Second, the Latin Hypercube with dependence (LHSD) method in Stage IV makes the best out of dependence and the corresponding joint distribution, since the scenarios or samples generated by LHSD usually are smaller in size than those produced by Monte Carlo sampling. The scenario generation with dependence tool offers several flexibilities. First, the dependence between concerned random variables may be linear or nonlinear. This dependence can be characterized by a proper choice of copula. This is one flexibility offered in Stage II of the proposed method. In addition, the LHSD employed here is quite different from the classical LHS (which usually is only applied to random vectors with independent components). Indeed, LHS is just a special case of LHSD, while by properly tuning $\eta_{k,N}^q$, LHSD can generate (asymptotically unbiased) samples for the independent and dependent structures, whose effectiveness is ensured by analytical studies [11]. So LHSD provides extra flexibility for modeling various (in-) dependent structures. On the application side, the proposed method can be seamlessly integrated into the studies on stability assessment and economic dispatch problems for large-scale power systems with high-penetration renewable energy.

Scenario generation with the dependence tool offers several flexibilities. First, the dependence between concerned random variables may be linear or nonlinear. This dependence can be characterized by a proper choice of copula. In addition, the LHSD employed here is quite different from the classical LHS (which usually is only applied to random vectors with independent components). Indeed, LHS is just a special case of LHSD, which can generate (asymptotically unbiased) samples for independent and dependent structures, and whose effectiveness is ensured by analytical studies [11]. So LHSD provides extra flexibility for modeling various (in-) dependent structures. We feel that the method can be seamlessly integrated into the studies on stability assessment and economic dispatch problems for large-scale power systems with high-penetration renewable energy.

The detailed procedure to generate the renewable contingency list proceeds as follows. The architecture of the scenario generation tool is shown in Fig. 19.3. In the figure, the arrows in gray boxes show how to choose the proper options to obtain the expected results.

A numerical example using the data of 6 wind farms in the evaluation system is provided in the sequel to show the performance of vine copulas and the Archimedean and elliptic copulas. The example is detailed below to show the step-by-step procedure.

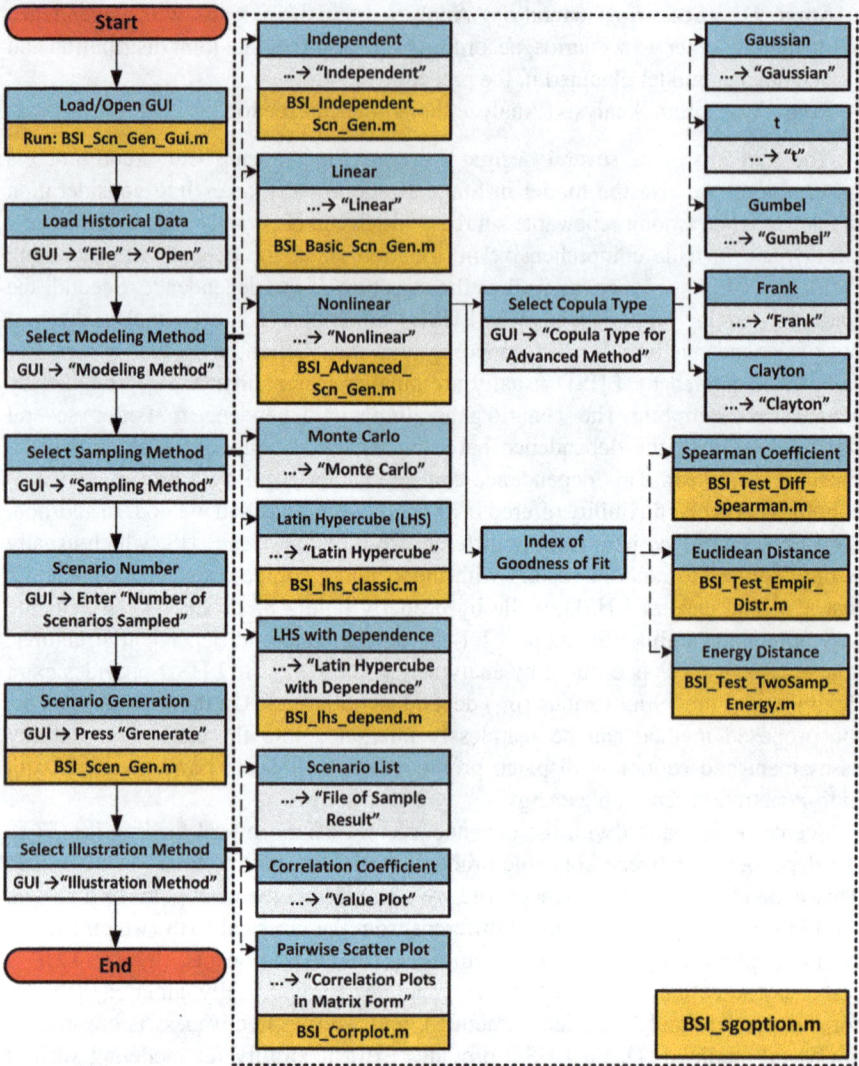

Fig. 19.3 The architecture of a scenario generation tool

Step 1: Load the data of 6 wind farms. Here, a set of parameters temporarily stores the data file names for the six wind farms, which includes the historical data of the kth wind farm with a list of observed actual power outputs and a list of forecast errors, respectively.

Step 2: Convert the forecast errors to the copula scale (i.e., unit scale) using the R function pobs() and store them. The pairwise Spearman's rank correlation coefficients can be calculated by the R function cor().

19 TEPCO-BCU for Transient Stability Assessment in Power System...

Step 3: Fit an RVine copula and a CVine copula and fit a DVine using the package CDVine. The fitted copulas are stored.

Step 4: Apply the fitted vine copulas. A set of 200 sample points is obtained for the RVine (and CVine) copula model. The same number of sample points is generated for the DVine copula model. Apply the newly developed function to the three sample sets and generate new sample sets by the Latin Hypercube sampling method with dependence.

Step 5: Compare the performance, then another five copulas are fitted and sampled by LHSD, namely, the Archimedean copulas (Clayton, Frank, Gumbel) and the elliptic copulas (Gaussian/normal, t).

Step 6: To select the best copula from the vine copulas, Archimedean copulas, and elliptic copulas, the indices for the goodness of fit, e.g., the geometric distance and the energy distance between the empirical distribution (of historical data) and the fitted copula (or the sample set), and the difference in correlation coefficients are calculated and tabulated below. The overall ranking value is given by the normalized average. The t copula attains the lowest overall ranking value and is selected as the best-fitted copula for the data set.

Step 7: Using the selected t copula, the set of samples for the forecast error is converted back to the scale of the historical data (instead of the unit scale), and then added to the forecast mean values for the power outputs of different wind farms, which produces the scenario list for the 6 wind farms. The results are illustrated using the different forms below.

Output 1 The pairwise correlation plots for the sample set of the forecast errors using the fitted t copula in the original scale.

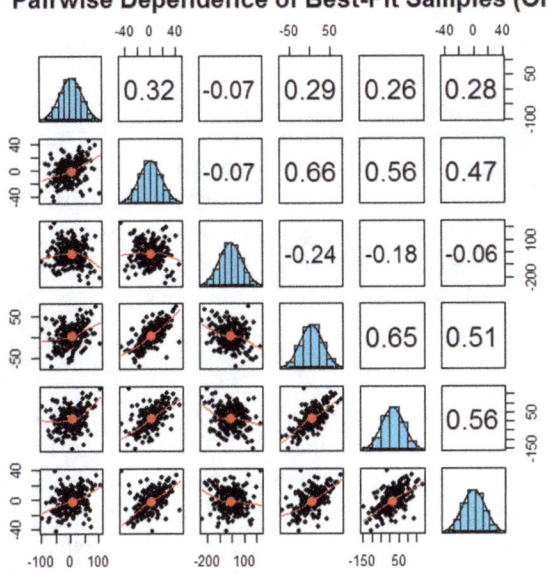

Output 2 A graph of 200 scenarios where the horizontal axis refers to the scenario number and the vertical axis indicates the power from a wind farm in the scenario.

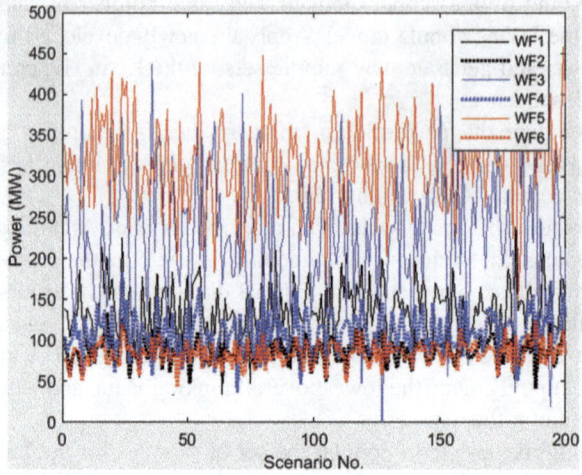

Output 3 Scenario List (only the first 10 scenarios of 200 in total are shown below)
Each row is a future scenario (power output, MW) generated for different wind farms (WFs)

No.	WF1	WF2	WF3	WF4	WF5	WF6
[1]	137.8590611	95.18818987	251.9441977	104.4338424	337.8578263	77.40986589
[2]	135.5854501	81.47163s567	277.8875807	84.85233395	288.4456309	73.72960408
[3]	103.169661	58.15795314	208.4977472	95.68192351	334.7009333	55.00841727
[4]	122.0070506	77.57971134	161.6312439	101.4039613	309.2193172	76.17110926
[5]	134.6752547	85.05847488	112.2816996	130.9489826	326.2792999	90.98218027
[6]	130.0931504	87.10991069	194.9441908	102.0858136	291.6703407	86.1931278
[7]	185.9233397	73.91066468	231.2431834	110.9672439	275.498279	85.27790787
[8]	128.2344564	59.96719602	393.0594875	55.20623886	259.1421988	51.57336101
[9]	102.4660014	99.40368793	185.4394097	130.0561256	294.8272337	65.29154376
[10]	139.6807426	101.3882466	233.3942111	106.7382808	342.7243469	80.51250824

We next demonstrate the application of TEPCO-BCU in performing TSA for planning purposes for which the contingency list is extensive to cover possible and yet credible contingencies including the network contingency list and the renewable contingency list. The size of the planning system is summarized below (Table 19.1).

Since the capabilities of the direct methods and the time-domain method should be used to complement each other, the current direction of applying these two tools is to combine a direct method and a fast time-domain method into an integrated

Table 19.1 Network data of the planning study

Buses	47,999	Branches	63,679
Loads	29,974	Exciter	3376
Generators	4776	Governor	2914
Transformers	16,604	Stabilizer	993
Compensator	288	Control	24

Table 19.2 The reliability measure of TEPCO-BCU

Total number of contingencies	Percentage of unstable contingencies captured by TEPCO-BCU
7850	100%

Table 19.3 The screening measure of TEPCO-BCU

Total number of contingencies	Percentage of stable contingencies screened out by TEPCO-BCU
7850	7821 (99.63%)

power system stability program to take advantage of the merits of both methods. To illustrate their merits, the outputs of both the time-domain step-by-step package and the TEPCO-BCU packages were written into two separate text files. Another set of software was developed to evaluate the integration results as follows:

- The stability status (i.e., stable/unstable) of each contingency
- The required CPU time by TEPCO-BCU software
- The reliability measure by TEPCO-BCU
- The screening measure by TEPCO-BCU

1. Reliability Measure

   To be useful as a fast screening and ranking tool, TEPCO-BCU needs to consistently give conservative stability assessments for each contingency and not give over-estimated stability assessments for any contingency. For a total of 7850 contingencies, TEPCO-BCU captures all the unstable contingencies (Table 19.2).

2. Screening Measure

   Depending on the loading conditions and network topologies, the screening rate may vary. The screening rate of TEPCO-BCU on this large planning system is 99.63%, meaning that TEPCO-BCU screens out 99.63 contingencies for every 100 stable contingencies, which are definitely stable (Table 19.3).

3. Speed Measure

   The total time required for the contingency list (a total of 7850 contingencies) by the time-domain is about 54.5 days while the total time required by the TEPCO-BCU package is the total time required for the 7850 contingencies, which is about 9 h, and the total time required by the time-domain simulation for the 29 (potentially) unstable contingencies, which is about 13.8 h. Hence, the total duration needed for TSA of the real power system model is noted in the following tables (Tables 19.4 and 19.5).

Table 19.4 Speed of TEPCO-BCU and the time-domain simulation package (with a 10-s simulation duration)

TEPCO-BCU package	4.12739 s per contingency (does not include the time-domain simulation)
Time-domain simulation package (commercial package)	600 s per contingency for a 10-s simulation time

Table 19.5 Overall TSA by the time-domain simulation and by TEPCO-BCU (Screening + Ranking + Time-domain simulation on (potentially) unstable contingencies)

Name of package	Total computation time	Computation time per contingency
TEPCO-BCU package + time-domain simulation on the identified potentially unstable contingencies	9 h for the screening and ranking stages and 4.8 h for detailed analysis by the time-domain package	6.328 s/contingency

19.5 Conclusion

TEPCO and BSI have jointly developed the TEPCO-BCU system to address the online DSA concern. After decades of research and development in direct methods, it has become clear that the time-domain method approach in stability analysis cannot be completely replaced. Instead, the capabilities of the direct methods and the time-domain method should be used to complement each other. TEPCO-BCU was developed under this direction by integrating the BCU method, improved BCU classifiers, and the BCU-guide time-domain method. TEPCO-BCU has been evaluated on several practical power system models. The evaluation results indicate that TEPCO-BCU works well with reliable transient stability assessment results for online transient stability assessment and for off-line planning study. Accurate energy margin calculations were obtained on several study power systems, including a 50,000-bus planning case system. A detailed evaluation of the integrated TEPCO-BCU and a time-domain package on the planning case favored the TEPCO-BCU package as a screening and ranking tool over the traditional time-domain approach as a detailed analysis tool for large-scale power system planning.

References

1. N. Balu, T. Bertram, A. Bose, V. Brandwajn, G. Cauley, D. Curtice, A. Fouad, L. Fink, M. G. Lauby, B. Wollenberg and J.N. Wrubel, On-line power system security analysis (Invited paper). *Proceedings of the IEEE*, 80(2), 262–280 (1992)
2. H.D. Chiang, C.S. Wang, H. Li, Development of BCU classifiers for on-line dynamic contingency screening of electric power systems. *IEEE Trans. Power Systems*, 14(2) (1999)

3. Y. Tada, A. Kurita, Y.C. Zhou, K. Koyanagi, H.D. Chiang, Y. Zheng, BCU-guided time-domain method for energy margin calculation to improve BCU-DSA system, in *IEEE/PES transmission and distribution conference and exhibition* 2002
4. Y. Tada, T. Takazawa, H.D. Chiang, H. Li, J. Tong, Transient stability evaluation of a 12,000-bus power system data using TEPCO-BCU, in *15th power system computation conference (PSCC)*, Belgium, August 24–28, 2005
5. H.D. Chiang, F.F. Wu, P.P. Varaiya, A BCU method for direct analysis of power system transient stability. IEEE Trans. Power Syst. **8**(3), 1194–1208 (1994)
6. H.D. Chiang, C.C. Chu, G. Cauley, Direct stability analysis of electric power systems using energy functions: theory, applications and perspective (Invited paper), *Proc. IEEE*, 83(11), 1497–1529 (1995)
7. P.W. Sauer, M.A. Pai, *Power system dynamics and stability* (Prentice-Hall, Englewood Cliffs, 1998)
8. H.D. Chiang, Y. Zheng, Y. Tada, H. Okamoto, K. Koyanagi, Y.C. Zhou, Development of an on-line BCU dynamic contingency classifiers for practical power systems, in *14th power system computation conference (PSCC)*, Spain, June 24–28 2002
9. U.S. Patent 6,868,311; Method and System for On-line Dynamical Screening of Electric Power System, Date of Patent, March 15, 2005 (Inventors: Hsiao-Dong Chiang, Atsushi Kurita, Hiroshi Okamoto, Ryuya Tanabe, Yasuyuki Tada, Kaoru Koyanagi, and Yicheng Zhou)
10. H.D. Chiang, *Direct methods for stability analysis of electric power systems: theoretical foundation, BCU methodologies, and applications* (Wiley, Hoboken, 2011)
11. N. Packham, W.M. Schmidt, Latin hypercube sampling with dependence and applications in finance. J. Computat. Financ. **13**(3), 81–111 (2010)
12. U.S. Patent 2006/0190227; Group-Based BCU Methods for On-Line Dynamical Security Assessments and Energy Margin Calculations of Practical Power Systems, Pub. Date, Aug. 24, 2006 (Inventors: Hsiao-Dong Chiang, Hua Li, Yasuyuki Tada, Tsuyoshi Takazawa, Takeshi Yamada, Atsushi Kurit, and Kaoru Koyanagi)

Chapter 20
Voltage Stability Assessment Using Synchrophasor Technology

Iknoor Singh, Ken Martin, Neeraj Nayak, Ian Dobson, Anthony Faris, and Atena Darvishi

20.1 Overview

Voltage stability is critical for grid operations and determining voltage stability in operations is a complicated problem. Relying only on the voltage magnitude to detect a voltage stability problem is insufficient, as it is possible to maintain voltages by reactive power controls while the system is becoming more susceptible to voltage instability [1]. The conventional methods for monitoring voltage stability problems rely upon state estimation or continuation power flow techniques which are iterative in nature, and hence may be too slow under the scenarios where several outages occur in a relatively short time span. The use of synchrophasor technology in grid operations is growing. High-resolution high-speed data from Phasor Measurement Units (PMUs) has found great use in detecting events such as electromechanical oscillations, islanding events, and other conditions. There have been efforts to develop voltage stability monitoring methods based on the

---

I. Singh (✉) · K. Martin · N. Nayak
Electric Power Group, Pasadena, CA, USA
e-mail: iknoor.singh@gmail.com; martin@electricpowergroup.com; Nayak@electricpowergroup.com

I. Dobson
Iowa State University, Ames, IA, USA
e-mail: dobson@iastate.edu

A. Faris
Bonneville Power Authority, Portland, OR, USA
e-mail: ajfaris@bpa.gov

A. Darvishi
New York Power Authority, White Plains, NY, USA
e-mail: Atena.Darvishi@nypa.gov

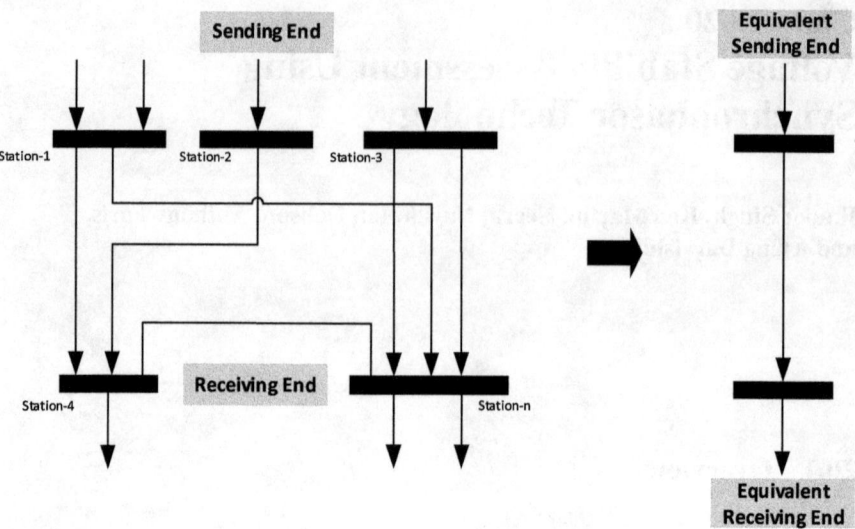

Fig. 20.1 Reduction of a transmission corridor to a single line equivalent

synchrophasor technology for use in real-time [2]. These methods can work independently of the existing state estimator and continuation power flow methods, and provide fast results, making them suitable for real-time operations. The initial methods measured voltage stability on essentially a single radial line [3]. More recently, this has been extended to measure the voltage stability of more elaborate power flow corridors, where corridors with multiple transmission lines feed the load centers. One such methodology is discussed here [4], which assesses the voltage stability of a predefined corridor by calculating a Voltage Stability Index (VSI) in real-time using synchrophasor data. The calculation uses the complex voltage and current measurements from the corridor boundary buses and can be performed at the speed of measurement (e.g., 30 s) since it does not require iterations. The approach has the effect of reducing a complicated transmission corridor (with multiple lines and several connection points) to a single line equivalent (Fig. 20.1).

This approach is particularly useful in quickly assessing the impact of multiple outages which are impractical to study systematically before they occur and may require emergency action to forestall subsequent cascading events (corridors with significant power transfers are susceptible to voltage stability issues due to multiple outages). Although voltage stability can also be routinely evaluated based on detailed state estimation [5], multiple outages are also a condition in which the state estimator is more likely to fail to converge. Thus, synchrophasor measurement-based VSI method can be seen as complementary to the conventional voltage stability monitoring methods. Since the reactive power supplied by generators is critical to maintain voltage stability, the method also considers the generator reactive power limits to provide a better assessment of voltage stability in real-time. The development of the method for use in real-time operations was part of a research project funded by the US Department of Energy and the method has been deployed at Bonneville Power Administration (BPA) and New York Power Authority (NYPA) for testing.

20.2 Methodology

The first step in assessing VSI is to define the corridor to be monitored. The corridor is defined so that the generation is at one end, feeding power into the corridor to the load centers at the other end. It works best if there are no significant branches across the corridor, but the method will accommodate several input and output taps.

Since VSI calculation requires synchrophasor data as input, it is important that the corridor is chosen such that a phasor measurement of the voltage and current of each line at the boundary of the corridor is available. In case some boundary buses have no direct PMU measurements, those buses should be observable through virtual PMUs obtained through Linear State Estimation (LSE). For that purpose, it is important that the desired virtual PMUs are observable from PMU measured buses. The LSE will also need the power system model information in the vicinity of boundary buses.

The next step is to reduce the transmission corridor with multiple transmission lines to a single line equivalent. Referring to Fig. 20.1, we reduce the problem using the following step-by-step process:

Notations:

$V_{s1}, \ldots V_{sn}$: Bus Voltage measurements at sending end
$I_{s1}, \ldots I_{sn}$: Line Current measurements at sending end
$V_{r1}, \ldots V_{rm}$: Bus Voltage measurements at receiving end
$I_{r1}, \ldots I_{rm}$: Line Current measurements at receiving end

(a) The complex currents and voltages ($I_{s1}, \ldots I_{sn}, I_{r1}, \ldots I_{rm}; V_{s1}, \ldots V_{sn}, V_{r1}, \ldots V_{rm}$) are obtained from the PMUs at all the buses that bound the transmission corridor. Then the sending and receiving complex power for each line is obtained from the measured complex currents and voltages:

$$S_{si} = V_{si} I_{si}^*, \; S_{ri} = V_{ri} I_{ri}^*$$

(b) The complete system will be reduced to a single line equivalent while preserving the complex powers entering and leaving the corridors. In other words, all or most of the power that is entering the transmission corridor is equal to the power that is leaving the equivalent system. The sending and receiving end powers are summed into overall power:

$$S_s = \sum_{i=1}^{n} S_{si}, \; S_r = \sum_{i=1}^{n} S_{ri}$$

(c) It is well known that voltage stability can be strongly affected by generators reaching their maximum reactive power output limits. Accordingly, the VSI calculation must monitor the generators providing significant reactive power to the corridor. If the limit is reached for a generator, it is treated as a negative load. Thus, the above equations need to be modified. For example, if the bus 2 generator reaches its reactive power limit, the modified equations are:

$$S_s = \sum_{i=1}^{n} S_{si} - S_{s2}, \; S_r = \sum_{i=1}^{n} S_{ri} + S_{s2}.$$

If the generator is not directly monitored by a PMU, its reactive power limit status could in some cases be estimated from nearby PMUs through LSE or by other signals.

(d) The total current at the sending and receiving ends can be calculated for the equivalent corridor as following:

$$I_s = \sum_{i=1}^{n} I_{si}, \; I_r = \sum_{i=1}^{n} I_{ri}$$

(e) If a generator reaches its reactive power limit, the modified current summation equations are (as above, assuming bus 2 generator reaches its reactive power limit):

$$I_s = \sum_{i=1}^{n} I_{si} - I_{s2}, \; I_r = \sum_{i=1}^{n} I_{ri} + I_{s2}$$

(f) Based upon the complex powers and complex currents of the transmission corridor and its equivalent, the voltages of the equivalent system are:

$$V_s = \frac{S_s}{I_s^*}, \; V_r = \frac{S_r}{I_r^*}$$

(g) Then the voltage across the equivalent is:

$$V_{sr} = V_s - V_r \text{ or } V_{sr} = \frac{S_s}{I_s^*} - \frac{S_r}{I_r^*}$$

(h) The voltage stability index (VSI) is then calculated as:

$$\text{VSI} = \frac{|V_{sr}|}{|V_r|} \times 100$$

The VSI is thus the ratio of the voltage across the system to the receiving end voltage expressed in percentage. This provides a figure of merit that directly relates to the stability of the voltage across the system.

VSIs can be calculated for various transmission corridors in the system which can then be displayed in real-time on the transmission operator's screen. An alarm is triggered if the calculated VSI exceeds a threshold. The thresholds are established by running studies using the utility's power flow and dynamic models for various loading scenarios and system conditions. This is done by gradually increasing the power flow through the corridor until voltage collapse and then finding the VSI threshold corresponding to the utility-defined emergency margin to voltage collapse. Additional alert thresholds can also be assigned at lower levels of stress thereby allowing the operator to take precautionary corrective steps.

Since this index is calculated very fast, at the speed of phasor measurements, it can respond to transients or noise in the measurement. A time delay is used to prevent spurious alarms. When the VSI exceeds a threshold, a timer starts and holds off an alarm. If the VSI is still beyond the threshold at timeout, the application issues the alarm. The user can set the time limit so that it is appropriate for the particular application. For an operator alarm, it could be several seconds. For a real-time control where the data is pre-processed to remove transients and noise, it could be a few cycles for very fast response

20.3 Case Study for Use at the Bonneville Power Administration (BPA)

BPA has deployed the VSI method for monitoring the voltage stability for a 500 kV transmission corridor through Oregon in the western US. The section from SUB-1 substation to SUB-2 and SUB-3 substations was chosen for the study and is shown in Fig. 20.2. Power flow is predominantly from SUB-1 to SUB-2 and SUB-3. A high load study case has 4308 MW entering the corridor at SUB-1 substation and 3808 MW leaving the corridor at SUB-2 and SUB-3 substations (with the difference leaving the corridor at SUB-4 and SUB-5 substations). Since most of the power is being received at the receiving end substations, the VSI method can be applied to this corridor. The VSI is calculated using the power flow output (voltages and currents derived from MW/MVAR flows) at SUB-1, SUB-2, and SUB-3.

The study procedure involved running multiple severe contingency scenarios to assess the voltage stability and to establish VSI alarm thresholds accordingly. One such severe scenario is the loss of generator units at Palo Verde Nuclear Generating Station (Palo Verde is the largest power plant in the USA and has three units, each rated at about 1310 MW). The simultaneous loss of two of the generating units is considered the largest credible loss based on the system configuration. Figure 20.3 shows how the VSI and SUB-3 voltage change upon the loss of two Palo Verde units for a corridor loading of 4667 MW. The VSI is seen to increase from 16.27 to 21.20

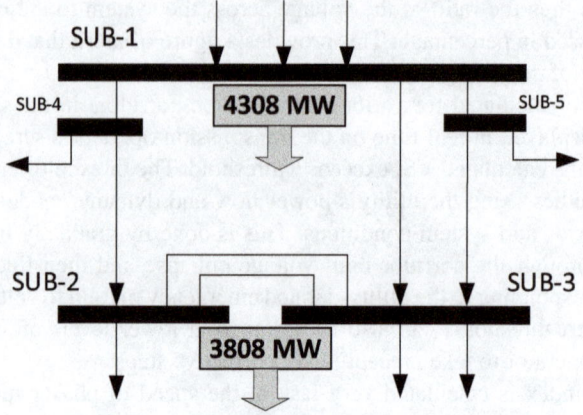

Fig. 20.2 Schematic layout of BPA study corridor

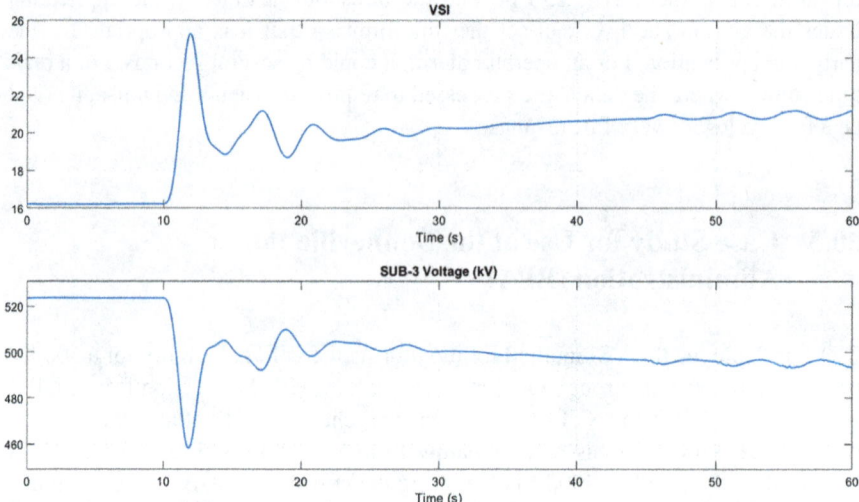

Fig. 20.3 VSI and SUB-3 Voltage plots—Loss of 2 Palo Verde units for corridor loading of 4667 MW

after the event and SUB-3 voltage sharply decreases from 523.71 to 493.79 kV. This illustrates the capability of the VSI method for quick indication of the voltage stress conditions.

Several severe contingencies were studied to establish the VSI alarm thresholds. For each of the contingencies, the VSI was calculated for various increased loading conditions on the corridor. Table 20.1 shows the study results for the loss of two Palo Verde units. The VSI worsens as the corridor loading is increased, indicating increased voltage stress on the corridor. The deterioration can also be seen in the decrease in voltage with increased corridor loading.

Table 20.1 Result summary—loss of two Palo Verde units

	Corridor loading	SUB-3 voltage	VSI
1	Base Case (3808 MW)	514.8 kV	16.61
2	4353 MW	504.8 kV	19.11
3	4667 MW	493.7 kV	21.21
4	4917 MW	480.2 kV	23.38
5	5125 MW	472.5 kV	24.87
6	5150 MW	471.5 kV	25.03
7	5160 MW	466.6 kV	25.93
8	5205 MW	NA	Case diverges

Table 20.2 Result summary—loss of three Palo Verde units

	Corridor loading	SUB-3 voltage	VSI
1	Base Case (3808 MW)	497.3 kV	19.45
2	4110 MW	488.4 kV	21.12
3	4234 MW	483.8 kV	21.98
4	4293 MW	483.3 kV	22.20
5	4402 MW	479.8 kV	22.79
6	4423 MW	483.4 kV	22.54
7	4475 MW	NA	Case diverges

Similarly, Table 20.2 shows the study results for the loss of three Palo Verde units. VSI is observed to be much more severe for the loss of three Palo Verde units for similar loadings, which is also indicated in much depressed voltages.

Based upon the VSI values and the corresponding voltages from Tables 20.1 and 20.2, the following VSI alert and alarm thresholds were established:

- Alert: 19
- Alarm: 23

Note that the above thresholds indicate the VSI values with significantly depressed voltages. The concept of VSI can also be utilized to suggest the necessary mitigation actions. For example, for the contingency in Fig. 20.3, inserting shunt capacitors at the receiving end substations (SUB-2 and SUB-3) would improve the voltage profile, which is also reflected in decrease in VSI as shown in Fig. 20.4.

Figure 20.5 shows an example real-time display for the loss of two Palo Verde units, with an initial loading of 4667 MW on the corridor. As can be seen, the VSI increases to 21, exceeding the Alert threshold of 19, triggering a real-time indication to the operator. The decline in the corridor bus voltages is also clear. The alarming delay can be set to avoid alarming on the first transient, in this case the first swing in VSI to 25. Therefore, here the delay could be about 3 s.

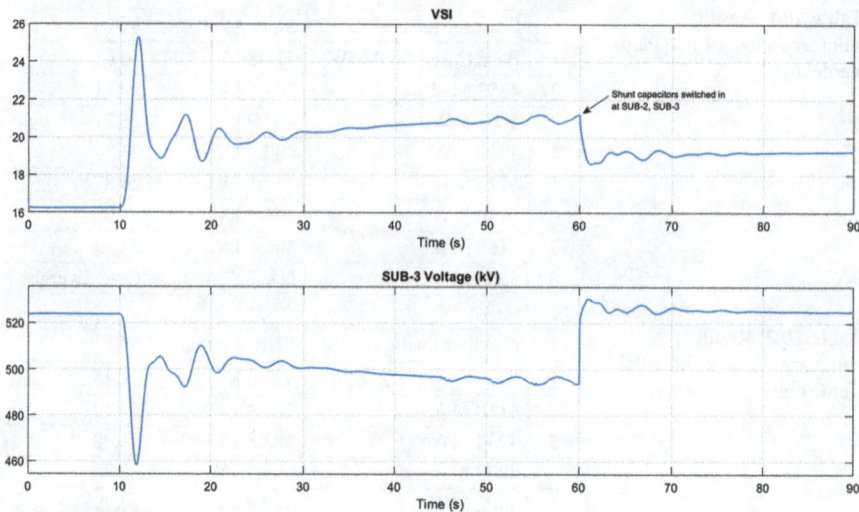

Fig. 20.4 VSI Corrective Action Example

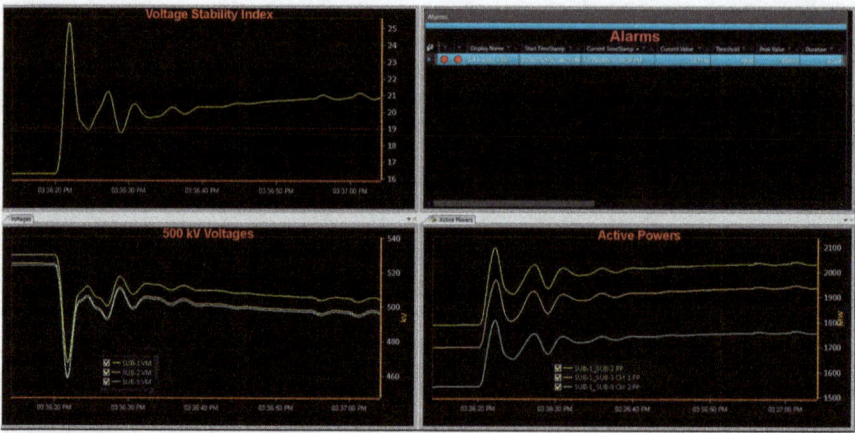

Fig. 20.5 VSI visualization in real-time for a severe event

20.4 Case Study for Use at the New York Power Authority (NYPA)

NYPA has deployed this VSI method to monitor a few transmission corridors, one of which is a 230 kV corridor from SUB-1 and SUB-2 substations to SUB-8 substation. Figure 20.6 shows a one-line diagram for the corridor, with the generators at SUB-1 and SUB-2 as the sending end and SUB-8 as the receiving end substation. The power flow is predominantly in one direction—from SUB-1 and SUB-2 to SUB-8. The lines being monitored by the PMUs are shown by the blue arrows at the two

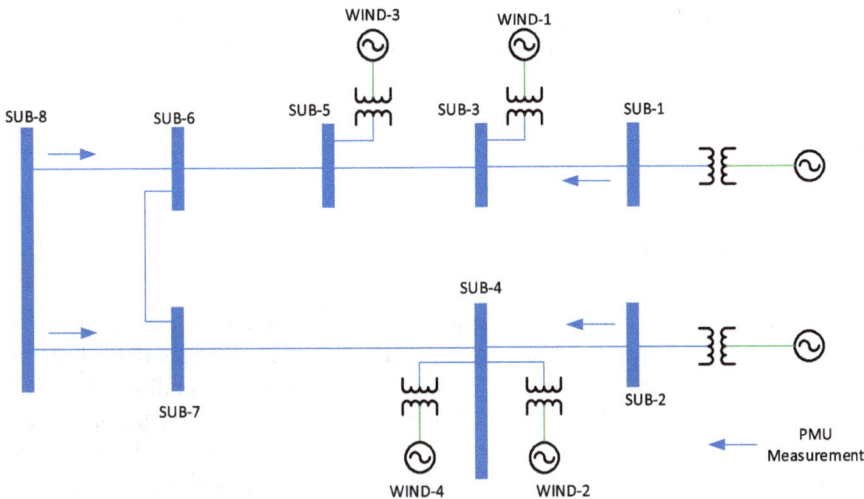

Fig. 20.6 Schematic layout of NYPA study corridor

Table 20.3 Result summary—various loadings for NYPA study corridor

	Case name	Power flow from SUB-1, 2	SUB-1 VM	Power flow at SUB-8	SUB-8 VM	VSI
1	Case-1	122 MW	239 kV	68 MW	234.6 kV	4.96
2	Case-2	159 MW	237.9 kV	105 MW	234.6 kV	6.74
3	Case-3	198 MW	237.4 kV	145 MW	234.6 kV	8.77
4	Case-4	224 MW	236.9 kV	172 MW	234.6 kV	10.16
5	Case-5	276 MW	235.8 kV	224 MW	234.6 kV	12.96
6	Case-6	315 MW	234.4 kV	262 MW	234.6 kV	15.12
7	Case-7	352 MW	227.2 kV	298 MW	233.5 kV	18.05
8	Case-8	375 MW	219.9 kV	320 MW	232 kV	20.60
9	Case-9	373 MW	217.8 kV	314 MW	230.6 kV	20.77

ends of the corridor. A high summer loading base case was used for establishing the alarm thresholds. Note that there are several intermediate wind generating stations which can add power through the corridor. However, to simplify the study, wind generation was assumed to be zero and the generators at SUB-1 and SUB-2 were ramped up to increase the power injection into the corridor thereby increasing the corridor stress.

Table 20.3 shows the VSI for increased loadings on the corridor. As can be seen, the VSI worsens as the corridor loading is increased, indicating increased voltage stress on the corridor. Decrease in voltage can also be seen with increased corridor loading.

The generators at SUB-1 and SUB-2 hit their reactive power limits for Case-7 and the active power limits for Case-8. Additionally, the voltages start to decrease significantly for Cases-8 and 9 at SUB-1 and SUB-2. Hence, the alert and alarm thresholds for VSI can be established as follows for the corridor:

- Alert: 18.00
- Alarm: 20.50

20.5 Conclusion

The Voltage Stability Assessment method discussed here combines multiple synchrophasor measurements at the generation and load ends of a predefined power flow corridor with multiple lines to form a real-time voltage stability index. The index is derived by reducing the complex corridor to a single line equivalent with the same power and current injections at the ends. Reactive power limits of the generators can be accounted for in the index calculation. The index provides a fast indication of a voltage instability condition in real-time grid operations. Since the index is fast and works well for multiple contingencies when the state estimator is less likely to converge, the approach complements the traditional voltage stability analysis calculations based on state estimation. The voltage stability assessment has been practically applied to power flow corridors in two utilities, and we discuss the choice of corridors, setting thresholds for recommending operator actions, and confirm the behavior of the index under contingencies.

References

1. C.A. Canizares, F.L. Alvarado, Point of collapse and continuation methods for large AC/DC systems. IEEE Trans. Power Syst., 1–8 (1993)
2. M. Glavic, T. Van Cutsem, A short survey of methods for voltage instability detection, in *IEEE PES General Meeting* (2011)
3. K. Vu, M. Begovic, D. Novosel, M. Saha, Use of local measurements to estimate voltage-stability margin, in *Power Industry Computer Applications Conference (PICA)* (1997)
4. L. Ramirez, I. Dobson, Monitoring voltage collapse margin with synchrophasors across transmission corridors with multiple lines and multiple contingencies, in *IEEE Power and Energy Society General Meeting*, Denver, USA (2015)
5. T. Van Cutsem, C. Vournas, *Voltage stability of electric power systems* (Kluwer, Boston, MA, 1998)

Index

A
Angle stability, 26, 79, 201
Auxiliary controllers, 91
   excitation system, 90
   PSS, 90

B
Bonneville power administration (BPA), 389–392
British Columbia Hydro (BC Hydro) and power authority, 103, 108, 109, 315–317
   EMS, 317–318
   integrated RAS system, 318–319
   RTAT integrated with EMS, 319, 320
   VSAT and TSAT, 320–323

C
California independent system operator (CAISO), 112, 113, 273–275, 325, 347, 348
   Bigwood Tool, 129–130
   changes in assumptions/methodology, 344–347
   NERC-registered functions, 328
   NERC, WECC, peak reliability and RC West, 329
   real-time voltage and transient stability analysis, 276–277
   RTVSA monitoring, 277–283
   SDG&E's relationship, 328–329
   SOL, 275–276
   transmission control agreement, 327
Colstrip ATR RAS monitoring, 110–111, 114
Consistency, 164, 347
Constraint equations, 202, 205
Contingency analysis, 9, 11, 13–14
   dynamic simulations, 141
   EMS, 254
   periodic sequence of actions, 136
   RTCA, 288–289, 354
   stability limits, 20–21
   steady-state, 155
Continuation power flow, 31–33, 40, 149, 385
Control center
   medium- and long-term planning, 2
   PMU applications, 19
   prototype tools, 182
   regional dispatching, 243
   SCADA platforms, 2
   TVA, 311–313
Control room information
   AEMO, 204
   Carolina power and light, 5
   decision factors, 355–358
   integration platform, 183
   mitigating transient insecurity, 151–152
   OSIsoft PI, 344
   real-time monitoring, 114
   transient security assessment, 141
Co-simulation
   offline studies, 52
   T&D, 39, 41
   VSM assessment, 42–47

D

Data integration, 230, 232–234, 236
Delayed voltage recovery, 53–55, 66, 72
DERs, *see* Distributed energy resources (DERs)
DEV, *see* Dominion Energy Virginia (DEV)
Direct methods, 222, 360, 372, 380, 382
Distributed computing, 232
Distributed energy resources (DERs)
    distribution system, 45
    grid-edge technologies, 37
    influence, 45–47
    net-load unbalance, 41
    New England 39-bus system case study, 72, 73
    phasor representation, 72
    power system components, 26
Dominion energy system
    decision factors
        generation retirement, 356–357
        increased system stress and upgrade delays, 358
        PJM results, 357
        regulatory considerations, 357
    overview, 351–354
    stability tool requirements, 358–359
    Virginia system operations center, 354–355
Dominion Energy Virginia (DEV)
    data acquisition, 362–363
        stability, 363–364
        stability program setup and use, 364–368
    stability analysis
        small signal, 361
        transient, 360
        voltage, 360–361
DSA, *see* Dynamic security assessment (DSA)
Dynamic composite load model
    key parameters for the composite, 63
    single-phase induction motor, 60–62
    static and electronic, 58
    substation and feeder model, 57–58
    three-phase induction motor, 58–60
Dynamic contingency screening, 360, 372–374
Dynamic security assessment (DSA), 204, 205, 235, 260, 284, 297–299, 301, 320–322, 371, 374
    add-ons, 9
    issues, 208
    MISO real-time, 301
    Powertech, 104–105
    transient stability assessment, 205–206
Dynamic stability, 142, 145, 224, 231, 233, 276

E

Electric Reliability Council of Texas Inc. (ERCOT)
    industrial and urban load centers, 135
    planning/ad hoc studies, 136
    transient stability, 140–143
    voltage stability, 136–139
Electro-magnetic transient (EMT) simulation, 83, 84, 95–97, 139, 212, 214, 215
EMS, *see* Energy management systems (EMS)
EMT simulation, *see* Electro-magnetic transient (EMT) simulation
Energy management systems (EMS)
    BC Hydro's, 317–320
    CIP, 359
    ERCOT, 141
    ICCP server, 102
    integrated RAS system, 318–319
    network
        model *vs.* planning network model, 260–261
        topology, 260
    process setup and integration, 252
    real-time stability tool, 204
    SCADA (*see* SCADA/EMS)
    snapshot, 266
    software, 104
    system architecture, 119–120
    transient event metrics, 141
    WSM-TSAT system architecture, 103
ERCOT, *see* Electric Reliability Council of Texas Inc. (ERCOT)

F

Fast computation, 8, 9, 11, 16–18, 20, 373

G

Grid operations
    real-time transient stability monitoring
        frequency response estimation, 286–288
        inertia tracking, 288
        model validation, 289, 290
        small signal stability analysis, 290–291
        transient stability limit monitoring, 286
        transient stability results, 288–289
    RTVSA
        congestion management, 280–281
        RAS threshold determination, 280
        supplementing seasonal studies, 282–283

Index

total transfer capability calculations, 281–282
voltage stability limit monitoring, 279
security assessment (*see* Security assessment)

I

Induction motor (IM), 53, 54
  load models
    single-phase, 60–62
    three-phase, 58–60
  model, 87–89
Integrated RAS System (iRAS)
  BC Hydro, 318–319
  BC Hydro's EMS, 317–318
Interconnection reliability operating limit (IROL), 103, 112, 258, 328, 344, 348
  limits, 276
  MISO, 300
  Northwest
    Net Export, 128–129
    Washington load, 124–126
  Oregon Net Export, 113, 114
  PEAK ROSE with CAISO's Bigwood Tool, 129–131
  real-time assessment, 118
  SDGE, 126–128
  SDGE/CFE import, 112–113, 126–128
  TTC numbers, 282
  VSA basic assumptions, 123–124
IROL, *see* Interconnection reliability operating limit (IROL)

K

Key data inputs
  additional impedance data, 156
  contingency list, 155–156
  control system status data, 157
  dynamic model data, 156
  power flow modification file, 155
  state estimator case, 154–155

L

Limiting contingency, 128, 149, 204, 258, 278, 286, 307, 332–333, 335, 345, 348
Load-flow, 1, 6, 15, 20, 154
Load models
  dynamic
    generic dynamic load model, 87
    inclusion, transient stability program, 89
    induction motor model, 87–89
  static, 32–33, 86
  WECC
    key parameters, 63
    single-phase induction motor, 60–62
    static and electronic, 58
    substation and feeder model, 57–58
    three-phase induction motor, 58–60
Long-term voltage stability
  data-driven methods
    centralized Thevenin equivalent-based methods, 48
    distribution network in Thevenin index, 51–53
    local Thevenin equivalent-based methods, 47
    sensitivity-based Thevenin index, 48–51
  maximum loading (P-V curve and Q-V curve), 27–29

M

Market dispatch, 297, 302
Midcontinent Independent System Operator (MISO)
  base power flow model development process, 295–296
  footprint, 293–295
  operational planning stability assessment tool, 296–297
  real-time
    DSA architecture, 301
    dynamic security assessment, 298–299, 301–302
    stability assessment tool, 297–298
    transient stability results, 299–300
Mitigation actions, 290, 391

N

National Electricity Market (NEM)
  AEMO, 199
  NEM AC transmission system, 199
  power system
    events, 211–213
    stability, 200–203
  real-time stability assessment
    DSA, 205–206
    scenario definition, 205
    stability criterion, 206–208
    typical DSA issues, 208

National Electricity Market (NEM) (*cont.*)
  transformation, 214–215
  VSAT, 208–210
National Grid UK
  government renewable energy, 162
  identify unknown stability problems solution, 163–164
  OFSA, 170–173
  OSA, 167–170
  power system analysis, 161
  SQSS, 165–167
  stability assessment, 161, 162
  time simulation approach, 164–165
New England power system, 247–249
New York Power Authority (NYPA), 392–394

O

Offline stability assessment (OFSA), 170–173
  accuracy of, 177–178
  TA, 178
OFSA, *see* Offline stability assessment (OFSA)
OLTSA, *see* Online transient security assessment (OLTSA)
Online assessment
  stability
    computation accuracy, 226
    computation time, 224
    constraints on modeling, 224–226
    transient stability algorithm
      simulation process description, 222–223
      simultaneous solution of differential algebraic equations, 223–224
    voltage
      assessment algorithm, 226–228
      assessment issues, 228
Online stability assessment (OSA), 167–170
  accuracy, 174–175
  problems identified, 175–177
  robustness, 173
  statistics, 175
Online stability assessment system design
  framework and functions, 229–231
  hardware architecture, 235–237
  implementation and case studies
    demonstration, 239–241
    online voltage demonstration, 241–244
    unified dispatching and control system D5000, 237–239
  process, 231–233
  software architecture, 233–235
Online transient security assessment (OLTSA), 317, 320–323

Online transient stability assessment
  dynamic modeling
    equivalent of external systems, 266–267
    model integration, 264–265
    PMU-based power plant model verification, 268
  network modeling
    EMS *vs.* planning, 260–261
    mapping and modification, 261–264
  online TSA program, 269–270
Online transient stability monitoring, 184–187
Online voltage stability assessment, 249–250
  ISO-NE operations, 256
  online VSAT architecture, 252–256
  process setup and integration, EMS, 252
  recommendation for future improvement, 256–257
  2D transfer analysis, 250–251
Online voltage stability monitoring, 187–193
Operations planning, 151, 158, 276, 284, 286, 319, 355, 357, 365, 366, 369
OSA, *see* Online stability assessment (OSA)
Oscillatory stability, 201, 204

P

PEAK ROSE RT-VSA, 118–120, 129
PEAK RT-VSA tool, 113, 118, 121, 130, 131
  key software improvements, 121
  other tool features, 122–123
  two stage RAS logic implementation, 121
Performance
  BCU classifiers, 374
  CPU, 13
  dynamic, 110
  enhancement, 255–256
  online implementation, 167
  OSA, 175
  security, 321
  TSAT simulation, 114
  voltage stability, 228
  VSA solution quality, 127
Phasor measurement units (PMUs), 19, 181, 204, 385
  deployment, 50
  measurements, 48
  micro-PMUs, 55
  postmortem analysis, 19
  power plant model validation, 268
  robust configuration, 194
  Thevenin methods, 47
  US-Canada separation event playback, 108
  WAMS Platform, 183

PJM tools
  implementation, 148–150
  interfaces, 150
  inter-regional operating limits, 147–148
  transient stability analysis, 150–158
  voltage stability, 147–148
    and instability, 146–147
PMUs, *see* Phasor measurement units (PMUs)
Power oscillation monitoring, 184, 185
Power system
  network elements, 10
  pre-fault, 85
  PSS (*see* Power system stabilizer (PSS))
  reliability, 357
  security, 3
  stability
    classification, 26
    instability, 25
  time domain simulators, 57
Power system events
  multiple NEM region separation event, 212–213
  South Australian black system incident, 211–212
Power system operation, 3, 17, 26, 200–202, 231, 256, 308
Power system stability
  managing stability issues, NEM, 200–202
  NEM stability issues, 202–203
Power system stabilizer (PSS), 90
  classification, 26
  EMT-based and RMS-based models, 212
  excitation systems, 91
  NEM, 200–203
  oscillations, 92
  PSCAD-based simulation, 211
  reference voltage, 81
  synchronous machines, 79
Power transfer limit, 178, 235
PSS, *see* Power system stabilizer (PSS)
PV analysis, 208
PV-curves, 331, 334–335, 340

Q
QV-curves, 331
  dV/dQ calculation, 336
  VSA import limit, 335

R
RAS, *see* Remedial action scheme (RAS)
Reactive power
  demand, 27
  generation units, 303
  implementation and use cases
    reactive power management, 306–313
    reactive reserve zone definition, 305–306
  measured voltage, 220
  PMU, 388
  TVA, 304–305
  voltage and current phasors, 185
  voltage stability, 34
Reactive power management
  online assessment, 306–307
  online mitigation, 307–308
  online VMS application, 311–313
  testing to production, 308–311
Reactive reserve zones, 305–306, 313
Real-time
  calculations, 2
  MISO, 297–298
  OSA display, 170
  Pacific Southwest blackout, 118
  software enhancements in RT-VSA, 120–123
  stability assessment, 203–208
  vs. study-mode processes, 6–8
  tool validation, 123–130
  transient stability monitoring, 284–292
  TVA, 303–313
  voltage
    stability analysis, 118
    stability monitoring, 277–283
    and transient stability analysis, 276–277
  WECC system, 117
  See also Remedial action scheme (RAS)
Real-time assessment tools (RTAT), 317, 319–321
Real-time contingency analysis (RTCA), 100, 101, 109–111, 118, 120, 280, 288–289, 317, 320, 354, 357
Real-time monitoring, 114, 238, 275, 276
Real-time operations, 139, 152, 167, 175, 282, 294, 298–299, 322, 368
Real-time stability assessment
  DSA, 205–206
  MISO, 297–298
  scenario definition, 205
  stability criterion, 206–208
  typical DSA issues, 208
Real-time transient stability monitoring
  challenges, 291–292
  grid operations, 286–291
Real-time voltage stability assessment (RTVSA)
  HVDC frequency modulation, 210

Real-time voltage stability assessment (RTVSA) (cont.)
  islanding issues, 210
  monitoring
    challenges, 283
    grid operations, 279–283
  scaling issues, 210
  VSAT limit, 209, 210
Remedial action scheme (RAS)
  backup RT-VSA for transfer capability analysis
    Oregon Net Export IROL, 113, 114
    SDGE/CFE import IROL calculation, 112–113
  BC Hydro's EMS, 317–320
  Colstrip ATR RAS Monitoring, 110–111
  feedback, 289
  IFRO measure calculation, 111, 112
  logic implementation, 122
  model validation and impact study
    April 2017 PDCI loss event, 106–107
    October 2017 US-Canada separation event, 107–110
  software enhancement implementation
    EMS software changes, 104
    Powertech DSA manager changes-part A, 104
    Powertech DSA PI interface code changes-part D, 105
    Powertech DSA service code changes-part B, 105
    Powertech TSAT code changes-part C, 105
  threshold determination, 280
ROSE software
  CAISO's Bigwood Tool, 129–130
  PEAK ROSE, 118–120
  SDGE, 332–340
Rotor angle, 26, 30, 79, 81, 164, 165, 207, 208, 285
RTAT, *see* Real-time assessment tools (RTAT)
RTCA, *see* Real-time contingency analysis (RTCA)
RTVSA, *see* Real-time voltage stability assessment (RTVSA)

S
San Diego Gas and electric (SDG&E) system
  metered boundaries, 325
  transmission system and system changes, 326–327
  voltage stability constrained utility, 326

SCADA/EMS
  execution modes, 11
  modeling requirements, 9–10
  non-standard network analysis applications
    fast computation of the risk of blackout, 16, 17
    optimal power flow, 17–18
  purpose and scope, 9, 10
  service routines, 15–16
  standard network analysis applications
    contingency analysis, 13–14
    dispatcher's power flow, 14–15
    state estimation, 12, 13
SDG&E, *see* San Diego Gas and electric (SDG&E) system
Secure operation, 196, 200, 204
Security assessment
  context, 3–4
  DSA (*see* Dynamic security assessment (DSA))
  ERCOT (*see* Electric Reliability Council of Texas Inc. (ERCOT))
  interactive computation paradigm, 4–6
  MISO, 298–299
  network analysis subsystem, 9–18
  The Northeast blackout of 1965, 2
  phasor measurements impacts, 19
  RAS (*see* Remedial action scheme (RAS))
  real-time
    calculations, 2
    vs. study-mode processes, 6–8
    stability limits in contingency analysis, 20–21
  static *vs.* dynamic, 8
  TEPCO-BCU, 374
  vertically integrated utility, 1
  See also Transient stability assessment (TSA)
Service routines
  bus-load forecast, 15
  transmission losses penalty factors, 16
SE, *see* State estimation (SE)
Short-term voltage stability
  data-driven methods
    Lyapunov exponent computation, 64–65
    simulation results, 65–68
    time series data, 64–65
  DERs, 71–73
  fault induced delayed voltage recovery, 53–55
  FIDVR events observed, 55, 56
  transient voltage criteria, 56–57
Simulation
  contingency, 6

Index

Lyapunov exponent, 65–66
transient stability, 85
VSM assessment, 42–47
Situational awareness, 100, 114, 120, 131, 179, 195, 244, 247, 255, 260, 280, 286, 348
Small signal stability, 8, 229, 230, 239, 290–291, 361, 362, 367
SOC, *see* System operations center (SOC)
Stability assessment
   MISO, 297–299
   operational planning
     base power flow model development process, 295–296
     MISO footprint, 293–295
     operational planning stability assessment tool, 296–297
     transient stability results, 299–300
State estimation (SE), 12, 13
   and power system security assessment, 3
   computational accuracy and convergence, 226
   iEMS system, 167, 168, 174
   load-flow, 6
   powerflow, 288
   real-time voltage, 276
   transient stability analyses, 276
Steady-state stability, 8, 331, 332, 334, 343
Synchronous machine, 20, 79, 82, 83, 90, 165, 168, 177
Synchrophasor application
   generators' dynamic models, 268
   SCADA/EMS solutions, 182
   system disturbances, 185
   voltage and current, 189
Synchrophasor technology
   BPA, 389–392
   methodology, 387–389
   NYPA, 392–394
   power transfers, 386
   transmission corridor, 386
   voltage stability, 385
System model validation, 109
System operating limits (SOL) Methodology, 101, 111, 122, 129, 258, 275–276, 299, 300, 310
System operations center (SOC), 351, 354–355, 358, 359
System strength
   applicability, 139
   constraints, 201
   The Texas Panhandle, 138–139
   voltage/power oscillations, 214

T

T&D co-simulation
   influence of DER, 45–47
   long-term voltage stability margin assessment, 44–45
   steady-state and quasi-steady-state studies, 42–44
   VSM assessment, 41
Tennessee Valley Authority (TVA)
   online assessment, 306–307
   online mitigation, 307–308
   online VMS application, 311–313
   testing to production, 308–311
TEPCO-BCU, 371–372
   architecture, 373–375
   dynamic contingency screening, 372–373
   planning under uncertainty, 375–382
Thevenin equivalent, 47, 48, 51, 53, 190
TOGs, *see* Transmission operating guides (TOGs)
Transfer analysis (TA), 118, 121, 123, 128, 170, 178, 252, 270, 276, 299, 333, 334
Transfer limit, 102, 104, 150, 161, 170, 178, 206, 207, 235, 277, 281, 283, 297, 322
Transient event metrics, 141–142
Transient security/stability assessment tool (TSAT)
   ERCOT operations, 140
   grid operations, 320–323
   PJM tools, 145–158
   real-time voltage stability, 361
   WSM-TSAT, 101–105
Transient stability
   advanced computing techniques, 142
   angular acceleration, 80
   auxiliary controllers, 90
   case study, 92–95
   electrical system of the synchronous machine
     network and synchronous generators, 84–85
     simulation, 85
   electromagnetic torque, 81
   load models, 85–89
   modelling limitations, 95–96
   new developments, 96–97
   other dynamic devices
     HVDC line, 91–92
     non-linearities, 92
   PJM tools
     cycle execution process, 157

Transient stability (*cont.*)
    evolution of use, 152–153
    key data inputs, 154–157
    key outputs and integrations, 157
    mitigating transient insecurity process, 151–152
    study mode, 157
    training and job responsibilities, 152
  power system stability, 79
  shaft system dynamics, 81–82
  simultaneous event simulation, 143
  synchronous generator, 79
  three-phase short circuit, 80
  transient event metrics, 141–142
  W–N interface, 140–141
Transient stability assessment (TSA)
  data preparation and model update, 269–270
  IEEE papers, 100
  implementation and configuration, 270
  NEM (*see* National Electricity Market (NEM))
  PEAK's WSM-TSAT project, 101–102
  Powertech Labs, 100–101
  RAS model validation and RAS impact study, 106
  system architecture, 269
  WECC system, 99
  WSM RAS models, 100
  WSM-TSAT (*see* WSM-TSAT)
Transmission operating guides (TOGs)
  future work to formalize transmission, 259
  stability guides, 258
TSAT, *see* Transient security/stability assessment tool (TSAT)
TVA, *see* Tennessee Valley Authority (TVA)

U
Unknown constraints, 163–164

V
V&R PEAK ROSE RT-VSA tool
  PEAK WSM facts, 120
  project milestones, 120
  system architecture, 119–120
VAR management, 40, 57, 154, 249, 304, 306–308, 313, 328, 329
Voltage stability
  assessment in ERCOT operations, 136–137
  China's power supply, 217
  HVDC/UHVDC/UHVAC Interconnection, 217, 218
  hydro power exporting network, 217, 218
  instability, 221
  MMC transmission power, 219
  multi infeed HVDC/UHVDC power grid, 221
  NEM (*see* National Electricity Market (NEM))
  online assessment technique, 222–228
  online stability assessment system design, 229–237
  PJM tools (*see* PJM tools)
  PMUs, 181
  reactive voltage regulation, 220
  system strength
    applicability, 139
    The Texas Panhandle, 138–139
  UHVDC projects, 219
Voltage stability analysis (VSA)
  import limit alarms, 337
  methodology of computations
    reverse power transfer, 333–334
    VSA import limit and limiting contingency, 332–333
  off-line studies, 330
  online, 330
  PV-curve analysis, 334–335
  QV-curve analysis, 335–336
  scenario-based analysis, 340–341
  SDGE-ROSE operation, 337–340
  stressing the interface
    contingencies and associated RAS, 342–343
    maximum transfer level, 343
    sink area, 342
    source area, 342
    transfer increment, 343
Voltage stability assessment, 26–27
  dynamic composite load model, 57–63
  ERCOT operations, 136–137
  FIDVR, 66, 68–71
  long-term voltage stability, 27–29
  parameter sensitivity
    effect of transformer taps, 36–37
    generator limits, 35, 36
    network impedance, 33–35
    static load models, 32–33
  PJM tools, 145–158
  power flow divergence and instability
    voltage stability margin, 31–32
  power system classification, 26
  short-term voltage stability, 53–57
  T&D interactions
    co-simulation and its application, 42

Index

influence of load unbalance, 39–41
modeling distribution networks, 38, 39
Voltage stability assessment tool (VSAT)
  HVDC frequency modulation, 210
  HVDC/UHVDC/UHVAC Interconnection, 217, 218
  hydro power exporting network, 217, 218
  islanding issues, 210
  limit display, 208, 209
  online assessment technique, 222–228
  online stability assessment
    implementation and case studies, 237–244
    system design, 229–237
  reactive voltage regulation capability, 219, 220
  scaling issues, 210
  SD local load level, 220
  transient angle instability, 219
Voltage stability index (VSI)
  distribution network, 52
  PMU measurements, 47
  power flow output, 389
Voltage stability limit, 113, 201–204, 299, 322, 323, 326, 330, 359, 361
Voltage stability margin, 28, 31, 32, 40, 44–45, 226, 228, 242, 365
VSA Import Limit, 332–336, 343
VSAT, *see* Voltage stability assessment tool (VSAT)
VSI, *see* Voltage stability index (VSI)

W

WAMS, *see* Wide area measurement systems (WAMS)
WECC, *see* Western electricity coordinating council (WECC)

Weighted Short Circuit Ratio (WSCR), 137–139
Western electricity coordinating council (WECC), 329, 332, 335
  dynamic composite load model, 57–63
  individual company level, 111
  system operating paradigms, 100
  transient voltage criteria, 56
Wide area measurement systems (WAMS), 9, 181, 232
  application development
    online transient stability monitoring, 184–187
    online voltage stability monitoring, 187–193
  data and information, 233
  other examples
    access to voltage phase angle, 194, 195
    HVDC monitoring, 195–196
    robust configuration, 194
  platform, 182–183
WSCR, *see* Weighted Short Circuit Ratio (WSCR)
WSM-TSAT, 101
  custom software enhancements, 104
  model scales, 103
  RAS software enhancement implementation
    EMS software changes, 104
    Powertech DSA manager changes-part A, 104
    Powertech DSA PI interface code changes-part D, 105
    Powertech DSA service code changes-part B, 105
    Powertech TSAT code changes-part C, 105
  system architecture, 102–103

CPSIA information can be obtained
at www.ICGtesting.com
Printed in the USA
LVHW080950290123
738055LV00024BA/41